Metagenomics for Legume Crop Improvement

Metagenomics for Legume Crop Improvement offers a comprehensive and timely exploration of the enormous potential hidden within legume-associated microbial communities. Legumes play a crucial role in global food security and sustainable agriculture due to their natural symbiosis with nitrogen-fixing and plant growth–promoting microbes. Yet, despite their significance, the microbiomes surrounding legumes remain poorly understood. This book aims to bridge that gap by bringing together cutting-edge research, analytical tools, and practical applications of metagenomics, a transformative approach that allows the direct study of entire microbial communities without the need for cultivation. The book provides an interdisciplinary perspective, integrating plant science, microbiology, biotechnology, and bioinformatics to reveal how metagenomic insights can enhance legume productivity, resilience, soil health, and long-term sustainability. It uniquely combines foundational concepts with advanced methodologies and real-world agricultural relevance, making it stand out as a vital reference for emerging research in plant–microbe interactions.

Key Features:

- Provides a holistic overview of legume-associated microbiomes, including microbial ecology, nutrient cycling, and stress responses
- Highlights advanced metagenomic techniques, sequencing platforms, and bioinformatics tools tailored for legume research
- Explores how environmental stresses such as extreme pH, nutrient deficiencies, fertilisers, and climate stressors shape legume microbiomes
- Introduces integrative "omics" approaches (metagenomics–metabolomics) to uncover microbial functions and plant interactions

Written for researchers, graduate students, academicians, and agricultural practitioners, this book serves as both an educational resource and a practical guide. By connecting scientific discovery with field-level applications, *Metagenomics for Legume Crop Improvement* provides the knowledge needed to harness beneficial microbes for sustainable farming and future-ready legume cultivation. It is an essential addition to the libraries of professionals working in plant science, microbiome research, and agricultural biotechnology.

Metagenomics for Legume Crop Improvement

Edited by Anil Kumar Singh and Alok Ranjan

CRC Press
Taylor & Francis Group
Boca Raton London New York

CRC Press is an imprint of the
Taylor & Francis Group, an **informa** business

Designed cover image: Getty

First edition published 2027
by CRC Press
2385 NW Executive Center Drive, Suite 320, Boca Raton FL 33431

and by CRC Press
4 Park Square, Milton Park, Abingdon, Oxon, OX14 4RN

CRC Press is an imprint of Taylor & Francis Group, LLC

ISBN: 978-1-041-22649-9 (hbk)
ISBN: 978-1-041-22648-2 (pbk)
ISBN: 978-1-003-73332-4 (ebk)

DOI: 10.1201/9781003733324

Typeset in Times
by Apex CoVantage, LLC

Contents

Foreword

Legumes occupy a vital position in global agriculture due to their remarkable capacity to associate with diverse microbial communities that enhance soil fertility, support biological nitrogen fixation, and promote sustainable crop production. As modern agriculture faces increasing challenges from climate change, soil degradation, and the need for higher productivity with fewer inputs, understanding the legume microbiome has become more important than ever. Metagenomics, an advanced tool for studying complex microbial ecosystems, provides transformative insights into the composition, functions, and interactions of legume-associated microbes, offering new opportunities for improving crop resilience and sustainability. *Metagenomics for Legume Crop Improvement* presents a comprehensive overview of the current landscape of legume microbiome research. The book begins by outlining the relevance and potential of metagenomic approaches in understanding legume biology, followed by detailed discussions on experimental designs, analytical strategies, and advanced methodologies that have propelled microbiome research forward. It explores the dynamic and diverse microbial ecology of the legume rhizosphere, emphasising the essential roles of both symbiotic and non-symbiotic microbes in nutrient cycling, soil health, and plant productivity. The text further delves into the responses of legume-associated microbial communities under various environmental stresses, including biotic and abiotic pressures, extreme soil pH conditions, nutrient deficiencies, and the impact of chemical fertilisers. These insights provide a deeper understanding of how microbiomes adapt to adverse conditions and how these interactions can be leveraged to enhance crop resilience. Integrative approaches combining metagenomics with metabolomics and other "omics" technologies are highlighted for their potential to unravel functional complexities within the legume microbiome. Additionally, the book addresses practical applications, industrial relevance, methodological protocols, and essential bioinformatics tools required for metagenomic analysis. This volume serves as a valuable resource for students, researchers, and agricultural professionals. By uniting foundational knowledge with advanced research and practical applications, it offers powerful insights to harness legume-associated microbiomes for sustainable agriculture and global food security.

Preface

Legumes have long been recognised as nature's allies in sustainable agriculture. Their unique ability to form symbiotic relationships with nitrogen-fixing bacteria and other beneficial microbes makes them critical players in enhancing soil fertility, reducing the need for chemical fertilisers, and supporting global food security. Yet, despite their ecological and agricultural significance, the intricate microbial ecosystems that contribute to legume productivity remain vastly underexplored. This book, *Metagenomics for Legume Crop Improvement*, seeks to bridge that knowledge gap by delving into the emerging field of metagenomics and its transformative potential for legume agriculture. Metagenomics has revolutionised our understanding of microbial communities, enabling researchers to study the genetic potential of complex microbiomes without the need for culturing. For legumes, this approach has opened doors to uncovering the diversity, composition, and functional roles of their associated microbiomes, from rhizosphere microbes that promote nutrient cycling to endophytes that enhance stress resilience. By harnessing these insights, we can pave the way for innovative strategies to improve legume crop yields, enhance resilience to climate change, and promote sustainable farming practices. This book is designed to serve as a comprehensive resource for researchers, students, and agricultural practitioners. It brings together foundational concepts, cutting-edge methodologies, and real-world applications of metagenomics in legume crop improvement. The chapters span topics such as the microbial ecology of legume rhizospheres, the role of plant growth–promoting microorganisms, advances in sequencing technologies, and case studies highlighting successful applications in sustainable agriculture. As global challenges such as population growth, climate change, and soil degradation intensify, the need for sustainable and efficient agricultural solutions becomes ever more urgent. Legume crops, with their natural ability to enrich soils and support human nutrition, stand as a beacon of hope. By unlocking the power of their microbiomes through metagenomics, we can usher in a new era of agricultural innovation that benefits farmers, ecosystems, and society at large. We extend my deepest gratitude to the contributors, whose expertise and dedication have shaped this book, and to the global community of scientists working tirelessly to advance our understanding of metagenomics and its applications. It is our hope that this book not only serves as an informative guide but also inspires further research and collaboration in this exciting field.

Anil Kumar Singh

Alok Ranjan

About the Editors

Dr. Anil Kumar Singh is Principal Scientist at the ICAR-National Institute for Plant Biotechnology, New Delhi, India. He has been engaged in plant molecular biology and biotechnology research for more than two decades. His research group has extensively characterised genomes and transcriptomes of several crop plants and commercially important microbes, generating valuable gene resources for crop improvement. Many of these genes have been functionally validated in model plants as well as in economically important crops such as rice and potato through the gene pyramiding approach. Dr. Singh has also worked as a visiting scientist at the University of Minnesota, USA, where he worked on novel strategies for multiplexed genome editing and approaches to overcome low transformation efficiency in recalcitrant legume crops, thereby advancing precision breeding tools for underutilised and technically challenging species.

He is currently leading an ambitious project aimed at deciphering the influence of soil pH on the rhizosphere microbiome of legume crops, with the goal of developing microbe-based mitigation strategies for acidic soil stress. Dr. Singh has published more than 90 research articles in peer-reviewed international journals with a cumulative Impact Factor of ~ 300, over 4,700 citations, and an h-index of 37. He has also edited two books and authored 15 book chapters and delivered invited or keynote lectures at more than 40 national and international conferences in India and abroad. He serves on the editorial board of several reputed journals, including *Frontiers in Plant Science*, *PLOS ONE*, *BMC Research Notes*, and *Phyton-International Journal of Experimental Botany*. Furthermore, he has guest-edited special issues in *Antioxidants, Genes, Tree Physiology and Physiologia Plantarum*. In recognition of his significant contributions to plant molecular biology research, Dr. Singh has been elected as a member of the National Academy of Sciences, India (NASI), and the Plant Tissue Culture Association–India.

Dr. Alok Ranjan is a plant biotechnologist and molecular biologist with over 15 years of research experience and more than seven years of teaching in modern plant sciences and biotechnology. He is currently an assistant professor in the Department of Biotechnology at Patna Women's College, Patna University, Bihar. Dr. Ranjan completed his PhD in biotechnology at CSIR–National Botanical Research Institute, Lucknow, in collaboration with Banaras Hindu University, Varanasi. He possesses over eight years of international postdoctoral research experience from renowned institutions, including the Umeå Plant Science Centre, Umeå University (Sweden),

Oklahoma State University (USA), Meharry Medical College (USA), and Université libre de Bruxelles (Belgium). His research expertise includes plant stress physiology, hormone signalling, adventitious root development, and multi-omics approaches. Dr. Ranjan has published more than 40 peer-reviewed articles and several book chapters and has served as a reviewer for leading scientific journals. His research output has garnered over 1,800 citations, with an h-index of 22, and he holds one Indian patent. He has been actively engaged in root biology, metagenomics, and plant stress physiology, and his recent work on legume metagenomics aligns closely with the objectives of this book.

Contributors

Aarti Yadav
Department of Botany
University of Delhi, India

Abhishek Kumar Mishra
Department of Biotechnology, School of Biosciences and Technology (SBST), Vellore Institute of Technology
Tamil Nadu, India

Adarsh Kumar
ICAR-Indian Institute of Pulses Research
Kanpur, India

Alok Ranjan
Department of Biotechnology
Patna Women's College, Patna University
Bihar, India

Anil Kumar Singh
ICAR-National Institute for Plant Biotechnology
New Delhi, India

Ankit Singh Negi
Amity Institute of Biotechnology, Amity University,
Uttar Pradesh, India

Ankita Pradhan
Plant Molecular Biology Laboratory
School of Biotechnology, Kalinga Institute of Industrial Technology (KIIT) Deemed-to-Be University
Odisha, India

Ashwani Kumar
ICAR-Central Soil Salinity Research Institute
Karnal, India

Deepak Kumar
Department of Botany, Institute of Science
Banaras Hindu University
Varanasi, India

Dhruvi Vaish
University School of Biotechnology
Guru Gobind Singh Indraprastha University
New Delhi, India

Ekta Narwal
ICAR-Indian Agricultural Research Institute
New Delhi, India

Geetika Sirhindi
Department of Botany
Punjabi University
Patiala, Punjab, India

Gopal Chowdhary
Plant Molecular Biology Laboratory, School of Biotechnology
KIIT Deemed-to-Be University
Odisha, India

Gunja Gond
Zoology Department
Rashtriya PG College
Uttar Pradesh, India

Ishita Samanta
Plant Molecular Biology Laboratory, School of Biotechnology
KIIT Deemed-to-Be University
Odisha, India.

Jairam Choudhary
ICAR-Indian Agricultural Research Institute
New Delhi, India

Jamilur Rahman
Department of Genetics and Plant Breeding Sher-e-Bangla Agricultural University Dhaka
Bangladesh

Jyotsana Tilgam
ICAR-National Bureau of Agriculturally Important Microorganisms, India

Kreeti Kajal
Institute of Cellular and Organismic Biology, Academia Sinica
Taipei, Taiwan

Kumar M.
ICAR-National Bureau of Agriculturally Important Microorganisms, India

Kumari Deepali
Amity Institute of Biotechnology, Amity University
Uttar Pradesh, India

Lopamudra Ray
School of Biotechnology, KIIT Deemed-to-Be University
Odisha, India

Madan Kumar
ICAR-Indian Institute of Agricultural Biotechnology
Jharkhand, India

Md. Sajedur Rahman
Department of Genetics and Plant Breeding
Sher-e-Bangla Agricultural University
Dhaka, Bangladesh

Md. Zahidur Rahman
Department of Genetics and Plant Breeding
Sher-e-Bangla Agricultural University
Dhaka, Bangladesh

Meghna Thapa
ICAR-National Institute for Plant Biotechnology
New Delhi, India

Neeraj Kumar Dubey
Botany Department
Rashtriya PG College
Jaunpur, Uttar Pradesh, India

Neeshma Jaiswal
Department of Zoology, School of Life Sciences
Babasaheb Bhimrao Ambedkar University
Uttar Pradesh, India

Nikhil Kumar Singh
ICAR-Indian Institute of Agricultural Biotechnology
Jharkhand, India

Nikita Srivastava
Plant Molecular Biology Laboratory, Department of Botany
Dayanand Anglo-Vedic (PG) College
Chhatrapati Shahu Ji Maharaj University
Kanpur, India

Niloy Gain
Department of Genetics and Plant Breeding
Sher-e-Bangla Agricultural University
Dhaka, Bangladesh

Nishant
School of Agriculture and Rural Development
Ramkrishna Mission Vivekananda Educational and Research Institute (RKMVERI), India

Parichita Priyadarshini
ICAR-Indian Grassland and Fodder Research Institute
Jhansi, India

Pooja Sharma
Plant Molecular Biology Laboratory, Department of Botany
Dayanand Anglo-Vedic (PG) College
Chhatrapati Shahu Ji Maharaj University
Kanpur, India

Pradeep Kumar Yadav
Botany Department
University of Allahabad, Prayagraj
Uttar Pradesh, India

Ragini Sinha
Department of Life Science
Central University of Jharkhand
Cheri-Manatu, Ranchi, Jharkhand, India

Rakesh Srivastava
Department of Biotechnology, School of Sciences, Noida
International University, Greater Noida, Gautam Budh Nagar, UP, India

Ram Singh Purty
University School of Biotechnology
Guru Gobind Singh Indraprastha University
New Delhi, India

Ran Vijay Singh
Zoology Department, Rashtriya PG College
Uttar Pradesh, India

Rashmi Srivastava
Department of Zoology, School of Life Sciences
Babasaheb Bhimrao Ambedkar University
Uttar Pradesh, India

Ritesh Pattnaik
Applied Research Laboratory, School of Biotechnology
KIIT Deemed-to-Be University
Odisha, India

Ritu
University School of Biotechnology
Guru Gobind Singh Indraprastha University
New Delhi, India

Riya Mary Cherian
Department of Life Science
Central University of Jharkhand
Jharkhand, India

Ruchi Yadav
Amity Institute of Biotechnology
Amity University
Uttar Pradesh, India

Sakshi Suman
Department of Botany
University of Delhi
India

Sandeep Jaiswal
ICAR-Research Complex for NEH Region
Umiam, Meghalaya, India

Sangeeta Singh
ICAR-Indian Grassland and Fodder Research Institute
Jhansi, India

Sasmita Mishra
Department of Biology
Kean University
USA.

Satendra Pal Singh
Department of Botany
Y.D. (PG) College
Lakhimpur, Uttar Pradesh, India

Seemon Giri
School of Biotechnology
KIIT Deemed-to-Be University
Odisha, India

Shobit Thapa
ICAR-Indian Institute of Pulses Research
Kanpur, India

Shreyasi Bhowmik Tonton
Department of Genetics and Plant Breeding Sher-e-Bangla Agricultural University Dhaka
Bangladesh

Shruti Kaushik
ICAR-Central Soil Salinity Research Institute
Karnal, India

Smaranika Pradha
School of Biotechnology
KIIT Deemed-to-Be University
Odisha, India

Sohan Lal
Zoology Department
Rashtriya PG College
Uttar Pradesh, India

Sourav Ranjan Parida
School of Biotechnology
KIIT Deemed-to-Be University
Odisha, India

Subhransu Sekhar Behera
School of Biotechnology
KIIT Deemed-to-Be University
Odisha, India

Sujeet Pratap Singh
Amity Institute of Biotechnology
Amity University
Uttar Pradesh, India

Sujit Kumar Bishi
ICAR-Indian Institute of Agricultural Biotechnology
Jharkhand, India

Supriti Chowdhury
Plant Molecular Biology Laboratory, School of Biotechnology
KIIT Deemed-to-Be University
Odisha, India

Surendra Pratap Singh
Plant Molecular Biology Laboratory, Department of Botany
Dayanand Anglo-Vedic (PG) College
Chhatrapati Shahu Ji Maharaj University
Kanpur, India

Tikshana Yadav
Plant Molecular Biology Laboratory, Department of Botany, Dayanand Anglo-Vedic (PG) College
Chhatrapati Shahu Ji Maharaj University
Kanpur, India

Varsha Singh
Plant Molecular Biology Laboratory, Department of Botany, Dayanand Anglo-Vedic (PG) College
Chhatrapati Shahu Ji Maharaj University
Kanpur, India

Vijai Pal Bhadana
ICAR-Indian Institute of Agricultural Biotechnology
Jharkhand, India

Yamini Agrawal
Department of Botany, University of Delhi
India

Yashika Sarangle
University School of Biotechnology, Guru Gobind Singh Indraprastha University
New Delhi, India

Zahra Parwez
School of Biotechnology, KIIT Deemed-to-Be University
Odisha, India

Introduction

Metagenomics for Legume Crop Improvement presents a comprehensive examination of the immense yet underutilised potential of legume-associated microbiomes through the advanced lens of metagenomics. Legumes play a crucial role in sustainable agriculture due to their inherent ability to form symbiotic associations with nitrogen-fixing bacteria and a variety of other beneficial microbes. These interactions enrich soil fertility, reduce dependency on synthetic fertilisers, and support environmentally friendly farming systems. Despite their importance, the complex microbial communities linked with legumes remain insufficiently explored, and their functional roles in enhancing crop productivity, stress resilience, and soil health are still not fully understood. This book seeks to bridge this critical knowledge gap. Metagenomics, a transformative approach for studying whole microbial communities without the need for cultivation, has opened new avenues for unravelling the structure, diversity, and functional dynamics of legume microbiomes. This volume elucidates the latest research advancements, analytical tools, and practical applications in this rapidly expanding field. It covers essential themes, including microbial ecology, nutrient cycling, plant–microbe interactions, biotic and abiotic stress tolerance, and integrative multi-omics approaches such as metabolomics. Additionally, this book discusses the influence of environmental stresses, soil conditions, and chemical fertilisers on the legume microbiome, offering insights into strategies for strengthening crop resilience and promoting sustainable cultivation. Its interdisciplinary perspective spanning plant science, microbiology, bioinformatics, and agricultural biotechnology ensures both scientific depth and practical relevance. Beyond documenting cutting-edge discoveries, this book highlights real-world applications, including the development of biofertilisers, biostimulants, and microbial solutions for greener agriculture. In an era marked by escalating concerns over food security, climate change, and soil degradation, the subject matter is timely and highly impactful.

Key Features

- *Comprehensive Insight into Legume Microbiomes*
  The book provides an in-depth understanding of the diversity, structure, and functional roles of microbial communities associated with legumes, with a special focus on the rhizosphere.
- *Advanced Metagenomic Tools and Techniques*
  It highlights cutting-edge metagenomic approaches, experimental methods, sequencing technologies, and bioinformatics tools essential for studying and analysing legume-associated microbiomes.
- *Microbial Contributions to Stress Tolerance and Soil Health*
  The content emphasises how microbial communities contribute to nutrient cycling, biotic and abiotic stress tolerance, and overall soil fertility, offering solutions for climate-resilient legume cultivation.

DOI: 10.1201/9781003733324-1

- *Impact of Environmental Factors and Farming Practices*
  The book examines how soil pH, nutrient deficiencies, fertilisers, and environmental stressors influence legume microbiomes, providing strategies to mitigate negative impacts on crop performance.
- *Applications for Sustainable and Industrial Agriculture*
  It explores practical applications, including biofertilisers, biostimulants, bioremediation, and industrial uses of legume-associated microbes, bridging research insights with real-world agricultural innovation.

By connecting metagenomic research with practical agricultural innovations, *Metagenomics for Legume Crop Improvement* aims to serve as an essential reference for researchers, students, and practitioners committed to advancing sustainable and resilient legume cultivation.

1 Legume Metagenomics
Status and Scope for Crop Improvement

Abhishek Kumar Mishra, Kumari Deepali, Sujeet Pratap Singh, and Ruchi Yadav

1.1 INTRODUCTION

Legumes have become integral to worldwide farming, which is in great part not only due their nutritional values and food security but also due to the fact that they improve the health of the soil. Legumes have the ability to fix nitrogen from the air, which they put back into the soil at a rate of 100 to 300 kg/hectare annually. Consequently, it lowers chemical fertiliser use, curbing greenhouse gas emissions from intensive agriculture. Studies show legumes, compared to non-nitrogen-fixing crops, are eco-friendly, support sustainable yields, and maintain soil health (Kebede 2021). There are various legume cultivation challenges that threaten both yield targets and the sustainability of the environment. Extreme harsh conditions such as drought, excessive heat, and soil salinity, waterlogging, and even heavy metal contamination interrupt plant growth and cause loss of yield. Under stresses, the legume-rhizobia partnership is one of the first to suffer as the nitrogen-fixing nodules weaken and lose their functions in a regular manner (Anjum 2016). Biotic stresses such as pests and diseases can affect entire fields at once, sometimes resulting in total crop failure, particularly in Asian and African regions with favourable conditions for disease outbreaks.

The nutrient deficiencies, for example, low phosphorus levels, reduce nodulation and the nitrogen fixation process, which in turn reduces the yield (Jha et al. 2024). Plant-associated microbiomes, forming a "holobiont" with their host, influence legume growth, adaptation, and survival. The rhizosphere microbiome bacteria, fungi, archaea, and other microbes enhance nutrient uptake, stress tolerance, disease resistance, and growth, making it a key factor for healthier, more resilient plants (Levy and Lubell 2017). Metagenomics, which is an advanced technology, now allows for unparalleled exploration of the concealed microbial world. Figure 1.1 shows a microbiome-centred approach in legumes, demonstrating plant–microbe interactions and the role of metagenomics in sustainable crop improvement. Genetic analysis of soil or root samples allows scientists to study unculturable microbes, uncover functional populations, and design microbiomes for sustainable agriculture (Nwachukwu and Babalola 2022).

DOI: 10.1201/9781003733324-2

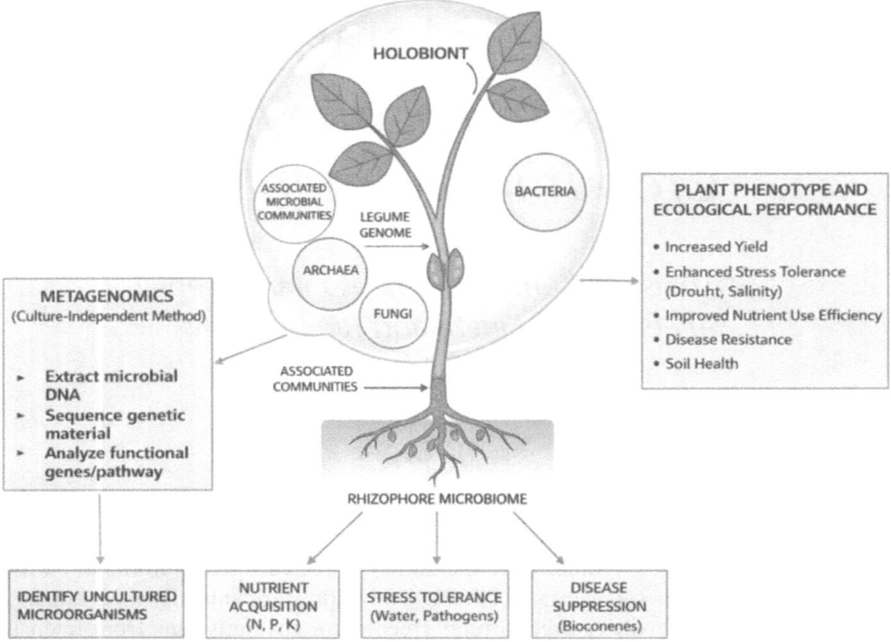

FIGURE 1.1 Microbiome-centred approach in legumes showing plant–microbe interactions and the role of metagenomics in sustainable crop improvement.

1.2 METAGENOMICS: CONCEPT AND RELEVANCE IN AGRICULTURE

Metagenomics allows researchers to study the genomes and functions of all microbes in an environment, bypassing culturing limits. By sequencing entire communities, it provides a comprehensive view of microbial consortia and their ecological roles, which traditional methods capture only partially (Nwachukwu and Babalola 2022). Figure 1.2 illustrates the general workflow for a metagenomic study applied to crop improvement, comprising six stages in total. Samples of soil or plant material are collected, and DNA is extracted to obtain the metagenome. This DNA is sequenced via next-generation sequencing (NGS), and bioinformatics reconstructs fragments, identifies genes, and predicts functions using databases such as NCBI or KEGG. The results guide crop improvement and agricultural innovation.

Collateral benefits of innovation in the data economy in the fields of rapid sequencing and data analysis, particularly in sequencing and metagenomics, have broadened opportunities to address even more informational and biological challenges in agricultural research (Fatma Şeyma Gökdemir et al. 2022). Environmental metagenomics sequences and analyses large DNA pools efficiently, revealing microbial community structure, functions, and interactions. Combined with bioinformatics,

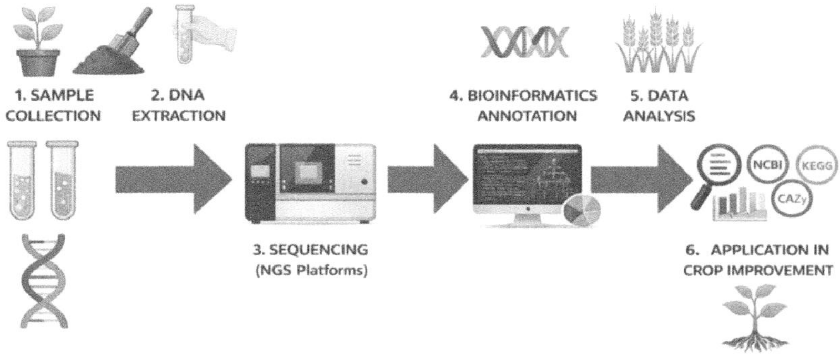

FIGURE 1.2 An overview of the metagenomic workflow showing the sequential process from sample collection and sequencing to bioinformatics-based taxonomic and functional analysis, leading to agricultural applications.

it identifies metabolic pathways, gene functions, and uncharacterised traits, aiding crop improvement, plant resilience, and ecosystem understanding (Mir Asif Iquebal et al. 2022). Scientists are reporting that they are putting more focus on the microbes which associate with legumes via the use of metagenomic tools. These microbes, which were previously unidentified in the plant's root system, are present in the rhizosphere, which is the soil which clings to roots, in the endosphere, which is the tissue within the root, and in the nodules, where the action of nitrogen fixation is seen (Xiao et al. 2017). Each plant niche hosts distinct microbes shaped by soil, environment, or plant genotype. This structure allows researchers to study legume–microbe interactions, including nitrogen fixation, nutrient uptake, and stress tolerance (Schaedel et al. 2021). Unlike traditional culturing, metagenomics overcomes lab limitations, revealing microbial diversity and functions by sequencing DNA, including rare and unculturable species. Metagenomics reveals which microbes are present, their gene functions, and their roles, helping to clarify complex plant–microbe relationships.

1.3 ADVANCES IN SEQUENCING TECHNOLOGIES AND ANALYTICAL METHODS

"Next-generation sequencing has transformed metagenomics, producing large-scale data faster and cheaper. Short, accurate Illumina reads, complemented by longer sequences, allow detailed study of microbes across soil, roots, and plants, accelerating gene discovery and crop trait improvement (Gökdemir et al. 2022). For comparison, Table 1.1 outlines key features, including read length, precision, output volume, and relevance to legume microbiomes for widely used sequencing platforms."

Bioinformatics workflows now handle large metagenomic datasets, using databases and models (Pfam, KEGG, BLAST) for taxonomic and functional annotation. New tools such as DeepFRI, MetaErg, and MEDUSA provide detailed gene functions and streamline analysis from reads to visualisation (Dong and Strous 2019). Table 1.2 lists common software and frameworks currently utilised in this field.

TABLE 1.1
Comparative Overview of Major Sequencing Technologies Used in Agricultural Metagenomics

Sequencing Platform	Read Length	Accuracy	Throughput	Advantages	Limitations	Reference
Illumina (HiSeq, MiSeq)	150–300 bp	High (>99%)	High	Cost-effective, high accuracy	Short reads, assembly challenges	(Gökdemir et al. 2022)
Oxford Nanopore	>10 kb	Moderate	Variable	Real-time sequencing, long reads	Higher error rates	(Gökdemir et al. 2022)
PacBio SMRT	10–25 kb	High	Moderate	Long reads, reduced bias	High cost per base	(Gökdemir et al. 2022)
Ion Torrent	200–400 bp	Moderate	High	Fast run time	Homopolymer errors	(Gökdemir et al. 2022)

TABLE 1.2
Bioinformatics Tools and Pipelines for Metagenomic Data Analysis

Analytical Step	Common Tools	Key Features	Reference Databases	Remarks	Reference
Quality Control	FastQC, Trimmomatic	Read filtering and trimming	–	Essential for error-free downstream analysis	(Dong and Strous 2019)
Assembly	MEGAHIT, MetaSPAdes	De novo metagenome assembly	–	Handles complex microbial communities	(Dong and Strous 2019)
Taxonomic Profiling	Kraken2, MetaPhlAn, QIIME2	Assign reads to taxa	NCBI, GTDB	Enables community structure analysis	(Dong and Strous 2019)
Functional Annotation	Prokka, EggNOG-mapper, DeepFRI	Predict gene function	KEGG, Pfam, SwissProt	Combines homology and machine learning	(Dong and Strous 2019)

Such tools cover sequence checking, genome reconstruction, species identification, or gene function prediction.

Data integration tools such as MicrobiomeAnalyst, MetaGENassist, and Mian allow multi-level analysis of microbial communities, providing statistics, visualisations, and functional insights from amplicon and shotgun data. Advanced systems combine DNA, transcripts, proteins, and metabolites. Challenges remain due to massive data, microbial diversity, sequencing errors, incomplete reference databases, and inconsistent analysis practices, which can bias gene prediction and hinder cross-study comparisons (Bharti and Grimm 2019).

1.4 DIVERSITY AND FUNCTIONAL ROLES OF LEGUME-ASSOCIATED MICROBIOMES

Legume-associated microbes also include bacteria, archaea, fungi, and actinobacteria, but most are bacteria, especially from the classes (or phyla) of Proteobacteria, Actinobacteria, Firmicutes, and Bacteroidetes. Root nodules contain clusters of both symbiotic and nitrogen-fixing microorganisms. Primary symbiotic rhizobia are group- specific – for example, *Ensifer* (*Sinorhizobium*) with *alfalfa* and *Medicago*, *Bradyrhizobium* and *Ensifer* with soybeans, and *Mesorhizobium* with chickpeas. Apart from atmospheric N_2 fixation, other internal *Pseudomonas*, *Bacillus*, and *Niastella*, *Shewanella* and co-acting microbes also enhance the host plant's health (Hassen et al. 2025). Archaea, even in low numbers, aid nutrient transformations in the rhizosphere, especially near legumes. Fungi include mycorrhizal species for phosphorus uptake, pathogens affecting crops, and saprophytes decomposing organic matter. Table 1.3 shows microbes associated with major legumes, highlighting nitrogen-fixers and beneficial colonisers, with their roles in nitrogen fixation, nutrient uptake, and stress tolerance

Disease-suppressing genes produce antimicrobials, enzymes, and siderophores to protect legumes, while growth-promoting genes make phytohormones that boost roots and nutrient uptake (Timofeeva, Galyamova, and Sedykh 2023). Metagenomics reveals genes driving legume–microbe functions. Nitrogen fixation genes such as nifH track microbial nitrogenase activity, while phosphorus solubilisation genes produce enzymes and acids to make phosphate available. Stress-related genes, including otsA and Usp, enhance microbial and plant resilience. Disease-suppressing genes

TABLE 1.3
Diversity of Microbial Taxa Associated with Major Legume Species

Legume Species	Dominant Symbionts	Other Endophytes/ Rhizosphere Taxa	Functional Roles
Soybean (*Glycine max*)	*Bradyrhizobium*, *Ensifer*	*Pseudomonas*, *Bacillus*	Nitrogen fixation, pathogen suppression
Chickpea (*Cicer arietinum*)	*Mesorhizobium cicero*	*Bacillus*, *Niastella*	Phosphorus solubilisation, growth promotion
Lentil (*Lens culinaris*)	*Rhizobium leguminosarum*	*Burkholderia*, *Streptomyces*	Disease resistance, stress tolerance
Groundnut (*Arachis hypogaea*)	*Bradyrhizobium* spp.	*Shewanella*, *Enterobacter*	Co-inoculation benefits, growth promotion

TABLE 1.4
Functional Genes Identified in Legume-Associated Metagenomes

Functional Category	Representative Genes	Function	Example Microbes	Agricultural Relevance	Reference
Nitrogen Fixation	*nifH*, *nifD*, *nifK*	Nitrogenase complex	*Bradyrhizobium*, *Mesorhizobium*	Biological nitrogen fixation	(Timofeeva, Galyamova, and Sedykh 2023)
Phosphate Solubilisation	*gcd*, *pqqC*	Gluconic acid production	*Pseudomonas*, *Bacillus*	Enhanced P uptake	(Timofeeva, Galyamova, and Sedykh 2023)
Stress Tolerance	*otsA*, *Usp*, *Auc*	Osmoprotection, drought tolerance	*Rhizobium*, *Ensifer*	Abiotic stress adaptation	(Timofeeva, Galyamova, and Sedykh 2023)
Disease Suppression	*sid*, *prn*, *phlD*	Antimicrobial and siderophore production	*Pseudomonas*, *Bacillus*	Biocontrol of pathogens	(Timofeeva, Galyamova, and Sedykh 2023)

produce antimicrobials and enzymes, and growth-promoting genes generate phytohormones that boost root growth and nutrient uptake (Timofeeva, Galyamova, and Sedykh 2023). Table 1.4 shows the major functional genes in legume-associated metagenomes and their roles in plant growth, nutrient uptake, and stress tolerance.

Lentil genotypes have both shared and genotype-specific bacterial communities, with some landraces showing greater fungal disease tolerance. Groundnut nodules host diverse endophytes, and co-inoculation with rhizobia and passenger bacteria boosts growth. Soybeans interact with Bradyrhizobium and Ensifer, guided by root flavonoids that trigger nodule formation. These findings highlight the value of crop-specific microbiomes for legume improvement (Brown et al. 2020). Chickpeas host diverse microbes, including Mesorhizobium ciceri. Higher rhizobia levels in soil correlate with greater flower and fruit production, linking microbial community structure to plant productivity (Brescia et al. 2023). In lentils we see a stable set of prokaryotes, which is true across many types of the plant, but also each variety has its own specific bacteria which also traditional strains tend to be more resistant to fungal infection than modern ones. *Arachis hypogaea* is home to many bacteria in its nodules, which include primary and secondary players. We see that the use of local rhizobia in combination with helper strains outperforms the use of rhizobia by itself. Also, we note that when it comes to improving legume crop performance, what works best is a very specific microbial treatment.

1.5 CULTIVAR-SPECIFIC MICROBIAL RECRUITMENT AND HOST–MICROBE INTERACTIONS

Plant genotypes shape microbial selection through root exudates, which vary by species, genotype, and development stage, promoting the growth of specific microbial taxa (Upadhyay et al. 2022). Legume genetic diversity shapes nodule microbiomes, with endophyte composition and gene functions varying by genotype despite similar diversity metrics. Studies in Medicago show soil primarily shapes the rhizosphere, while internal compartments reflect host genetics, selectively enriching beneficial microbes from bulk soil (Burns, Epstein, and Burghardt 2021). Root exudates, including flavonoids, organic acids, sugars, and amino acids, shape microbial activity. In legumes, flavonoids trigger rhizobia nodulation factors, enhancing mutualism. Soil properties such as pH, nutrients, and texture interact with plant traits to determine microbiome composition. In studies of certain plant types, we see that genetics play a role in microbiome change. In Cucurbita pepo, we note the development of distinct microbial communities within the roots and in the soil that which surrounds them (Gopal and Gupta 2016). Heirloom and disease-resistant varieties produce distinct root exudates that shape bacterial communities, while fungal responses are also influenced but less predictable (Anderson et al. 2024). In *Medicago* and *alfalfa*, plant genetics influence internal microbes, nutrient- and stress-related genes, and soil microbial communities, highlighting the potential of combining breeding with tailored microbiomes to develop climate-resilient legumes (Wu et al. 2023).

1.6 FUNCTIONAL METAGENOMICS: DISCOVERY OF NOVEL GENES AND PATHWAYS

Expression-based screens in hosts such as *E. coli* detect microbial activities, while sequence-based methods use conserved primers to find biosynthetic genes. Combining both approaches leverages their strengths and overcomes their limitations for innovative metagenomic discovery. Functional metagenomics seeks to discover novel genes that can be utilised in agriculture by employing diverse methodologies to access the natural environment. In the construction of gene libraries in foreign hosts, such as *Escherichia coli* or *R. metallidurans*, scientists use function-based methodologies to screen for traits, such as the production of certain enzymes or antibiotics. DNA-targeted approaches capture biosynthetic genes via conserved primers, though PCR biases remain. Combining functional assays with sequencing improves discovery, enabling eco-friendly farming advances, including enzymes such as glycoside hydrolases, cold-active lipases, esterases, and oxidoreductases (Gonçalves, Creevey, and Santana 2023). While some support biofuel development, others aid pollution cleanup or industrial processes. Instead of chemical pesticides, natural antimicrobials and cell-lysing proteins from microbial DNA may serve as biological controls (Iqbal, Craig, and Brady 2014). Microbial stress-relief compounds help plants tolerate harsh conditions. Functional metagenomics identifies microbes for biofertilisers and biopesticides, enabling nitrogen fixation, phosphate solubilisation, and hormone production. Combining metagenomics, screening, and synthetic biology improves biofertiliser performance, boosting yields and reducing chemical use, while biopesticides target pathogens via antibiosis, competition, and induced resistance, optimised with smart farming tools (Neeraj et al. 2024). Table 1.5 shows the key applications of functional metagenomics in sustainable agriculture, including enzyme discovery, biocontrol agents, biofertiliser development, and stress-resilient crops, along with their microbial sources and practical outcomes.

TABLE 1.5
Applications of Functional Metagenomics in Sustainable Agriculture

Application Area	Key Discoveries	Microbial Source	Practical Outcome
Enzyme Discovery	Glycoside hydrolases, lipases	Soil DNA libraries	Biomass conversion, biofuel production
Biocontrol Agents	Antimicrobial enzymes	Rhizosphere metagenomes	Natural pesticide development
Biofertiliser Development	Nitrogen-fixing and P-solubilising genes	Rhizobial consortia	Reduced chemical fertiliser use
Stress-Resilient Crops	Genes for osmoprotectants, antioxidants	Plant-associated microbes	Enhanced drought and heat tolerance

1.7 INTEGRATING METAGENOMICS WITH BREEDING AND BIOTECHNOLOGY

Microbiome-assisted breeding integrates plant-associated microbes into conventional and modern breeding, treating plants and microbiomes as co-evolving holobionts. Understanding these interactions helps develop crops that thrive through combined genetic and microbial strengths. (Kusstatscher et al. 2021). Rather than relying solely on DNA markers or GWAS, new methods track which plant types naturally attract positive microbes. Strategies for the co-selection of plant traits and beneficial microbe partners use many complementary approaches. In the case of reciprocal soil transplantation, we see that it is a method to study plant genotype performance in different soil microbiome settings, which in turn identifies genotypes that do well and also attract beneficial microbes in a wide range of environments (Chen et al., 2025). Artificial ecosystem selection uses synthetic microbiomes to enhance plant–microbe compatibility, improving plant performance. Genetic screens identify microbes influencing traits, and models incorporating genes, microbiomes, and environment guide in silico design of optimal plant–microbiome pairs before field testing (Cui et al. 2025). Combining breeding techniques improves legumes. In chickpeas, varieties retaining more soil rhizobia produced more flowers and fruits, while in maize, integrating microbes with genomic predictions enhanced biomass forecasting. Depending on the nitrogen, *Massilia putida* and *Pseudomonas parafulva* were fixed. In the sorghum-based synthetic community, the defined microbial communities positively affected the plant traits in the field and in the controlled environments. There was evidence of changes in the pathways of lignin biosynthesis, and microbial inoculation stimulated the pathways of plant immunity. These examples give evidence of practical crop breeding that exploit plant–microbe interactions. Metagenomics enables precision agriculture by linking soil microbial genetics with satellite and on-site data. Farmers can tailor practices adding microbes or adjusting water and nutrients based on real-time microbial activity, optimising crops according to local soil communities (Ambaru, Manvitha, and Madas 2025).

1.8 CHALLENGES AND LIMITATIONS

Metagenomic breeding faces computational challenges: massive, diverse datasets require deep sequencing; short, error-prone reads complicate assembly; memory limits hinder genome reconstruction; and lack of standardised workflows, metadata, and testing protocols hampers tool comparison and cross-study analyses. Field performance of microbial inoculants is highly variable. Native soil microbes often outperform commercial strains, and lab results rarely match field outcomes due to soil, environmental factors, and complex microbial communities, highlighting the need for location-specific solutions (Hamed, Azarbad, and Junker 2024). Factors that are out of your control will delay implementing this information in agriculture. The information that could be obtained from sequencing data has the potential to be of great use, whether that be from taxonomy. The problem lies in the fact that without an active inference, we are not able to determine which of these microbes are present in or expressing the genes we find in the soil. In most of the databases we are using to match sequences, many microbes are either absent or misidentified altogether due to the lack of representation of less studied microbes, particularly environmental isolates. The insufficient data leads to the recruitment of many genes having a hypothetical function. The missing function is a result of similarity predictions over experimentation. (Maranga et al. 2023). Linking plants, microbes, and environment requires integrating sequencing with field data, transcriptomics, proteomics, and metabolomics, but these resource-intensive approaches remain technically challenging. Figure 1.3 shows the current challenges and emerging strategies shaping the future of legume metagenomics research.

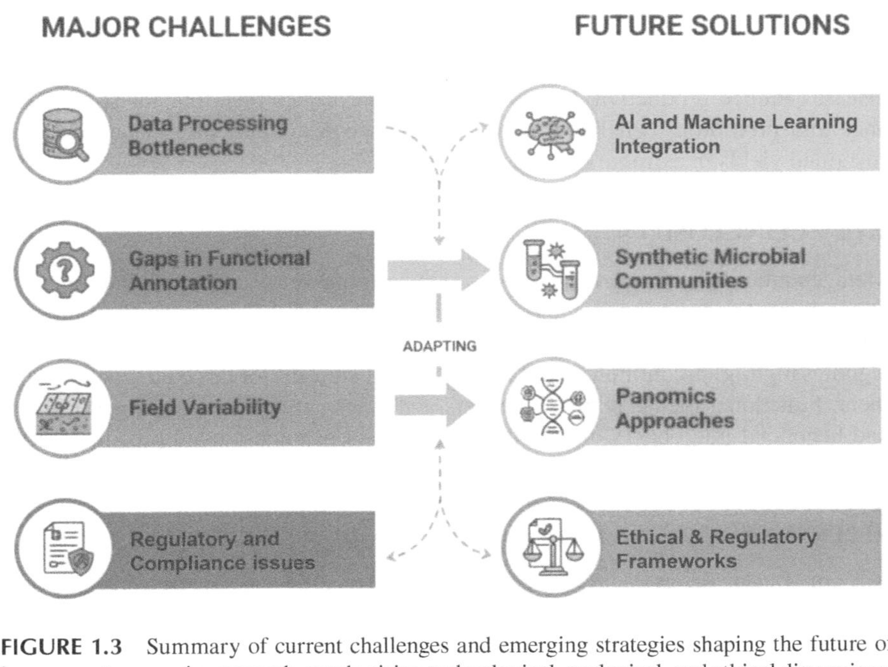

FIGURE 1.3 Summary of current challenges and emerging strategies shaping the future of legume metagenomics research, emphasising technological, ecological, and ethical dimensions.

Ethics, ecology, and regulations are crucial in microbiome design. Releasing engineered microbes may involve multiple authorities and poses ecological risks, such as loss of native microbes and horizontal gene transfer, which remain poorly understood (Thakur et al. 2023). Questions remain as to who owns what rights to microbiome tools, and this also makes it difficult to sustain supportive creativity and inequitable access to beneficial developments. Engaging the public, explaining the role of engineered microbes in food systems, helps gain their trust and reflect on the common purpose of food production.

1.9 FUTURE DIRECTIONS

Synthetic microbial communities (SynComs) enhance crop growth using a reduced but relevant set of species (Yamazaki et al. 2025). Top-down approaches survey microbial roles via genetics, while bottom-up assembly tests individual strains for beneficial traits. SynComs alter root-zone communities, promoting helpful species and stabilising ecosystems (Kaur et al. 2022). Genome-based metabolic simulations design small-scale communities that deliver core plant benefits. Combining metagenomics with panomics and systems biology allows multi-level analysis from genomics to metabolomics, identifying key microbes for drought resilience, nitrogen cycling, and yield (Ghatak et al. 2022). Functional gene tags linked to pangenomes reveal genetic commonalities across strains, clarifying microbe roles (Kimotho and Maina 2023). AI and machine learning, including autoencoders, CNNs, RNNs, random forests, SVMs, and boosted trees, improve DNA sorting, gene function annotation, species identification, and predict plant performance by integrating microbial, plant, and climate data, guiding model-based crop improvement (Roy et al. 2024).

Improving legumes under climate change requires combining genetics, cultivation, and microbiome management. Droughts, floods, temperature extremes, and diseases reduce productivity, but synergistic approaches breeding for stress tolerance and promoting beneficial microbes enhance resilience, nutrient uptake, and sustained yields in extreme climates

1.10 CONCLUSION

Metagenomics helps improve legume growth sustainably by analysing plant–microbe interactions. DNA sequencing and bioinformatics identify genes involved in nitrogen fixation, phosphorus use, stress tolerance, and disease resistance. Research across legumes highlights common and unique microbial patterns for targeted crop improvement. Functional metagenomics uncovers genes for natural fertilisers, pest control, and beneficial microbes. Combining this data with genome-guided breeding, precision farming, and AI supports resilient, low-input crops and eco-friendly cultivation.

REFERENCES

Ambaru, Bindu, Reena Manvitha, and Rajini Madas. 2025. "Synergistic Integration of Remote Sensing and Soil Metagenomics Data: Advancing Precision Agriculture through Interdisciplinary Approaches." *Frontiers in Sustainable Food Systems* 8 (January). Frontiers Media SA. doi: https://doi.org/10.3389/fsufs.2024.1499973.

Anderson, Hannah M., Grace A. Cagle, Erica L.-W. Majumder, Erin Silva, and Zachary B. Freedman. 2024. "Root Exudation and Rhizosphere Microbial Recruitment Are Influenced by Novel Plant Trait Diversity in Carrot Genotypes." March 12. doi: https://doi.org/10.1101/2024.03.12.583384.

Anjum, Naser A. 2016. "Book Review: Legumes under Environmental Stress: Yield, Improvement and Adaptations." *Frontiers in Plant Science* 7 (June). Frontiers Media SA. doi: https://doi.org/10.3389/fpls.2016.00798.

Bharti, Richa, and Dominik G. Grimm. 2019. "Current Challenges and Best-Practice Protocols for Microbiome Analysis." *Briefings in Bioinformatics* 22 (1). Oxford University Press (OUP): 178–93. doi: https://doi.org/10.1093/bib/bbz155.

Brescia, Francesca, Fabiano Sillo, Elisabetta Franchi, Ilaria Pietrini, Vincenzo Montesano, Giovanni Marino, Matthew Haworth, et al. 2023. "The 'Microbiome Counterattack': Insights on the Soil and Root-Associated Microbiome in Diverse Chickpea and Lentil Genotypes after an Erratic Rainfall Event." *Environmental Microbiology Reports* 15 (6). Wiley: 459–83. doi: https://doi.org/10.1111/1758-2229.13167.

Brown, Shawn P., Michael A. Grillo, Justin C. Podowski, and Katy D. Heath. 2020. "Soil Origin and Plant Genotype Structure Distinct Microbiome Compartments in the Model Legume Medicago Truncatula." *Microbiome* 8 (1). Springer Science and Business Media LLC. doi: https://doi.org/10.1186/s40168-020-00915-9.

Burns, Mannix, Brendan Epstein, and Liana T. Burghardt. 2021. "Comparison of Nodule Endophyte Composition, Diversity, and Gene Content between *Medicago Truncatula* Genotypes." *Phytobiomes Journal* 5 (4). American Phytopathological Society: 400–407. doi: https://doi.org/10.1094/pbiomes-10-20-0077-r.

Chen, W., B. Zhu, C. Wei, Y. He, F. Fan, X. Liu, J. Yang, E. Siemann, and X. Lu. 2025. "Soil Transplantation: Complex Interplay Between Soil Microbes and the Local Environment." *New Phytologist* 247 (6): 2903–2913. doi: https://doi.org/10.1111/nph.70393.

Cui, Yunhe, Yue Tao, Yang Luo, Rui Sun, Siyue Han, Yang Yang, Zhongmin Dai, and Ying Zhang. 2025. "Harnessing Synthetic Microbial Communities for Sustainable Agriculture: Enhancing Soil Health and Crop Yields." *Critical Reviews in Environmental Science and Technology* 55 (19). Taylor & Francis: 1479–505. doi: https://doi.org/10.1080/10643389.2025.2543793.

Dong, Xiaoli, and Marc Strous. 2019. "An Integrated Pipeline for Annotation and Visualization of Metagenomic Contigs." *Frontiers in Genetics* 10 (October). Frontiers Media: 999. doi: https://doi.org/10.3389/fgene.2019.00999.

Ghatak, Arindam, Palak Chaturvedi, Steffen Waldherr, Guntur Venkata Subbarao, and Wolfram Weckwerth. 2022. "PANOMICS at the Interface of Root–Soil Microbiome and BNI." *Trends in Plant Science* 28 (1). Elsevier BV: 106–22. doi: https://doi.org/10.1016/j.tplants.2022.08.016.

Gökdemir, Fatma Şeyma, Özlem Darcansoy İşeri, Abhishek Sharma, Premila N. Achar, and Füsun Eyidoğan. 2022. "Metagenomics Next Generation Sequencing (MNGS): An Exciting Tool for Early and Accurate Diagnosis of Fungal Pathogens in Plants." *Journal of Fungi* 8 (11). Multidisciplinary Digital Publishing Institute: 1195. doi: https://doi.org/10.3390/jof8111195.

Gonçalves, Osiel S., Christopher J. Creevey, and Mateus F. Santana. 2023. "Designing a Synthetic Microbial Community through Genome Metabolic Modeling to Enhance Plant–Microbe Interaction." *Environmental Microbiome* 18 (1). BioMed Central: 81. doi: https://doi.org/10.1186/s40793-023-00536-3.

Gopal, Murali, and Alka Gupta. 2016. "Microbiome Selection Could Spur Next-Generation Plant Breeding Strategies." *Frontiers in Microbiology* 7 (December). Frontiers Media SA. doi: https://doi.org/10.3389/fmicb.2016.01971.

Hamed, Azarbad, and Robert R. Junker. 2024. "Biological and Experimental Factors That Define the Effectiveness of Microbial Inoculation on Plant Traits: A Meta-Analysis." *ISME Communications* 4 (1). Springer Nature: ycae122. doi: https://doi.org/10.1093/ismeco/ycae122.

Hassen, Ahmed Idris, Esther K. Muema, Mamonokane O. Diale, Tiisetso Mpai, and Francina L. Bopape. 2025. "Non-Rhizobial Endophytes (NREs) of the Nodule Microbiome Have Synergistic Roles in Beneficial Tripartite Plant–Microbe Interactions." *Microorganisms* 13 (3). MDPI AG: 518. doi: https://doi.org/10.3390/microorganisms13030518.

Iqbal, Hala A., Jeffrey W. Craig, and Sean F. Brady. 2014. "Antibacterial Enzymes from the Functional Screening of Metagenomic Libraries Hosted in*Ralstonia Metallidurans*." *FEMS Microbiology Letters* 354 (1). Oxford University Press (OUP): 19–26. doi: https://doi.org/10.1111/1574-6968.12431.

Iquebal, M. A., J. Jagannadham, S. Jaiswal, R. Prabha, A. Rai, and D. Kumar. 2022. "Potential Use of Microbial Community Genomes in Various Dimensions of Agriculture Productivity and Its Management: A Review." *Frontiers in Microbiology* 13, 2–7. doi: https://doi.org/10.3389/fmicb.2022.708335.

Jha, Uday Chand, Manu Priya, Yogesh Dashrath Naik, Harsh Nayyar, Mahendar Thudi, Somashekhar M. Punnuri, Kadambot H.M. Siddique, and P.V. Vara Prasad. 2024. "Major Abiotic Stresses on Quality Parameters in Grain Legumes: Impacts and Various Strategies for Improving Quality Traits." *Environmental and Experimental Botany* 228 (December). Elsevier BV: 105978. doi: https://doi.org/10.1016/j.envexpbot.2024.105978.

Kaur, Simranjit, Eleonora Egidi, Zhiguang Qiu, Catriona A Macdonald, Jay Prakash Verma, Pankaj Trivedi, Juntao Wang, Hongwei Liu, and Brajesh K Singh. 2022. "Synthetic Community Improves Crop Performance and Alters Rhizosphere Microbial Communities." *Journal of Sustainable Agriculture and Environment* 1 (2). Wiley: 118–31. doi: https://doi.org/10.1002/sae2.12017.

Kebede, Erana. 2021. "Contribution, Utilization, and Improvement of Legumes-Driven Biological Nitrogen Fixation in Agricultural Systems." *Frontiers in Sustainable Food Systems* 5 (November). Frontiers Media SA. doi: https://doi.org/10.3389/fsufs.2021.767998.

Kimotho, Roy Njoroge, and Solomon Maina. 2023. "Unraveling Plant–Microbe Interactions: Can Integrated Omics Approaches Offer Concrete Answers?" *Journal of Experimental Botany* 75 (5). Oxford University Press: 1289–313. doi: https://doi.org/10.1093/jxb/erad448.

Kusstatscher, Peter, Eveline Adam, Wisnu Adi Wicaksono, Maria Bernhart, Expedito Olimi, Henry Müller, and Gabriele Berg. 2021. "Microbiome-Assisted Breeding to Understand Cultivar-Dependent Assembly in Cucurbita Pepo." *Frontiers in Plant Science* 12 (April). Frontiers Media: 642027. doi: https://doi.org/10.3389/fpls.2021.642027.

Levy, Michael A., and Mark N. Lubell. 2017. "Innovation, Cooperation, and the Structure of Three Regional Sustainable Agriculture Networks in California." *Regional Environmental Change* 18 (4). Springer Science and Business Media LLC: 1235–46. doi: https://doi.org/10.1007/s10113-017-1258-6.

Maranga, Mary, Pawel Szczerbiak, Valentyn Bezshapkin, Vladimir Gligorijevic, Chris Chandler, Richard Bonneau, Ramnik J. Xavier, Tommi Vatanen, and Tomasz Kosciolek. 2023. "Comprehensive Functional Annotation of Metagenomes and Microbial Genomes Using a Deep Learning-based Method." *MSystems* 8 (2). American Society for Microbiology. doi: https://doi.org/10.1128/msystems.01178-22.

Neeraj, Neha Singh, Grace Angel, Nishant, Neeta Shweta Kerketta, Rashmi Mohapatra, V. Siva Jyothi, and A. K. Bhandari. 2024. "Developing Biofertilizers for Sustainable Agriculture: A Metagenomic Approach." *International Journal of Research in Agronomy* 7 (12): 993–97. doi: https://doi.org/10.33545/2618060x.2024.v7.i12l.2410.

Nwachukwu, B. C., and O. O. Babalola. 2022. "Metagenomics: A Tool for Exploring Key Microbiome With the Potentials for Improving Sustainable Agriculture." *Frontiers in Sustainable Food Systems* 6:886987, 1–3. https://doi.org/10.3389/fsufs.2022.886987.

Roy, Gaspar, Edi Prifti, Eugeni Belda, and Jean-Daniel Zucker. 2024. "Deep Learning Methods in Metagenomics: A Review." *Microbial Genomics* 10 (4). Microbiology Society. doi: https://doi.org/10.1099/mgen.0.001231.

Schaedel, M., G. Hidrobo, and J. Grossman. 2021. "From Microns to Meters: Exploring Advances in Legume Microbiome Diversity for Agroecosystem Benefits." *Frontiers in Sustainable Food Systems* 5:668195, 1–3. https://doi.org/10.3389/fsufs.2021.668195.

Thakur, Nitika, Mohit Nigam, Neharika Ann Mann, Shivendra Gupta, Chaudhery Mustansar Hussain, Sudheesh K. Shukla, Anis Ali Shah, Ryan Casini, Hosam O. Elansary, and Sher Aslam Khan. 2023. "Host-Mediated Gene Engineering and Microbiome-based Technology Optimization for Sustainable Agriculture and Environment." *Functional & Integrative Genomics* 23 (1). Springer Science and Business Media LLC. doi: https://doi.org/10.1007/s10142-023-00982-9.

Timofeeva, Anna M., Maria R. Galyamova, and Sergey E. Sedykh. 2023. "Plant Growth-Promoting Soil Bacteria: Nitrogen Fixation, Phosphate Solubilization, Siderophore Production, and Other Biological Activities." *Plants* 12 (24). Multidisciplinary Digital Publishing Institute: 4074. doi: https://doi.org/10.3390/plants12244074.

Upadhyay, Sudhir K., Abhishek K. Srivastava, Vishnu D. Rajput, Prabhat K. Chauhan, Ali Asger Bhojiya, Devendra Jain, Gyaneshwer Chaubey, Padmanabh Dwivedi, Bechan Sharma, and Tatiana Minkina. 2022. "Root Exudates: Mechanistic Insight of Plant Growth Promoting Rhizobacteria for Sustainable Crop Production." *Frontiers in Microbiology* 13 (July). Frontiers Media: 916488. doi: https://doi.org/10.3389/fmicb.2022.916488.

Wu, Linkun, Leslie A. Weston, Shusheng Zhu, and Xingang Zhou. 2023. "Editorial: Rhizosphere Interactions: Root Exudates and the Rhizosphere Microbiome." *Frontiers in Plant Science* 14 (September). Frontiers Media: 1281010. doi: https://doi.org/10.3389/fpls.2023.1281010.

Xiao, X., W. Chen, L. Zong, J. Yang, S. Jiao, Y. Lin, E. Wang, and G. Wei. 2017. "Two Cultivated Legume Plants Reveal the Enrichment Process of the Microbiome in the Rhizocompartments." *Molecular Ecology* 26 (6): 1641–1651. doi: https://doi.org/10.1111/mec.14027.

Yamazaki, S., M. Nakayasu, K. Kanai, R. Mizuno, R. Kaida, S. Masuda, A. Shibata, K. Shirasu, A. J. Nagano, Y. Fujii, A. Sugiyama, and Y. Aoki. 2025. "From Model to Field: Predictive Design of Microbial Communities for Resilient Crop Performance." *bioRxiv* 7–8. https://doi.org/10.1101/2025.10.21.683643.

2 Metagenomics Approaches in Legume Biology

Ankit Singh Negi, Sujeet Pratap Singh, and Ruchi Yadav

2.1 INTRODUCTION

Legumes, referring to the plants or the fruits/seeds of species belonging to the Fabaceae (Leguminosae) family (Chatzivassiliou, 2021), including ~800 genera and 20,000 species, were recently classified into six subfamilies (Egan et al., 2019; Stagnari et al., 2017). They form symbioses with nitrogen-fixing bacteria (Rhizobium, Bradyrhizobium, Sinorhizobium, Mesorhizobium), reducing fertiliser use and enhancing soil health. Legumes also host diverse microbiota – bacteria, fungi, archaea, and viruses – that aid nutrient uptake, stress tolerance, and disease resistance (Rana et al., 2023; Wahab et al., 2023). Metagenomics, leveraging next-generation sequencing (NGS), enables culture-independent profiling of microbial communities, including rhizosphere and endosphere microbiomes, uncovering taxonomic and functional genes related to nitrogen fixation, phosphorus cycling, pathogen suppression, and stress resilience (Usyk et al., 2023; Jagadesh et al., 2024; Sivaprakasam et al., 2024). Amplicon-based approaches target marker genes, while shotgun sequencing provides broader functional insights. Advanced bioinformatics reconstructs metabolic pathways, identifies novel genes, and links microbial diversity to ecosystem functions. These insights support microbiome-informed innovations – biofertilisers, biocontrol agents, and synthetic consortia – enabling sustainable legume cultivation and improved agricultural productivity

2.2 METAGENOMIC ANALYSIS–BASED WORKFLOW REPRESENTATION FOR LEGUME RESEARCH

Metagenomic research has as its first step the collection of high-quality environmental samples, which is done in a very careful manner, as the properness and the representativeness of the sample have a direct impact on the accuracy of the microbial community analysis conducted further. In studying legumes, commonly the sources of samples are rhizosphere soil, root nodules, phyllosphere, and seeds, from an ecosystem standpoint, each of these sources has its own unique microbial community, as illustrated in Figure 2.1.

DOI: 10.1201/9781003733324-3

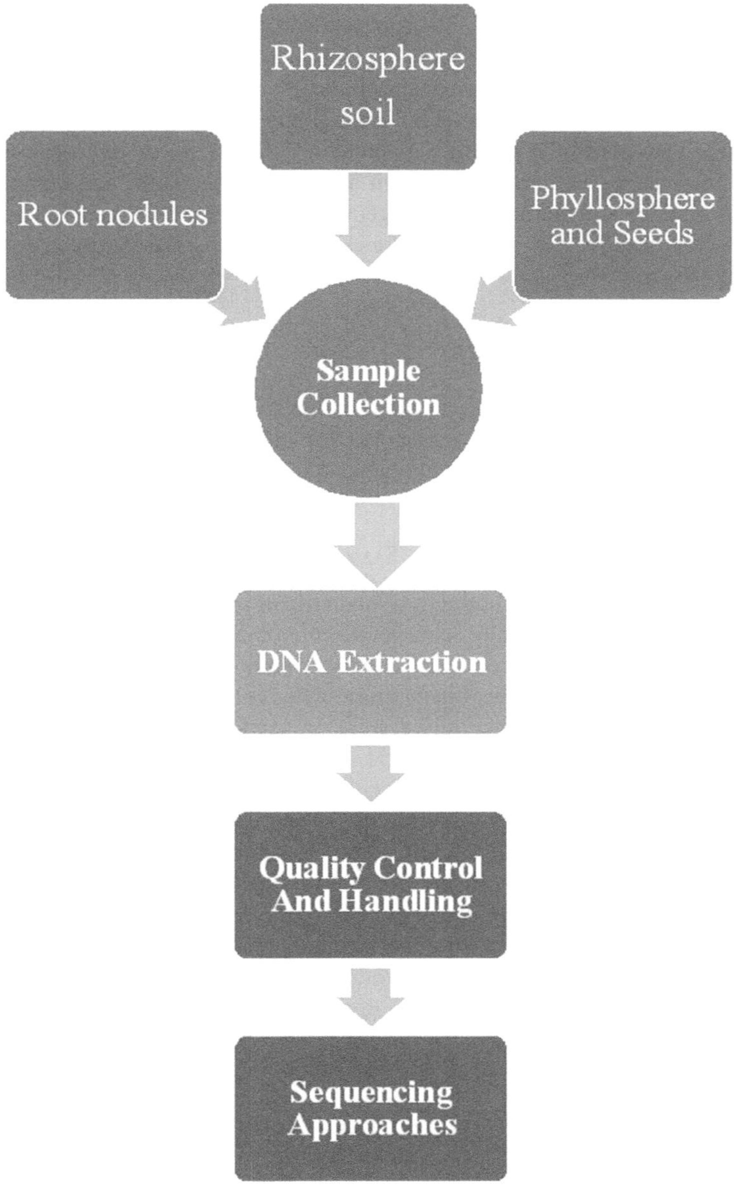

FIGURE 2.1 The linear and sequential wet-lab methods applied in legume metagenomic studies, inclusive of sample collection, DNA extraction, library preparation, quality control and handling and sequencing approaches.

2.2.1 Rhizosphere Soil Samples

The rhizosphere is a layer of soil that is composed of microorganisms and is the thickest around plant roots. Those microorganisms carry out critical methods, such as nutrient biking, making vegetation lives simpler, and offering them resilience to stresses. The vicinity is generally populated with the aid of symbiotic nitrogen-solving bacteria, particularly Rhizobium and Bradyrhizobium, and different bacteria that make a contribution to the vegetation's well-being in specific approaches together with the aid of generating hormones, making phosphorus reachable, or combating illnesses.

2.2.2 Root Nodule Samples

The structures are particular to the plant and develop during symbiosis, and they serve as a completely unique microhabitat for nitrogen-fixing microorganisms in order that direct observation of the energetic symbionts can be initiated.

2.2.3 The Phyllosphere and Seed Sample

The phyllosphere is the aerial surfaces of plant life or seeds above the floor, and those aerial surfaces may additionally incorporate microbial communities, which can be recognised to affect plant health, resistance to the pathogen, and vertical transmission of beneficial microbes to the next generation. After collection of samples, high-yield extraction of quality DNA is required so as to carry out successful metagenomic profiling. Quality DNA means high yield, intact, and free from any contaminants. Major contaminants are humic acids and secondary metabolites, since their presence will inhibit the reactions of polymerase chain amplification as well as sequencing reactions. When getting the DNA out, think about a few things: don't break it up too much by machines, keep other stuff from mixing in, and watch how cold-chain handling can make the DNA go bad or get dirty with a tilt towards any later looks that might be amplicon sequencing, shotgun metagenomics, or functional metagenomic work. Also, write down where you got your samples from, which site, what stage of plant growth, and what the weather was like to help with strong comparison checks and doing it again the right way.

2.3 METAGENOMICS SEQUENCING APPROACHES

Metagenomic studies are meant to throw some light on the composition, characteristics, and behaviour of microbial communities from a diversity of environments such as soil, rhizosphere, plant tissues, and many ecological niches. The choice of sequencing method largely depends on the purpose kept in mind – whether it is taxonomic profiling, functional characterisation, or an attempt at understanding microbial activity. On a broader level, metagenomic sequencing falls under amplicon sequencing, shotgun metagenomics, and metatranscriptomics – each bringing with them sure advantages plus views.

2.3.1 Amplicon Sequencing (16S/ITS/18S rRNA)

Amplicon sequencing is the primary molecular technique not only to analyse the most conserved genetic markers in microbial genomes but also to do the detailed taxonomic profiling and even the assessment of microbial diversity. The 16S ribosomal RNA (rRNA) gene for bacteria (Hugerth et al., 2017), internal transcribed spacer (ITS) regions for fungi, and 18S rRNA for other eukaryotes are the most common target markers (National Research Council et al., 2007). The procedure typically involves the PCR amplification of the designated marker genes followed by library preparation and the high-throughput sequencing on platforms such as Illumina MiSeq or NovaSeq (Cruaud et al., 2017). The bioinformatics analysis afterwards includes the quality control of the raw sequences, clustering into operational taxonomic units (OTUs) or amplicon sequence variants (ASVs) (Bharti et al., 2021), and taxonomic classification against the curated reference databases SILVA, Greengenes, or UNITE. This method allows for microbial community composition and structure insights to be obtained at high resolution, thus making it particularly useful for comparative research studies involving different sampling, environmental conditions, or treatment groups. Amplicon sequencing, however, is still limited because it only allows taxonomic analysis, which is a major drawback, since the whole potential of functional traits cannot be assessed, nor can the genes beyond the targeted regions be detected. Nevertheless, it continues to be a popular, cost-effective, and reliable method for examining microbial diversity, ecological interactions, and community dynamics shifts in various environments, such as the soil, plant-associated microbiomes, and even clinical samples.

2.3.2 Shotgun Metagenomics

Shotgun metagenomics sequences all DNA in a sample, capturing bacteria, archaea, fungi, viruses, and other microbes. Unlike targeted amplicon sequencing, it allows simultaneous taxonomic and functional profiling, genome reconstruction, novel gene discovery, and metabolic pathway analysis. Using high-throughput platforms (Illumina, Nanopore) and tools such as MEGAHIT, MetaPhlAn, and HUMAnN, it reveals microbial interactions, rare or unculturable species, and ecosystem dynamics, providing a holistic view of microbiomes despite high resource demands

2.3.3 Metatranscriptomics

Metatranscriptomics shifts focus from genetic potential to active gene expression by sequencing total RNA from environmental or host-associated samples (Butowski et al., 2025). It reveals microbial metabolism, stress responses, and ecological interactions in real time. After RNA extraction, rRNA removal, cDNA synthesis, and library prep, high-throughput sequencing (commonly Illumina) is followed by quality control, read mapping, differential expression, and functional annotation (KEGG, COG, EggNOG). This approach captures active microbial functions, community responses, and adaptations, providing insights into plant–microbe interactions, microbial ecology, and ecosystem processes

2.4 BIOINFORMATICS ANALYSIS OF SEQUENCING DATA

Once the sequencing run is done, bioinformatics analysis has to be robust enough to convert raw data into biological information. The computational workflow, described in Figure 2.2, involves several consecutive steps that would help in proper understanding of microbial community composition and their functional potential as well as ecological significance. Normally, bioinformatics analysis contains the following steps: quality control, assembly and annotation, taxonomic profiling, and functional characterisation.

Bioinformatics analysis begins with strict quality control of raw reads to remove low-quality bases, adapters, and contaminants using tools like FastQC (Wingett et al., 2018) and Trimmomatic or Trim Galore! (Bolger et al., 2014; Krueger, 2015). High-quality reads are then assembled into contigs with MEGAHIT or MetaSPAdes and annotated using Prokka to map genes and functional elements. Taxonomic profiling uses QIIME2 for amplicons or Kraken2/MetaPhlAn for shotgun data to determine community composition, diversity, and relative abundances. Functional profiling assigns genes to metabolic pathways and protein families via databases such as KEGG, COG, and eggNOG, with pipelines such as MG-RAST enabling abundance estimates and metabolic reconstruction. This workflow reveals microbial ecology, host interactions, nutrient cycling, and environmental adaptations. Table 2.1 summarises key bioinformatics tools, URLs, and their metagenomic applications.

2.5 APPLICATIONS OF METAGENOMICS IN LEGUME BIOLOGY

Metagenomics has emerged as a powerful tool to unravel the complexity of microbial communities associated with legumes, enabling researchers to move beyond traditional culture-dependent methods. By providing insights into both the taxonomic composition and functional potential of microbial communities, metagenomics facilitates a deeper understanding of plant–microbe interactions that are crucial for legume growth, productivity, and stress resilience.

2.5.1 Rhizosphere Microbiome Characterisation

Soil microbial activity in the rhizosphere is vital for plant health and nutrient uptake. Metagenomics profiles bacteria, fungi, archaea, and viruses in pulses, such as soybean, chickpea, and Medicago, revealing key taxa such as Rhizobium, Bradyrhizobium, Bacillus, and Pseudomonas. These microbes support nitrogen fixation, phosphorus solubilisation, phytohormone production, and pest suppression. Functional mapping identifies enzymes and pathways, including indole-3-acetic acid synthesis, highlighting microbial roles in root growth and nutrient availability, enabling targeted rhizosphere management for legume productivity under changing environments

2.5.2 Root Nodule Metagenome

Symbiotic N_2 fixation in legume nodules involves rhizobia and non-rhizobial endophytes with key functions. Shotgun metagenomics reveals genera like Enterobacter,

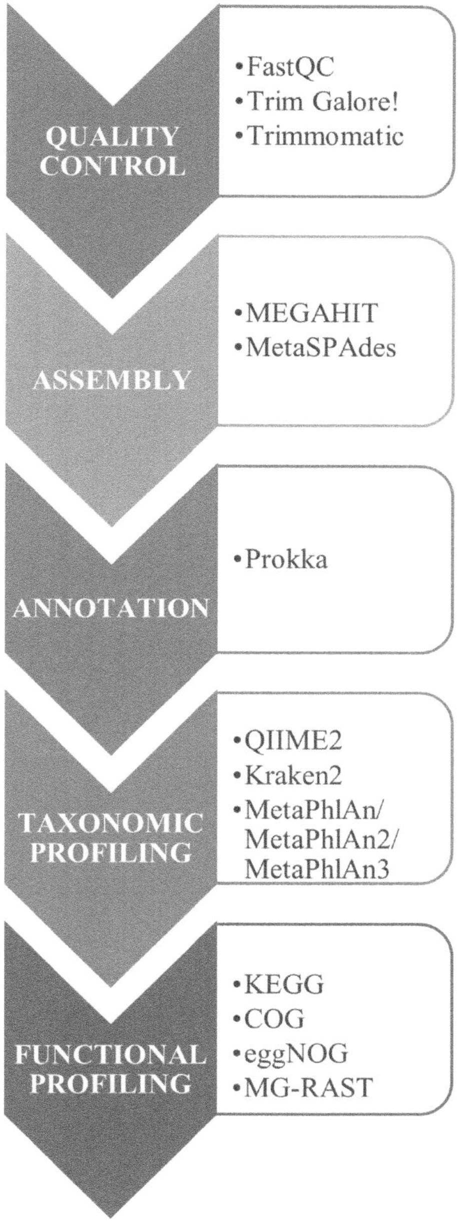

FIGURE 2.2 A schematic representation of bioinformatics tools for sequencing data analysis.

TABLE 2.1
Bioinformatics Tools and Databases for Legume Metagenomics

S.No.	Method	Tools/ Databases	URL	Description	References
1.	**Quality Control**	FastQC	https://toolshed.g2.bx.psu.edu/repository?repository_id=ca249a25748b71a3	Raw sequence data of high-throughput sequencing pipelines can be inspected for quality control using FastQC.	(Andrews, 2010)
		Trim Galore!	https://toolshed.g2.bx.psu.edu/repository?repository_id=5399c3e263b33bd5	A tool used to eliminate low quality nucleotide bases and also used for trimming the adapters.	(Krueger, 2021)
		Trimmomatic	https://toolshed.g2.bx.psu.edu/repository?repository_id=ef9e620e9ac844b3	Trimmomatic is a command-line program that trims adapters, discards low-quality bases, and deletes short reads from Illumina sequenced data	(Kumar et al., 2020)
2.	**Assembly and Annotation**	MEGAHIT	https://github.com/voutcn/megahit	A very quick and memory-efficient de novo assembler for short reads that is frequently utilised in metagenomic research to put together big, complicated sequencing datasets.	(Yang C et al., 2021)
		MetaSPAdes	https://github.com/ablab/spades	A metagenomic modification of the SPAdes assembler, MetaSPAdes is optimised to produce high-quality de novo assemblies from complicated microbial communities and was created especially for short read datasets.	(Yang C et al., 2021)
		Prokka	https://toolshed.g2.bx.psu.edu/repository?repository_id=9578f22fc7e9c8de	Prokka is a software device to rapidly annotate bacterial, archaeal and viral genomes, and bring output documents that require most effective minor tweaking to put up to GenBank/ENA/DDBJ.	(Seemann, 2014)

3.	**Taxonomic profiling**	QIIME2	https://github.com/qiime2/galaxy-tools or https://docs.qiime2.org/2024.10/install/native/ and https://docs.qiime2.org/2024.10/install/virtual/index.html	A tool that designed to analyse high-throughput microbiome marker gene amplicon sequencing data.	(Fung et al., 2021)
		Kraken2	https://toolshed.g2.bx.psu.edu/repository?repository_id=1004e042d582be95&changeset_revision=269ea4b78bbb	Kraken2 is a short and correct taxonomic classifier that assigns DNA reads to taxa via matching k-mers with a database of genomes the use of the lowest commonplace ancestor method.	(Wood et al., 2014)
		MetaPhlAn/ MetaPhlAn2/ MetaPhlAn3	https://toolshed.g2.bx.psu.edu/repository?repository_id=33e14d7c88421861	MetaPhlAn profiles microbial network composition from metagenomic shotgun records, outputting predicted taxon abundances.	(Truong et al., 2015)
4.	**Functional annotation**	KEGG	https://www.genome.jp/kegg/	High-level organic features on the cell, organismal, and surroundings levels are linked with molecular information through the KEGG database.	(Kanehisa et al., 2000)
		COG	https://www.ncbi.nlm.nih.gov/research/cog	A comparative genomics database that organisations proteins through evolutionary relationships to assign useful annotations.	(Galperin et al., 2025)
		eggNOG	https://eggnogdb.org/	The EMBL-hosted eggNOG database expands at the COGs idea by generating non-supervised orthologous groups from other taxa.	(Huerta-Cepas et al., 2016)
		MG-RAST	https://www.mg-rast.org/	It is an open web-based server for automated phylogenetic and functional metagenomic analysis.	(Field et al., 2011)

Burkholderia, Pseudomonas, and actinobacteria contributing to nutrient acquisition, phytohormone production, siderophores, ACC deaminase, and plant defence. Characterising the nodule metagenome identifies genes for N_2 fixation, iron uptake, and antimicrobial activity, guiding the design of microbial consortia to enhance symbiotic efficiency and legume growth under stress

2.5.3 Plant Growth–Promoting and Stress-Alleviating Microbes

Microorganisms that exist in legumes form a unique assemblage of functional attributes that provide services to the plant and enhance its fitness when subjected to challenging environmental conditions. Metagenomic studies have identified functional genes, including 1-aminocyclopropane-1-carboxylate (ACC) deaminase that reduces ethylene production when the plant is stressed, phosphate-solubilising enzymes that enhance phosphorus availability, and osmoregulatory pathways that improve tolerance to drought and salinity. For instance, chickpea germplasm that was drought tolerant yielded a higher abundance of rhizobial bacteria that optimise abiotic stress responses through osmotic protection, and soybean microflora that were salt tolerant exhibited the capacity to maintain ionic equilibria in plants subjected to high levels of salinity. These functional traits suggest the potential towards the rational design of microbial communities or bioinoculants to increase the resilience of legumes, mitigate productivity inputs of industrial fertilisers, and promote sustainable agriculture.

2.5.4 Soil Health and Nitrogen Cycling

Microbiomes associated with legumes play a major role in nitrogen cycling and hence nutrient dynamics of the soil. Nitrogen is considered as one of the primary drivers of productivity in ecosystems. Functional genes like amoA (ammonia oxidation), nosZ (denitrification), and nifH (nitrogen fixation), relating to nitrogen transformations, have been identified through metagenomics methodology together with contributions from different microbes responsible for various steps in the sequence of transformations from one form to another. This paper establishes nitrogen-use efficiency and soil nutrient dynamics under legume cropping influence. The soil microbiome metagenomic signature related to an alternative legume cropping system can act as an indicator of soil health and be used for making management decisions towards improved nutrient cycling but minimised nitrogen loss for long-term sustainable cropping systems.

2.6 CASE STUDIES

Advances in metagenomics enable detailed study of rhizosphere microbes and plant-root interactions. In intercropped peanuts, qPCR and shotgun metagenomics revealed links between root exudates, nitrogen-fixing microbial diversity, and functional genes for nitrogen and organic acid metabolism. Crop combinations altered exudation, enhancing genes supporting BNF. Optimising cropping systems and exudates can boost rhizosphere functions, though temporal dynamics and DNA-based inferences require confirmation via transcriptomics or enzyme assays (Qiao et al. 2024).

- Legume root nodules host complex microbial consortia beyond rhizobia. Deep sequencing and 16S/metagenomic surveys reveal cohabiting non-rhizobial taxa – Pseudomonas, Bacillus, Burkholderia, Methylobacterium – harbouring genes for plant growth, stress tolerance, and nitrogen cycling. These endophytes may enhance rhizobial symbiosis and host fitness, making nodules potential targets for bioinoculant development. DNA-based studies alone cannot confirm activity; cultivation and transcriptomics are needed to validate their ecological roles (Zheng et al., 2020; Pang et al., 2021).
- Seed metagenomics shows beneficial microbes can be vertically inherited across legume generations. Shotgun and amplicon sequencing of bean seeds reveal a heritable 'core' microbiome that supports germination, seedling vigour, and stress resistance. Seed microbiomes thus offer a natural source for bioinoculants and breeding strategies. Accurate characterisation requires robust surface sterilisation and infection control due to low biomass and challenges in assembling metagenome-assembled genomes (MAGs (Sulesky-Grieb et al., 2024).
- Advances in metagenomics enable direct recovery of metagenome-assembled genomes (MAGs) from legume seeds and rhizosphere microbiomes, reconstructing nearly complete microbial genomes (>90% completeness). Annotation reveals plant growth-promoting traits – nitrogen fixation, phytohormone production, siderophores, and stress tolerance. Previously uncharacterised microbes are now identified by functional potential. MAG-based insights guide isolation and synthetic consortia, though activity validation still requires transcriptomics or stable-isotope probing (Tshisekedi et al., 2024).

Table 2.2 represents multiple case studies that demonstrate how researchers applied metagenomic and omic methodologies to study legume species through their complete rhizosphere and endophytic microbiomes and their functional genes and stress response systems and their microbiome-based methods for improving legume production.

2.7 CHALLENGES AND FUTURE PROSPECTS

Metagenomic research in legumes has advanced understanding of plant–microbe interactions, but challenges remain. Soil and rhizosphere microbiomes are highly complex, with thousands of bacteria, fungi, archaea, and viruses interacting dynamically with plants. Short-read sequencing, gene annotation ambiguities, horizontal gene transfer, rare microbes, and limited legume-specific reference genomes hinder precise taxonomic and functional analyses. Integrating metagenomics with transcriptomics, proteomics, and metabolomics is technically difficult. Emerging approaches – synthetic microbiomes, single-cell genomics, and long-read sequencing (Oxford Nanopore, PacBio HiFi) – enable study of rare taxa, complete genomes, and microbial functions, offering pathways to improve legume growth, nitrogen fixation, and stress resilience.

TABLE 2.2
Case Studies of Metagenomics Approaches in Legume Biology

S.No.	Study Focus	Metagenomic Approach	Key Findings	References
1	Effect of rhizodeposition on nitrogen-fixing communities in peanut rhizosphere	Shotgun metagenomics of rhizosphere soils; functional gene profiling (N-cycle, C-metabolism)	Crop combinations altered rhizosphere gene profiles; enriched N-fixation and organic acid metabolism genes	(Qiao et al., 2024)
2	Diversity of non-rhizobial endophytes in legume root nodules	Full-length 16S and metagenomic sequencing of surface-sterilised nodules	Nodules host diverse bacterial taxa with genes for plant growth promotion and stress tolerance	(Zheng et al., 2020; Pang et al., 2021)
3	Vertical transmission of seed microbiota in legumes across generations	Shotgun metagenomics of seeds over multiple generations under normal and stress conditions	Stable heritable core seed microbiome; stress alters minor taxa composition	(Sulesky-Grieb et al., 2024)
4	Genome-level trait discovery in legume seed and rhizosphere microbiomes	Deep shotgun sequencing, MAG reconstruction, genome annotation	High-quality MAGs reveal novel PGP genes (nitrogenase, siderophores, ACC deaminase)	(Tshisekedi et al., 2024)
5	Functional activity of rhizosphere microbiota under drought stress	Metatranscriptomics of root/rhizosphere RNA; mapping to metagenome references	Increased transcription of osmoprotection, ROS detoxification, and stress-related genes	(Butowski et al., 2025)
6	Identification of marker bacteria across legume developmental stages	Multi-omics meta-analysis (16S, metagenome, transcriptome)	Consistent stage-specific bacterial markers correlated with growth and symbiosis genes	(Sahil et al., 2025)
7	Microbiome-based inoculant design for improved drought and salinity tolerance	Metagenomic discovery → isolation → greenhouse/field validation	Inoculants derived from -omics data enhance nodulation, N-fixation, and yield	(Kumar et al., 2020; Petrushin et al., 2023)

2.8 CONCLUSION

Metagenomics has transformed understanding of legume–microbe interactions, revealing diverse, often uncultivable bacteria, fungi, archaea, and viruses. Beyond taxonomy, it uncovers functional genes for nitrogen fixation, phosphorus solubilisation, stress tolerance, and secondary metabolite production, clarifying microbial roles in growth, nutrient uptake, disease resistance, and environmental adaptation. These insights guide next-generation biofertilisers, microbial inoculants, and precision agriculture, reducing chemical inputs and improving soil fertility. Advances in sequencing, bioinformatics, and legume-specific reference databases will further link taxa to functions, track microbial shifts, and enhance sustainable legume productivity and resilience.

REFERENCES

Andrews, S. *FastQC: A Quality Control Tool for High Throughput Sequence Data.* 2010 [2017]. Accessed n.d. http://www.bioinformatics.babraham.ac.uk/projects/fastqc/.

Bharti, Richa, and Dominik G. Grimm. "Current Challenges and Best-Practice Protocols for Microbiome Analysis." *Briefings in Bioinformatics* 22, no. 1 (2021): 178–193.

Bolger, Anthony M., Marc Lohse, and Bjoern Usadel. "Trimmomatic: A Flexible Trimmer for Illumina Sequence Data." *Bioinformatics* 30, no. 15 (2014): 2114–2120.

Butowski, Christina F., Yash Dixit, Marlon M. Reis, and Chunlong Mu. "Metatranscriptomics for Understanding the Microbiome in Food and Nutrition Science." *Metabolites* 15, no. 3 (2025): 185.

Chatzivassiliou, Elisavet K. "An Annotated List of Legume-Infecting Viruses in the Light of Metagenomics." *Plants* 10, no. 7 (2021): 1413.

Cruaud, Perrine, Jean-Yves Rasplus, Lillian Jennifer Rodriguez, and Astrid Cruaud. "High-Throughput Sequencing of Multiple Amplicons for Barcoding and Integrative Taxonomy." *Scientific Reports* 7, no. 1 (2017): 41948.

Egan, Ashley N., and Mohammad Vatanparast. "Advances in Legume Research in the Genomics Era." *Australian Systematic Botany* 32, no. 6 (2019): 459–483.

Field, Dawn, Linda Amaral-Zettler, Guy Cochrane, James R. Cole, Peter Dawyndt, George M. Garrity, Jack Gilbert, et al. "The Genomic Standards Consortium." *PLoS Biology* 9, no. 6 (2011): e1001088.

Fung, Calvin, Mathew Rusling, Thomas Lampeter, Charles Love, Anisha Karim, Christian Bongiorno, and Li-Lian Yuan. "Automation of QIIME2 Metagenomic Analysis Platform." *Current Protocols* 1, no. 9 (2021): e254.

Galperin, Michael Y., Roberto Vera Alvarez, Svetlana Karamycheva, Kira S. Makarova, Yuri I. Wolf, David Landsman, and Eugene V. Koonin. "COG Database Update 2024." *Nucleic Acids Research* 53, no. D1 (2025): D356–D363.

Huerta-Cepas, Jaime, Damian Szklarczyk, Kristoffer Forslund, Helen Cook, Davide Heller, Mathias C. Walter, Thomas Rattei, et al. "eggNOG 4.5: A Hierarchical Orthology Framework with Improved Functional Annotations for Eukaryotic, Prokaryotic and Viral Sequences." *Nucleic Acids Research* 44, no. D1 (2016): D286–D293.

Hugerth, Luisa W., and Anders F. Andersson. "Analysing Microbial Community Composition through Amplicon Sequencing: From Sampling to Hypothesis Testing." *Frontiers in Microbiology* 8 (2017): 1561.

Jagadesh, M., Munmun Dash, Aradhna Kumari, Santosh Kumar Singh, Krishan K. Verma, Prasann Kumar, Rajan Bhatt, and Satish Kumar Sharma. "Revealing the Hidden World of Soil Microbes: Metagenomic Insights into Plant, Bacteria, and Fungi Interactions for Sustainable Agriculture and Ecosystem Restoration." *Microbiological Research* 285 (2024): 127764.

Kanehisa, Minoru, and Susumu Goto. "KEGG: Kyoto Encyclopedia of Genes and Genomes." *Nucleic Acids Research* 28, no. 1 (2000): 27–30.

Krueger, Felix. *Trim Galore!: A Wrapper around Cutadapt and FastQC to Consistently Apply Adapter and Quality Trimming to FastQ Files, with Extra Functionality for RRBS Data.* Babraham Institute, 2015.

Krueger, Felix. *Trim Galore. GitHub repository. GitHub*, 2021. https://github.com/FelixKrueger/TrimGalore.

Kumar, Ashwani, and Anamika Dubey. "Rhizosphere Microbiome: Engineering Bacterial Competitiveness for Enhancing Crop Production." *Journal of Advanced Research* 24 (2020): 337–352.

National Research Council, Division on Earth, Board on Life Sciences, Committee on Metagenomics, and Functional Applications. *The New Science of Metagenomics: Revealing the Secrets of Our Microbial Planet.* Washington, DC: National Academies Press, 2007.

Pang, Jinfeng, Marike Palmer, Henry J. Sun, Cale O. Seymour, Ling Zhang, Brian P. Hedlund, and Fanjiang Zeng. "Diversity of Root Nodule-Associated Bacteria of Diverse Legumes along an Elevation Gradient in the Kunlun Mountains, China." *Frontiers in Microbiology* 12 (2021): 633141.

Petrushin, Ivan S., Ilia A. Vasilev, and Yulia A. Markova. "Drought Tolerance of Legumes: Physiology and the Role of the Microbiome." *Current Issues in Molecular Biology* 45, no. 8 (2023): 6311–6324.

Qiao, Mengjie, Ruibo Sun, Zixuan Wang, Kenneth Dumack, Xingguang Xie, Chuanchao Dai, Ertao Wang, et al. "Legume Rhizodeposition Promotes Nitrogen Fixation by Soil Microbiota under Crop Diversification." *Nature Communications* 15, no. 1 (2024): 2924.

Rana, Amrita Kumari, Pratibha Vyas, Sandeep Sharma, and Virender Sardana. "Groundnut Harbours Non-Nodulating Non-Rhizobial Plant Growth-Promoting Bacterial Endophytes." *3 Biotech* 13, no. 12 (2023): 420.

Sahil, Rishav, Vivek Pal, Arun S. Kharat, and Mukesh Jain. "A Multi-Omics Meta-Analysis of Rhizosphere Microbiome Reveals Growth-Promoting Marker Bacteria at Different Stages of Legume Development." *Plant, Cell & Environment* (2025): 1–16.

Seemann, Torsten. "Prokka: Rapid Prokaryotic Genome Annotation." *Bioinformatics* 30, no. 14 (2014): 2068–2069. https://academic.oup.com/bioinformatics/article/30/14/2068/2390517?login=true

Sivaprakasam, Navarasu, Sendhilvel Vaithiyanathan, Karthikeyan Gandhi, Swarnakumari Narayanan, P. S. Kavitha, Raghu Rajasekaran, and Raveendran Muthurajan. "Metagenomics Approaches in Unveiling the Dynamics of Plant Growth-Promoting Microorganisms (PGPM) vis-à-vis *Phytophthora* sp. Suppression in Various Crop Ecological Systems." *Research in Microbiology* 175, no. 7 (2024): 104217.

Stagnari, Fabio, Albino Maggio, Angelica Galieni, and Michele Pisante. "Multiple Benefits of Legumes for Agriculture Sustainability: An Overview." *Chemical and Biological Technologies in Agriculture* 4, no. 1 (2017): 2.

Sulesky-Grieb, Abby, Marie Simonin, A. Fina Bintarti, Brice Marolleau, Matthieu Barret, and Ashley Shade. "Stable, Multigenerational Transmission of the Bean Seed Microbiome Despite Abiotic Stress." *mSystems* 9, no. 11 (2024): 1–15. (https://doi.org/10.1128/msystems.00951-24)

Truong, D. T., E. A. Franzosa, T. L. Tickle, M. Scholz, G. Weingart, E. Pasolli, A. Tett, C. Huttenhower, and N. Segata. "MetaPhlAn2 for Enhanced Metagenomic Taxonomic Profiling." *Nature Methods* 12, no. 10 (2015): 902–903.

Tshisekedi, Kalonji A., Pieter De Maayer, and Angela Botes. "Metagenomic Sequencing and Reconstruction of 82 Microbial Genomes from Barley Seed Communities." *Scientific Data* 11, no. 1 (2024): 484.

Usyk, Mykhaylo, Brandilyn A. Peters, Smruthi Karthikeyan, Daniel McDonald, Christopher C. Sollecito, Yoshiki Vazquez-Baeza, Justin P. Shaffer, et al. "Comprehensive Evaluation of Shotgun Metagenomics, Amplicon Sequencing, and Harmonization of These Platforms for Epidemiological Studies." *Cell Reports Methods* 3, no. 1 (2023).

Wahab, Abdul, Murad Muhammad, Asma Munir, Gholamreza Abdi, Wajid Zaman, Asma Ayaz, Chandni Khizar, and Sneha Priya Pappula Reddy. "Role of Arbuscular Mycorrhizal Fungi in Regulating Growth, Enhancing Productivity, and Potentially Influencing Ecosystems under Abiotic and Biotic Stresses." *Plants* 12, no. 17 (2023): 3102.

Wingett, Steven W., and Simon Andrews. "FastQ Screen: A Tool for Multi-Genome Mapping and Quality Control." *F1000Research* 7 (2018): 1338.

Wood, Derrick E., and Steven L. Salzberg. "Kraken: Ultrafast Metagenomic Sequence Classification Using Exact Alignments." *Genome Biology* 15, no. 3 (2014): R46.

Yang, C., D. Chowdhury, Z. Zhang, W. K. Cheung, A. Lu, Z. Bian, and L. Zhang. "A Review of Computational Tools for Generating Metagenome-Assembled Genomes from Metagenomic Sequencing Data." *Computational and Structural Biotechnology Journal* 19 (2021): 6301–6314.

Zheng, Yanfen, Jing Liang, Dong-Lin Zhao, Chen Meng, Zong-Chang Xu, Zhi-Hong Xie, and Cheng-Sheng Zhang. "The Root Nodule Microbiome of Cultivated and Wild Halophytic Legumes Showed Similar Diversity but Distinct Community Structure in Yellow River Delta Saline Soils." *Microorganisms* 8, no. 2 (2020): 207.

3 Microbial Ecology of the Silent Workers in the Legume Rhizosphere

Exploring an Informed and Purposeful Community

Meghna Thapa, Deepak Kumar, and Anil Kumar Singh

3.1 INTRODUCTION

Plants are not independent biological entities in the soil, which itself is a highly complex and dynamic microbial ecosystem (Gans 2014). Below the ground, a silent yet dynamic life exists, shaping much of a plant's life cycle. Microbial communities influence nutrient availability, plant health, stress resistance, and the restoration of soil properties, while also developing symbiotic relationships that enable mutualistic interactions. Together, these microorganisms operate within highly complex functional networks defined by cooperation, competition, and commensalism. To explore this hidden world, we focus on the legume rhizosphere, which is a densely populated habitat where millions of microorganisms coexist and interact intimately with the plant.

The soil zones are divided into compartments that are dependent on the extent of influence a plant root can exert on the soil. Bulk soil is one that the root and its action cannot reach, and therefore, the plant has no bearing on it. The rhizosphere, on the contrary, is directly and closely affected by the plant root and its exudates first explained by Hiltner in 1904 (Okutani et al. 2019). Studying microbes in isolation reveals only a portion of a broader picture, because the visible plant phenotype is not solely the product of its own genome; it is, in fact, jointly shaped by the plant's genome and the diverse microbial community that coexists with it (Ravanbakhsh et al. 2019). When we reverse the idea, the microbial community itself is also shaped by the plant, resulting in dynamic feedback that intensifies in the rhizosphere. It is in this environment that the answers to most of the cause-and-effect relationships of microbial ecology lie. So, the bigger question is what exactly is happening in the rhizosphere and why should we understand it? This question inevitably brings us to a classic example of the partnership between legumes and nitrogen-fixing rhizobia, one of the best-studied symbiotic relationships in plant biology (Mathesius 2022). This interaction explains the molecular dialogue

DOI: 10.1201/9781003733324-4

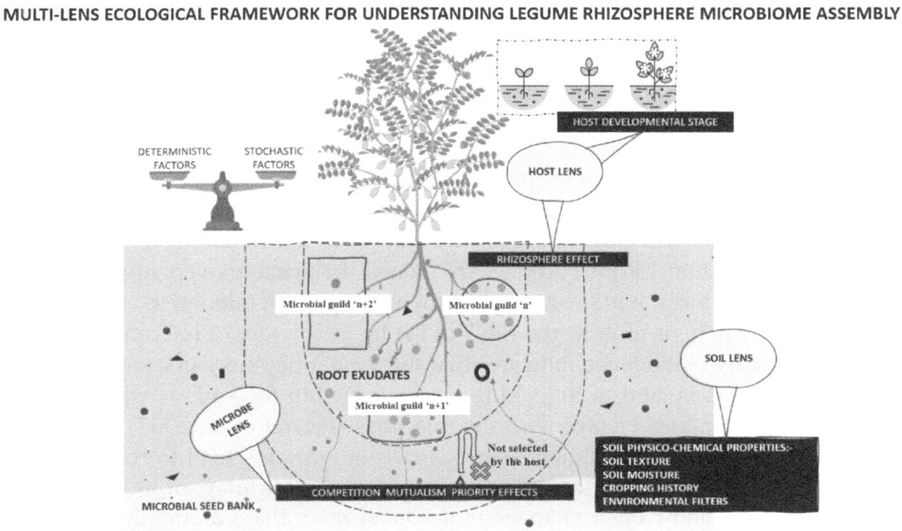

FIGURE 3.1 Multi-lens ecological framework for understanding legume rhizosphere microbiome assembly.

between plants and microbes and how it can shape ecosystems and inform future agricultural practices. It is necessary to mention that symbiosis is not friendship; it is a transaction that can only exist if the conditions are still mutually advantageous. To comprehensively understand rhizosphere microbial ecology via such relationships, we must examine it through multiple lenses-those of the plant (legumes), the microbes, and the soil, each of which contributes to shaping the community by influencing who inhabits the rhizosphere, and how they behave. Before we explore these perspectives in detail, it is essential to acknowledge that the rhizosphere arises from an interplay between deterministic and stochastic processes. Deterministic forces-such as root exudation patterns, soil physicochemical properties, and inter-microbial interactions shape which taxa are selectively recruited. In contrast, stochastic factors, including dispersal limitation, ecological drift, and priority effects, generate variability even under similar environments. Because these processes operate simultaneously and often overlap, understanding rhizosphere dynamics requires several ecological lenses to fully decipher the complexity of microbial community assembly. The present chapter attempts to compile existing knowledge on rhizosphere microbial ecology with particular emphasis on the legume-rhizobia interaction model, and to present a coherent, broad-based overview that serves as an accessible entry point for beginners in this field.

3.1.1 Through the Lens of the Legumes

This host-centric view naturally leads to the question of mechanism: how do legumes exert such control over their surrounding microbiome? A primary interface through which legumes shape rhizosphere assembly is root exudation, which mediates microbial recruitment, exclusion, and functional specialisation.

3.1.2 Microbial Selection by the Legume Root Exudation

The impact of root exudates on the microbial communities, known as the rhizosphere effects (Zhalnina et al. 2018) goes well beyond the previous effects of the legume associated with just nitrogen fixation, up to the functions of shaping the microbial ecosystems. Root exudates are indeed the prime movers of rhizosphere microbiome assembly, and the pattern of their mapping in relation to environmental stimuli can help us to draw the links of the bigger picture of microbial ecology. A growing body of evidence indicates that the exudation of organic substances by the plant roots is pivotal to unlocking nutrients within the soil, as well as determining the way plants engage with the microbes (Pierret et al. 2007). Roots that tend to develop and branch more in areas rich in nutrients tend to release larger quantities of exudates. The resulting increased exudation may, in its turn, affect nutrient cycling and trigger changes in the local microbial community (Paterson et al. 2004).

Exudates include flavonoids, sugars, amino acids, organic acids, fatty acids, and secondary metabolites (Badri and Vivanco 2009). The role of Flavonoids is central to legume rhizobium communication, which induces Nod factor synthesis and consequently determines the initial phases of symbiotic colonisation (Desbrosses and Stougaard 2011). Interestingly, nodulation can be stimulated by experimental addition of certain flavonoids, which induce the expression of nod genes in rhizobia (Liu and Murray 2016). Additionally, they are not just rhizobia-specific signal molecules. They can have antimicrobial properties and act as chemo-attractants for other taxa, other than rhizobia, to benefit from mutualist or helper bacteria (Hassan and Mathesius 2012). At this point, it is only natural to pose the following question: What controls the exudate profile of the plant? And, are those factors actually dictating plants to only attract a certain type of microbes? The exudates are normally influenced by plant genotype, development stage, and external conditions such as biotic and abiotic stressors (Zhalnina et al. 2018). (Chaparro, Badri, and Vivanco 2014) examined the impact of the developmental stage of the plant on the assembly of rhizosphere bacteria. They took Arabidopsis at four time points: seedling, vegetative, bolting, and flowering. Although there were no major differences in the community structure of bacteria, they found that the microbial community in the seedling stage was different from the others. Other significant phyla, such as acidobacteria, actinobacteria, bacteroidetes, and cyanobacteria, showed stage-specific patterns of distribution, which could be due to the selective recruitment of the phyla by the host via differentiating root exudation patterns. Additionally, comparative research on legumes such as soybeans, clover, and *Medicago* demonstrates cultivar- and genotype-specific variations in exudate chemistry, resulting in variations in rhizobial assemblages, total rhizosphere diversity, and nodule occupancy (Tkacz et al. 2022). These are plastic reactions, so exudation is not an unchangeable characteristic but a physical extended phenotype, both responsive to plant needs and to the context of the soil. It changes and fine-tunes over time to outsource the necessary functional activities (Sasse, Martinoia, and Northen 2017). Phosphorus and nitrogen status have a strong reprogramming effect on exudation patterns. Phosphate starvation enhances the release of organic acids and phosphatases, while an adequate supply of Nitrogen decreases the level of host investment in exudate pathways that attract nitrogen-fixing partners (Aurélie et al. 2025). Therefore, it's clear that the legumes do not exude without a strategic ecological plan. Their differential exudation patterns set the stage for density dependence, niche construction, and resource competition.

3.1.3 Root Exudates Dictate Density Dependence, Niche Construction, and Resource Competition

The population ecology of the rhizosphere is determined by its spatiotemporal and chemical dynamics, which govern the arrival, persistence, and competitive exclusion. Studies in Medicago and soybean showed that early colonisers rapidly consume flavonoids and organic acids, generating a classic density-dependent effects (Zhalnina et al. 2018). Early developmental roots release compounds that attract pioneering colonisers (priority effects), whereas later stages redirect resources to maintenance and defence, reorganising successional patterns in the rhizosphere (Steinauer et al. 2023). Such priority effects, where early-arriving microbes influence later community assembly, have been demonstrated in legume–rhizobium systems (Boyle et al. 2021). Beyond mediating microbial interactions, root exudates actively participate in metabolic niche construction alongside other environmental factors that determine microbial survival and reproduction (Olanrewaju 2019). In legumes, root-exuded flavonoids (e.g. in Medicago), and in some cases, isoflavones (e.g. in soybean) act as biochemical filters directing microbial recruitment and shaping rhizosphere community composition (Tkacz et al. 2022). However, the extent and consistency of this filtering across legume species remains to be experimentally validated. Within this framework, microbial niches in the rhizosphere are viewed as chemically defined microhabitats shaped by exudates, physicochemical gradients, and biotic interactions, consistent with modern extensions of Hutchinson's n-dimensional niche framework (Hutchinson 1957; Takola and Schielzeth 2022). Modern microbial ecology builds on this foundation, suggesting that assembly arises not solely from chance but from a balance between environmental filtering, biotic interactions, and stochastic colonisation (Jonathan, Chase, and Leibold 2004). Importantly, exudation-mediated selection is context dependent: an identical exudate cocktail can produce different assembly outcomes depending on soil properties and the resident microbial seed bank. Consequently, plant-driven recruitment acts on a pre-existing pool of microbes provided by soil legacy effects, highlighting the limitations of studying rhizosphere ecology in isolation from bulk soil processes (Jacoby et al. 2017). From a microbial perspective, the rhizosphere constitutes a highly competitive and chemically dynamic habitat characterised by patchy resources, where microbes actively detect, interpret, and exploit plant-derived signals to access carbon and nutrients. In this bottom-up view, microbes are not passive recruits but active agents that shape community structure, symbiotic trajectories, and their own evolutionary outcomes.

3.2 THROUGH THE LENS OF THE MICROBES

The rhizosphere becomes more of an ecosystem rather than a secretion zone when we change our perspective to a viewpoint of the microbial world. Exudates are perceived as an opportunity or limitation by microbes. The organic acids that are released when phosphorus is deficient nourish phosphate-solubilising bacteria; the phenolics released when iron is limited reward siderophore producers; and the amino acids released when nitrogen is limited nourish diazotrophs and nitrifiers Copiotrophs like Pseudomonas and oligotrophs thrive in situations where there are abundant sugars and amino acids, and complex organic acids, respectively, and then nutrients are mobilised or detoxified in the rhizosphere (Dutilh, Sánchez-gil, and Jonge 2023).

3.2.1 Microbial Competition, Cooperation, and Coexistence

However, the rhizosphere is not a peaceful environment; it is a busy, competitive, and constantly changing arena. These dynamics can be explained by using microbial ecology theories. Classical niche theory explains coexistence whereas the neutral theory of Hubbell (2009) emphasises stochasticity, treating species as functionally equivalent and governed by dispersal and drift. In the rhizosphere, these forces operate simultaneously. Deterministic filters (such as pH and exudate chemistry) constrain which taxa can persist (Philippot et al. 2013), while stochastic factors like colonisation timing and dispersal influences which microbes ultimately dominate (Philippot et al. 2024). Community assembly therefore reflects neither strict competition nor pure randomness, but an open system in which selection and chance intersect.

3.2.2 Strategies for Survival

Microbes must evolve mechanisms suited to this dynamic environment. R-strategists grow rapidly by, feeding on easily attainable exudates, whereas the K-strategists invest in persistence under their ability to endure scarcity. Biofilm formation provides community defence against predation and chemical stress, and dormancy in the form of spores or persister cells can serve as a microbial seed bank, where functional diversity can be maintained during unfavourable conditions. At the community level, microbial survival is highly influenced by collaboration or cooperation. Black Queen Hypothesis suggests that certain groups of organisms do not retain expensive functions, including detoxification or siderophore production, due to their ability to depend on the adjacent organisms that still have such features, leading to the formation of interdependent networks (Loss 2012). This interdependence is further enhanced by cross-feeding: the metabolic by-product of one microbe is used by another, creating metabolic consortia that are capable of completing nutrient-cycling steps which could not have been achieved by a single organism (Zhalnina et al. 2018). But this cooperation is also at risk; siderophores or extracellular enzymes that attract free-loaders who exploit shared resources without helping them to be made, which makes the maintenance of cooperation a game of constant bargaining between collaboration and exploitation (Figueiredo, Kramer, and Gerardo 2020).

However, only cooperation can support these systems that are accompanied by constant competition. Resource partitioning enables the microbes to be specialised on different types of exudates (Sasse, Martinoia, and Northen 2017). In cases of niche overlap, more vigorous strategies arise: exploitative competition is characterised by a high rate of uptake and metabolic efficiency, and interference competition is characterised by the production of antibiotics, Type VI secretion systems, bacteriocins, and toxins (Granato, Meiller-legrand, and Foster 2019). These ecological strategies direct the evolution of microbes, which is the generalism versus specialisation and rapid growth versus persistent existence (Dennis et al. 2011). This landscape is altered by plants: changing root exudation patterns with developmental stages transform microbial competition and influence rhizosphere community courses (Chaparro, Badri, and Vivanco 2014). These dynamics are reformed even by viruses and bacteriophages. The phages can regulate higher taxa using the top-down selection, inhibit competitive displacement, and maintain diversity in microbes through lytic selection (S. Wang et al. 2024). Their lysogenic processes influence the stability of the populations, and horizontal gene transfer spreads favourable characteristic traits like stress resistance or symbiosis efficiency (Koonin et al. 2002). These principles are reflected in the empirical cases. *Pseudomonas fluorescens* synthesises 2, 4-diacetylphloroglucinol, which prevents fungal and bacterial competitors; a typical example of interference competition (Weller, n.d.). The rhizobial strains interchange amino acids and signalling molecules with non-symbiotic neighbours, strengthening cooperation of micro-networks in the rhizosphere (Sciences and Road 1865). In the meantime, phage-based "viral shunts" stimulate microbial biomass turnover and maintain carbon and nutrient circulation (Pratama and Elsas 2017). In the end, the rhizosphere, viewed through the lens of a microbe, is not just soil being affected by roots but a place of a market, of carbon, of cooperation, and of competition. Plant carbon serves as a currency, which controls the abundance and the activity of microbes (Bais et al. 2010). This concept has its roots in the original 1904 insight made by Hiltner that the price of microbial abundance is always greater in the rhizosphere than in bulk soil (Badri and Vivanco 2009). Nonetheless, the microbial activities at the rhizosphere cannot be understood as intrinsic and fixed properties controlled by the plant exudates only. The microbial reactions depend on the history of soil, the degree of disturbance, the plant pasture left behind by vegetation, and microbial seed bank composition (Noah Fierer 2017). This means that varying microbial assembly results in different soils can be produced by the same exudate profile (Jacoby et al. 2017). Thus, rhizosphere behaviour emerges from the interaction between plant-driven selection and the ecological memory of the soil. The last rhizosphere puzzle piece comes in the recognition of soil as a historical and ecological system, one that keeps record of its past, succession marks, and disturbance markers. These legacy filters end up acting upon the assemblage of microbial communities as well as the interaction between plants and microbes.

3.3 THROUGH THE LENS OF THE SOIL

The rhizosphere is brought to life through the soil. The biophysical structure of the soil is an important architect of the diversity of microbes within the rhizosphere, the history of which still outlines the actions of its microbial occupants. The concept is similar to the original discoveries of Beijerinck and Winogradsky, who initially discovered the close connection between microbial life and the mosaic of the external environment depicted by soils. The unbelievable spatiotemporal difference in physicochemical properties of the soils has pushed the microbes to various mechanisms in order to survive, and this is what has modulated the organisation of the entire soil microbial ecology. In this regard, recent decades have been marked by significant attempts to define the soil qualities, chemical, physical, and historical, which contribute to the assembly and functioning of microbiomes. (Bakker et al. 2015) in their study revealed emerging evidence that soils act as ecological archives: past plant communities, pathogen outbreaks, and management decisions leave persistent microbial signatures that in turn influence which microbes colonise future roots. These "soil-borne legacies" can be protective (disease-suppressive assemblages) or detrimental (pathogen reservoirs), and they arise through reciprocal plant–microbe feedbacks: plants sculpt microbiomes via exudates and stress signalling, while microbes and their metabolites modify soil chemistry and nutrient cycling, reinforcing particular community states.

3.3.1 Effects of Soil Type and Properties on Microbiome

Previous studies by Bulgarelli et al. (n.d.) and Lundberg et al. (2012) established the effects that soil type has on the composition of bacterial communities in the rhizosphere of *Arabidopsis thaliana,* which was grown in different soils under the experimental greenhouse conditions. Kuramae et al. (2011) have also noted that several soil bacterial taxa had a high correlation with physicochemical soil properties. Further in agreement with this view, Smalla (2014) also demonstrated that even when soils have identical cropping history with the same agricultural management for a decade, and the same climatic regime for over thirty years, their underlying inherent physicochemical differences still resulted in distinct microbial communities. This has led to a basic question: how do soil physicochemical properties determine which microbes prosper, which do not prosper, or which do not compete in the rhizosphere? Many factors, such as climate, moisture availability, pore size, soil organic matter content, the total content of carbon and nitrogen, carbon–nitrogen ratio, and pH, make significant contributions to soil bacterial community diversity (Tecon and Or 2017). They also described the wet or aqueous phase of the soil as fragmented, and in dynamical cycles of wetting and drying – that is, moisture dynamics and its pore-space connectivity decisively determine the extent of dispersal of microbes, the diffusion of nutrients, and the existence of microhabitats. Such dynamics of moisture liberate the resources in the pulses, which can be more effectively considered in terms of the Birch Effect. The Birch effect, first characterised by H.F. Birch (1958), describes the sharp pulse of CO_2 release and nutrient mineralisation that occurs when a dry soil is rewetted. Rewetting dissolves accumulated solutes, ruptures desiccation-stressed cells, and increases the diffusivity of organic carbon, producing a sudden flush of labile substrates that favours fast-growing copiotroph taxa (N Fierer, Schimel, and Holden 2003; Imminger et al. 2024). Such pulses frequently cause short-term rises in inorganic nitrogen (ammonium and nitrate) and phosphorus, restructuring microbial competitive structures within a few hours to a few days (Miller and Cramer 2004). In legume systems, the Birch effect has some other consequences: the nutrient flush following rewetting may temporarily augment mineral nitrogen supply, put on hold the plant investment in nodulation, and postpone the development of rhizobial infection threads (Serraj, Sinclair, and Purcell 1999). Simultaneously, rewetting resumes the inactive rhizobial populations in micropores to enable them to recolonise root surfaces and compete for the chance of nodulation (Zahran 1999). The same property of the soil puts it in the middle stage of microbial ecology on its own. It is this feedback between the microbial activity and the soil structure that, in turn, maintains the coexistence of the microbes in the soil by forming their spatial niches, decelerating dominance, and developing resilience over time (Tecon and Or 2017).

Thus, understanding rhizosphere ecology "through the lens of the soil" brings us back to a plain yet weighty deduction: the legume root does not structure its microbiome all by itself. It does so in continuous dialogue with the soil's memory, structure, and constraints, and it is within this tightly interwoven soil-plant–microbe continuum that the true dynamics of the rhizosphere begin to emerge.

3.4 HOW CAN WE BENEFIT FROM THIS MULTIPLE LENS PERSPECTIVE IN UNDERSTANDING RHIZOSPHERE MICROBIAL ECOLOGY?

In a generic sense, metagenomics has traditionally been viewed as the means of providing answers to only two basic questions: "who are there in the microbial community? and what can they potentially do?" and rightly so. However, to figure out the microbial world, particularly the rhizosphere ecology, answering these questions is just the tip of the iceberg. Rhizosphere ecology is an area of science that cannot be comprehended without a solid scientific background, so it is an instinctive yet analytical process in most instances. The reason is that the rhizosphere patterns fail to arise out of a solitary source of influence. A multi-lens view is very critical in disclosing every tier of knowledge to improve the interpretation of the microbiome data. The ecological factors shaping it become more exoteric as we attempt to look deeper at the community composition. All of the above-discussed lenses explain the agency of all the ecological forces that coordinate the microbial community and provide us with a great foundation to understand any sequencing data. They assist us in observing whether the data is partially formed by the plant deliberately, or the trend in the data is by microbial survival strategies, or is predominantly determined by the soil heritage! The structure of the chapter is expected to assist the reader in detangling deterministic vs stochastic assembly of the microbiome. Ecological measures, such as the β-Nearest Taxon Index (β-NTI) with null model analysis, are used to measure which of them prevail in the community assembly. They identify deterministic signals by contrasting the observed phylogenetic turnover with the expectation value (randomisation)(Dini-andreote et al. 2015). It has been demonstrated by various studies in legumes that in the early stages of seedling development, the rhizosphere microbiome assembles more stochastically compared to the nodulation phase, when the community is highly selective or deterministic (Sugiyama et al. 2014; X.

Wang et al. 2023). By working together, these lenses enable us to design a comprehensive workflow that takes into consideration these ecological patterns, and thus, one will not miss any point, which is otherwise required to interpret the data accurately (Figure 3.2). Readers are expected to be able to design experiments having more profound hypotheses, answering questions such as, "Why are they there in the microbiome community? Which ecological processes chose them?" The responses to these questions allow obtaining a holistic and mechanistically significant picture of the rhizosphere microbial ecology. In the case of legumes, whose existence relies on the accuracy of microbial associations, the moral is obvious: the rhizosphere is an orchestrated ecological setup, and only a multi-perceptual explanation can show how purpose, choice, chemistry, and randomness are used to construct the community of associates that amass around their roots. It is maybe not the answers we make out in the rhizosphere microbiome research, but the lesson of looking at it in many different ways before the silent, unheard-of workers in the data make their voices heard. In the end, it is not really the goal of this chapter to offer conclusive solutions but to offer a viable interpretive framework. The framework allows the shift in focus between descriptive sequencing and causal, hypothesis-directed experimental studies by considering root-associated microbial communities as part of the modern ecological theory. Provided that future research utilises such lenses such as abiotic, biotic, and colonisation processes as structural instruments in the study design as well as data interpretation, the field will increasingly attain increased levels of predictability, reproducibility, and mechanistic knowledge. In this respect, the multi-lens approach can be taken as a strategic direction of the next phase of research on rhizosphere microbiomes.

3.5 FUTURE DIRECTIONS

The multi-lens ecological approach described in this chapter offers several explicit priorities in the development of the rhizosphere microbiome studies as follows:

- Multisite, field-scale experiments are needed to establish whether the observed rules of assembling microbiomes and beneficial microbiomes can be maintained in similar soils, climates, and management systems.
- A combination of multi-omics with experimental validation will go beyond functional potential to show function with transcriptomics, proteomics, metabolomics, and culture-based biochemical assays.
- Temporal stability in microbiome stability, particularly in response to seasonal variations, abiotic stress, and agro-disturbances, should be monitored over time to understand stability, succession, and collapse limits.
- Causal modelling/Systems Causal modelling at the systems-level, that relates microbiome composition, functional gene expression, nodulation, N-fixation activity, and plant phenotypes with structural equation modelling, network inference, and machine learning, is required. Further extension of the rhizosphere to include the endosphere, bulk soil, phyllosphere, and soil fauna will develop a comprehensive perspective of ecological relations that determine root-related microbiomes.

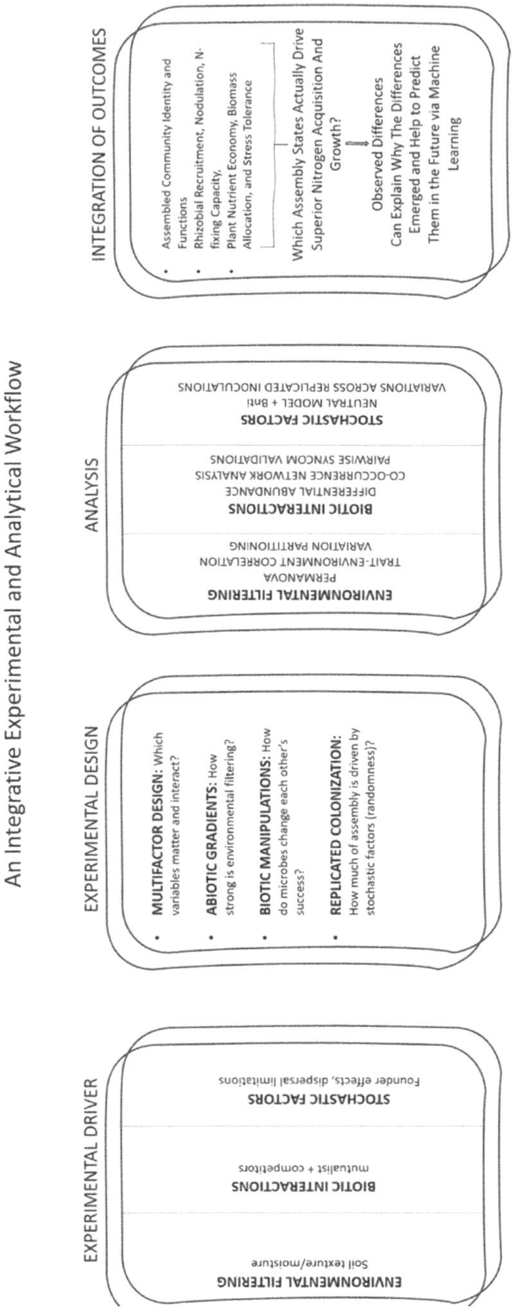

FIGURE 3.2 Decoding microbiome assembly in the legume rhizosphere.

- Converging on standardised methods and open data are essential towards enhancing comparability and reproducibility of microbiome studies and thus accelerating the rate of cumulative discovery. Agronomic translation and socio-economic feasibility should be secured in such a way that microbiome-based interventions, including inoculants and synthetic communities, can be reliably applied in fields controlled by farmers and be economically and logistically available.

3.6 ACKNOWLEDGEMENTS

The research in the laboratory of AKS is supported by the Indian Council of Agricultural Research (ICAR), Department of Agricultural Research and Education, Government of India; ICAR Network Project on Application of Microorganisms in Agriculture and Allied Sectors (AMAAS), and Department of Biotechnology, Government of India sponsored project National Network Project of Jawaharlal Nehru University under BTISNet.

REFERENCES

Aurélie, Christelle, Maud Robert, Paul Himmighofen, Sarah Mclaughlin, Tristan M Cofer, Sheharyar Ahmed Khan, Alexandra Siffert, and Joëlle Sasse. 2025. "Environmental and Biological Drivers of Root Exudation" 1: 317–39.

Badri, Dayakar V, and Jorge M Vivanco. 2009. "Regulation and Function of Root Exudates," 666–81. https://doi.org/10.1111/j.1365-3040.2009.01926.x.

Bais, Harsh P, Tiffany L Weir, Laura G Perry, Simon Gilroy, and Jorge M Vivanco. 2010. "The Role of Root Exudates in Rhizosphere Interactions with Plants and Other Organisms." https://doi.org/10.1146/annurev.arplant.57.032905.105159.

Bakker, Matthew G, Jacqueline M Chaparro, Daniel K Manter, and Jorge M Vivanco. 2015. "Impacts of Bulk Soil Microbial Community Structure on Rhizosphere Microbiomes of Zea Mays," 115–26. https://doi.org/10.1007/s11104-015-2446-0.

Boyle, Julia A, Anna K Simonsen, Megan E Frederickson, and John R Stinchcombe. 2021. "Priority Effects Alter Interaction Outcomes in a Legume – Rhizobium Mutualism," 1–8.

Bulgarelli, Davide, Klaus Schlaeppi, Stijn Spaepen, Emiel Ver, Loren Van Themaat, and Paul Schulze-lefert. n.d. "Structure and Functions of the Bacterial Microbiota of Plants." https://doi.org/10.1146/annurev-arplant-050312-120106

Chaparro, Jacqueline M, Dayakar V Badri, and Jorge M Vivanco. 2014. "Rhizosphere Microbiome Assemblage Is Affected by Plant Development," 790–803. https://doi.org/10.1038/ismej.2013.196.

Chen, Qingxia, Yingjie Song, Yuxing An, Yinglin Lu, and Guohua Zhong. 2024. "Soil Microorganisms: Their Role in Enhancing Crop Nutrition and Health," 1–26.

Dennis, Roger L H, Leonardo Dapporto, Simone Fattorini, and Laurence M Cook. 2011. "The Generalism – Specialism Debate: The Role of Generalists in the Life and Death of Species," 725–37.

Desbrosses, Guilhem J, and Jens Stougaard. 2011. "Root Nodulation: A Paradigm for How Plant-Microbe Symbiosis Influences Host Developmental Pathways." *Cell Host and Microbe* 10 (4): 348–58. https://doi.org/10.1016/j.chom.2011.09.005.

Dini-Andreote, Francisco, James C Stegen, Jan Dirk, Van Elsas, and Joana Falcão. 2015. "Disentangling Mechanisms That Mediate the Balance between Stochastic and Deterministic Processes in Microbial Succession." https://doi.org/10.1073/pnas.1414261112.

Dutilh, Bas E, Juan J Sánchez-gil, and Ronnie De Jonge. 2023. "Growth Rate Is a Dominant Factor Predicting the Rhizosphere Effect," no. May. https://doi.org/10.1038/s41396-023-01453-6.

Fierer, Noah. 2017. "Embracing the Unknown: Disentangling the Complexities of the Soil Microbiome." *Nature Publishing Group* 15 (10): 579–90. https://doi.org/10.1038/nrmicro.2017.87.

Fierer, N, J P Schimel, and P A Holden. 2003. "Influence of Drying – Rewetting Frequency on Soil Bacterial Community Structure," 63–71. https://doi.org/10.1007/s00248-002-1007-2.

Figueiredo, Alexandre R T, Jos Kramer, and Nicole Marie Gerardo. 2020. "Cooperation and Conflict Within the Microbiota and Their Effects On Animal Hosts," 8 (May): 1–15. https://doi.org/10.3389/fevo.2020.00132.

Gans, Jason. 2014. "Computational Improvements Reveal Great Bacterial Diversity and High Metal Toxicity in Soil," 1387 (2005). https://doi.org/10.1126/science.1112665.

Granato, Elisa T, Thomas A Meiller-legrand, and Kevin R Foster. 2019. "Review The Evolution and Ecology of Bacterial Warfare." *Current Biology* 29 (11): R521–37. https://doi.org/10.1016/j.cub.2019.04.024.

Hassan, Samira, and Ulrike Mathesius. 2012. "The Role of Flavonoids in Root – Rhizosphere Signalling: Opportunities and Challenges for Improving Plant – Microbe Interactions" 63 (9): 3429–44. https://doi.org/10.1093/jxb/err430.

Hubbell, Stephen P. 2009. "Neutral Theory and the Theory of Island Biogeography." In *The Theory of Island Biogeography Revisited*, edited by Jonathan B Losos and Robert E Ricklefs, 264–92. Princeton: Princeton University Press. https://doi.org/doi:10.1515/9781400831920.264

Hutchinson, G. Evelyn. 1957. "Concluding Remarks." Cold Spring Harbor Symposia on Quantitative Biology 22: 415–427. https://doi.org/10.1101/SQB.1957.022.01.039

Imminger, Stefanie, Dimitri V Meier, Arno Schintlmeister, Anton Legin, Jörg Schnecker, Andreas Richter, Osnat Gillor, Stephanie A Eichorst, and Dagmar Woebken. 2024. "Survival and Rapid Resuscitation Permit Limited Productivity in Desert Microbial Communities," https://doi.org/10.1038/s41467-024-46920-6.

Jacoby, Richard, Manuela Peukert, Antonella Succurro, and Anna Koprivova. 2017. "The Role of Soil Microorganisms in Plant Mineral Nutrition – Current Knowledge and Future Directions," 8 (September): 1–19. https://doi.org/10.3389/fpls.2017.01617.

Jonathan, Contemporary Approaches, M Chase, and Mathew A Leibold. 2004. "Book Review," 1791–93.

Koonin, Eugene V, Kira S Makarova, Yuri I Wolf, and L Aravind. 2002. "Horizontal Gene Transfer and Its Role in the Evolution of Prokaryotes," 277–305.

Kuramae, Eiko E, Etienne Yergeau, Lina C Wong, Agata S Pijl, Johannes A Van Veen, and George A Kowalchuk. 2011. "Soil Characteristics More Strongly Influence Soil Bacterial Communities than Land-Use Type" 2. https://doi.org/10.1111/j.1574-6941.2011.01192.x

Liu, Cheng-Wu, and Jeremy D. Murray. 2016. "The Role of Flavonoids in Nodulation Host-Range Specificity: An Update." Plants 5 (3): 33. https://doi.org/10.3390/plants503003

Loss, Adaptive Gene. 2012. "The Black Queen Hypothesis: Evolution of Dependencies Through," 3 (2): 1–7. https://doi.org/10.1128/mBio.00036-12.Copyright.

Lundberg, Derek S, Sarah L Lebeis, Sur Herrera Paredes, Scott Yourstone, Stephanie Malfatti, Julien Tremblay, Anna Engelbrektson, and Victor Kunin. 2012. "Defining the Core Arabidopsis Thaliana Root Microbiome," 488 (7409): 86–90.

Mathesius, Ulrike. 2022. "Are Legumes Different? Origins and Consequences of Evolving Nitrogen Fixing Symbioses." *Journal of Plant Physiology* 276: 153765. https://doi.org/10.1016/j.jplph.2022.153765.

Miller, A J, and M D Cramer. 2004. "Root Nitrogen Acquisition and Assimilation." https://doi.org/10.1007/s11104-004-0965-1.

Okutani, Fuki, Shoichiro Hamamoto, Yuichi Aoki, Masaru Nakayasu, Naoto Nihei, Taku Nishimura, Kazufumi Yazaki, and Akifumi Sugiyama. 2020. "Rhizosphere Modelling Reveals Spatiotemporal Distribution of Daidzein Shaping Soybean Rhizosphere Bacterial Community." *Plant, Cell & Environment* 43 (4): 1036–1046. https://doi.org/10.1111/pce.13708

Olanrewaju, Oluwaseyi Samuel. 2019. "Plant Health: Feedback Effect of Root Exudates-Rhizobiome Interactions," 1155–66.

Paterson, Eric, Thomas Gebbing, Claire Abel, Allan Sim, Gillian Telfer, and Eric Paterson. 2004. "Rhizodeposition Shapes Rhizosphere Microbial Community Structure in Organic Soil," 600–10.

Peters, N Kenr, John W Frost, and Sharon R Long. 2002. "A Plant Flavone, Luteolin, Induces Expression of Rhizobium Meliloti Nodulation Genes," 551 (1974).

Philippot, Laurent, Claire Chenu, Andreas Kappler, Matthias C Rillig, and Noah Fierer. 2024. "The Interplay between Microbial Communities and Soil Properties," 22 (April). https://doi.org/10.1038/s41579-023-00980-5.

Philippot, Laurent, Jos M Raaijmakers, Philippe Lemanceau, and Wim H Van Der Putten. 2013. "Going Back to the Roots: The Microbial Ecology of the Rhizosphere." *Nature Publishing Group*, no. September. https://doi.org/10.1038/nrmicro3109.

Pierret, Alain, Claude Doussan, Yvan Capowiez, François Bastardie, Loïc Pagès, S Pecial, and S Ection. 2007. "Root Functional Architecture: A Framework for Modeling the Interplay between Roots and Soil," 6 (2). https://doi.org/10.2136/vzj2006.0067.

Pratama, Akbar Adjie, and Jan Dirk Van Elsas. 2017. "The 'Neglected' Soil Virome – Potential Role and Impact." *Trends in Microbiology* xx: 1–14. https://doi.org/10.1016/j.tim.2017.12.004

Ravanbakhsh, Mohammadhossein, George A Kowalchuk, and Alexandre Jousset. 2019. "Root-Associated Microorganisms Reprogram Plant Life History along the Growth – Stress Resistance Tradeoff." *The ISME Journal*: 3093–3101. https://doi.org/10.1038/s41396-019-0501-1.

Sasse, Joelle, Enrico Martinoia, and Trent Northen. 2017. "Feed Your Friends: Do Plant Exudates Shape the Root Microbiome?" *Trends in Plant Science* 20: 1–17. https://doi.org/10.1016/j.tplants.2017.09.003.

Sciences, Plant, and South Parks Road. 1865. "Rachel M . Wheatley and Philip S . Poole," no. April 2018. https://doi.org/10.1093/femsre/fuy014/4975273

Serraj, Rachid, Thomas R Sinclair, and Larry C Purcell. 1999. "Symbiotic N Fixation Response to Drought" 50 (331): 143–55

Smalla, Kornelia. 2014. "Effect of the Soil Type on the Microbiome in the Rhizosphere of Field-Grown Lettuce," 5 (April): 1–13. https://doi.org/10.3389/fmicb.2014.00144.

Steinauer, Katja, Madhav P Thakur, S Emilia Hannula, Alexander Weinhold, Henriette Uthe, Nicole M Van Dam, and T Martijn Bezemer. 2023. "Root Exudates and Rhizosphere Microbiomes Jointly Determine Temporal Shifts in Plant–Soil Feedbacks," no. January: 1885–99. https://doi.org/10.1111/pce.14570.

Sugiyama, Akifumi, Yoshikatsu Ueda, Takahiro Zushi, Hisabumi Takase, and Kazufumi Yazaki. 2014. "Changes in the Bacterial Community of Soybean Rhizospheres during Growth in the Field," 9 (6): 1–9. https://doi.org/10.1371/journal.pone.0100709.

Takola, Elina, and Holger Schielzeth. 2022. "Hutchinson's Ecological Niche for Individuals," no. April 2021: 1–21.

Tecon, Robin, and Dani Or. 2017. "Biophysical Processes Supporting the Diversity of Microbial Life in Soil," no. November 2016: 599–623. https://doi.org/10.1093/femsre/fux039.

Tkacz, Andrzej, Raphael Ledermann, Anna Martyn, Sebastian Schornack, Giles E D Oldroyd, and Philip S Poole. 2022. "Nodulation and Nitrogen Fixation in Medicago Truncatula Strongly Alters the Abundance of Its Root Microbiota and Subtly Affects Its Structure," no. July: 5524–33. https://doi.org/10.1111/1462-2920.16164.

Wang, Shuo, Dong Zhu, Tida Ge, Yongfeng Wang, Ying Zhang, and Chao Liang. 2024. "Climate Smart Agriculture Unveiling the Top-Down Control of Soil Viruses over Microbial Communities and Soil Organic Carbon Cycling: A Review." *Climate Smart Agriculture* 1 (2): 100022. https://doi.org/10.1016/j.csag.2024.100022.

Wang, Xing, Shan-wen He, Qing He, Zhi-cheng Ju, Yi-nan Ma, Zhe Wang, Jia-cheng Han, et al. 2023. "Bacterial Communities across Rice Plant Growth Stages," 11 (5).

Weller, David M. n.d. "Pseudomonas Biocontrol Agents of Soilborne Pathogens: Looking Back Over 30 Years," (53).

Zhalnina, Kateryna, Katherine B Louie, Zhao Hao, Nasim Mansoori, Ulisses Nunes, Shengjing Shi, Heejung Cho, et al. 2018. "Substrate Preferences Drive Patterns in Rhizosphere Microbial Community Assembly," 3 (April). https://doi.org/10.1038/s41564-018-0129-3.

Zahran, Hamdi Hussein. 1999. "Rhizobium-Legume Symbiosis and Nitrogen Fixation under Severe Conditions and in an Arid Climate" 63 (4): 968–89.

4 The Crucial Role of Key Microbiome for Improving Sustainable Legume Cultivation

Supriti Chowdhury, Ankita Pradhan, Ishita Samanta, Yamini Agrawal, Sasmita Mishra, Ritesh Pattnaik, and Gopal Chowdhary

4.1 INTRODUCTION

According to a report by the Food and Agriculture Organisation (FAO), it is expected that the world population might reach 8.5 billion by the end of 2025. This will require a substantial increase in agricultural production. The global agricultural system will face immense pressure to supply food to such a rapidly growing population and will also require to mitigate the effects of climate change (Timmusk et al., 2017). In addition to food shortage, the extensive use of mineral fertilisers, pesticides, water, in intensive cropping to meet the world-wide food demand has led to the degradation of soil, pollution, as well as depletion of natural resources (Suman et al., 2022). In this regard, the aggregate of microorganisms in the soil environment has served as an essential component of the sustainable agricultural system (Schaedel et al., 2021). Legumes (Fabaceae) comprise of more than 20,000 species, are the third largest family of flowering plants and play a vital role in food production by generating plant-based proteins. They can intensively contribute to sustainable agricultural systems by enhancing soil fertility, reducing carbon footprints of extensive cropping and also by posing bioenergy security. The biological nitrogen fixation by legume-rhizobia symbiosis is crucial for ecosystem functioning as legumes are one of the main sources of nitrogen fixation in terrestrial soil (Ben Gaied et al., 2024). In intercropping systems, legumes may enable circulating the nutrients such as N, P, Fe, and Zn in the soil (Janati et al., 2021). Climate change has affected the productivity of legumes by inducing various biotic and abiotic stresses. Conventional research has mostly focused on bi-partite interactions like the legume-rhizobia interaction, but in nature and in agricultural fields, plants are associated with diverse communities of bacteria, fungi and viruses. Therefore, crop performance is not dictated by a single host genotype or a single introduced strain, but by a combined influence of a diverse microbiome (Li et al., 2022).

DOI: 10.1201/9781003733324-5

This creates the necessity of moving towards comprehensive microbial synergy instead of single-strain inoculants. Successful agricultural solutions are those that serve multiple purposes and that address co-existing environmental demands. For example, coordinating phosphorus solubilisation with nitrogen fixation, as biological nitrogen fixation is a phosphorus-requiring agronomic process (Janati et al., 2021). Meta-analysis states that the crop yield generated by the microbial inoculants improves plant nutrient availability and also alleviates abiotic/biotic stresses (Li et al., 2022). The objective of the current study is to investigate the role of the legume microbiome in establishing sustainable cultivation practices. This will be done by studying the basic microbiome and its molecular mechanisms by which it stores nutrients and resists stress. In this regard, next-generation sequencing, multi-omics analysis, synthetic biology, the incorporation of CRISPR/Cas9 techniques, RNA interference techniques and various other manipulation techniques appear to be a promising approach to enhance plant performance and thereby making sustainable agriculture possible (Joshi et al., 2025).

4.2 THE IMPERATIVE NEED FOR MICROBIAL SYNERGY IN LEGUME SUSTAINABILITY

Due to their unique ability to initiate symbiotic relationships, legumes are foundational to sustainable farming systems, offering nutritional, ecological, and economic benefits. Such a unique symbiotic relationship additionally allows legumes to thrive in nutrient-poor soils and reduce dependence on synthetic fertilisers. Microbial synergy refers to the cooperative interactions among multiple microbial species that collectively enhance plant health and soil function. Understanding and harnessing these interactions is essential for advancing legume sustainability. Some of the key roles of microbial synergy beyond nitrogen fixation include stress tolerance, disease resistance, improving nutrient acquisition, regulating plant hormones, and, most importantly, improving soil health and increasing microbial diversity.

Several studies have reported such role of microbes as mycorrhizal fungi and endophytes, helps legumes survive in drought, salinity, and nutrient-poor soils by improving water retention and nutrient absorption. Certain bacteria and fungi produce antimicrobial compounds or trigger plant immune responses, protecting legumes from pathogens and reducing reliance on chemical pesticides (Andrade-Domínguez et al., 2014). Although both the plant and soil microbiome have been studied over the decades, the efficiency of translating the laboratory and greenhouse findings to the field is largely dependent on the ability of the inoculants or beneficial microorganisms to colonise the soil and maintain stability in the ecosystem. As climate change and soil degradation from excessive fertiliser use continue to challenge global food systems, leveraging microbial synergy in legumes presents a sustainable and resilient pathway for future crop production.

4.3 THE SYMBIOTIC KEYSTONE: RHIZOBIA AND BIOLOGICAL NITROGEN FIXATION (BNF)

The most defined and ecologically significant symbiosis involving legumes is with a group of soil-dwelling nitrogen-fixing bacteria, referred to as rhizobia. This mutualistic relationship facilitates legumes to convert atmospheric nitrogen (N_2) into ammonia (NH_3), as a usable form for plant uptake, thereby reducing dependence on the application of synthetic nitrogen fertilisers. Biological nitrogen fixation (BNF) is the primary route for nitrogen (N) input into the ecosystem. It is estimated that BNF produces around 200 teragrams (Tg) of organic N per year; therefore, it may significantly contribute to climate conservation and sustainable farming. The molecular insights about symbiotic microbial interactions and biochemistry of symbiotic N_2-fixation emphasise the role and involvement of plant legumes in biological nitrogen fixation. The nitrogen-fixing prokaryotes convert atmospheric dinitrogen into ammonia using a relatively conserved nitrogenase complex that comprises dinitrogen synthase, dinitrogen synthase reductase, and metal cofactors. Dinitrogen synthases are categorised into three types according to the metal ions they bind at their active sites: molybdenum (Mo) nitrogenase (the predominant nitrogenase in N-fixing bacteria and archaea), vanadium nitrogenase, and iron (Fe) nitrogenase (Guo et al., 2023).

Both legumes and rhizobial bacteria exhibit significant phylogenetic diversity. No single strain of rhizobia is capable of forming symbiotic relationships with all legumes, and the reverse is also true. Specificity manifests at both the species and genotypic levels. This specificity can occur early in the interaction, enabling the same bacterial strains to infect and nodulate one host plant while failing to do so in another (Fan et al., 2017). Incompatibility often arises at later stages of nodule development, causing significant variations in nitrogen-fixing efficiency among different plant-bacteria combinations. The specificity of symbiosis is a result of the evolving signals from both the host and bacterial sides; consequently, various recognition mechanisms have developed during the co-adaptation process. Table 4.1 highlights the genes (rhizobial and legume) that are primarily involved in symbiotic nitrogen fixation.

4.4 ECOLOGICAL SYNERGY: INTERRELATION BETWEEN RHIZOBIA, PGPB, AND AMF

Sustainable agriculture is crucial for addressing global food needs while protecting the environment and our natural resources. One effective approach is to utilise beneficial microorganisms like arbuscular mycorrhizal fungi (AMF) and bacteria associated with plants to improve agricultural sustainability. AMF establish mutualistic relationships with the roots of plants, aiding in the absorption of nutrients, particularly phosphorus and micronutrients. At the same time, bacteria associated with plants enhance growth, resilience to stress, and resistance to diseases. The combined benefits of these microorganisms are thought to exceed the effects of each one acting alone. In agricultural ecosystems, plants coexist with a diverse community of beneficial microorganisms in the rhizosphere including rhizobia, plant growth–promoting bacteria (PGPB), and arbuscular mycorrhizal fungi (AMF) form an intricate tripartite symbiotic network influencing plant growth and development (Hassani et al., 2018).

Recent developments in high-throughput genome sequencing for plant and microbial populations have created opportunities to gain insights into the molecular mechanisms responsible for microbe detection, the establishment of symbiosis, signal

TABLE 4.1
Highlighting the Involvement of Various Genes in Symbiotic Nitrogen Fixation

Category	Gene	Organism	Function in Symbiotic Nitrogen Fixation
Bacterial (Rhizobial) Genes	nodA, nodB, nodC	*Rhizobium leguminosarum*, *Bradyrhizobium japonicum*	Synthesis of Nod factors (lipochitooligosaccharides) that initiate nodule formation in host roots
	Nod	*R. leguminosarum*	Regulatory gene; activates other *nod* genes in response to plant flavonoids
	nifH, nifD, nifK	*B. japonicum*, *Sinorhizobium meliloti*	Encode nitrogenase enzyme components responsible for $N_2 \rightarrow NH_3$ reduction
	fixA, fixB, fixC, fixJ, fixK	*B. japonicum*	Regulate and maintain microaerobic conditions necessary for nitrogenase activity in nodules
	exoA, exoB	*S. meliloti*	Involved in exopolysaccharide synthesis required for infection thread formation
Plant (Legume) Genes	NFR1, NFR5 (LysM receptors)	*Lotus japonicus*, *Medicago truncatula*	Nod factor receptors; perceive rhizobial Nod signals to initiate nodulation
	NSP1, NSP2 (GRAS transcription factors)	*M. truncatula*	Transcriptional regulators in the nodulations

transduction, and the plant's defence responses (Afridi et al., 2022). The identification of resistance gene analogs (RGAs), including nucleotide binding site-leucine rich repeats (NBS-LRRs) and receptor-like kinases (RLKs), offers promise for the development of tolerance genotypes. Additionally, research has concentrated on assessing the copy numbers and distributions of symbiosis genes to enhance our understanding of legume functional and comparative genomics.

4.5 MICROBIOME-MEDIATED STRESS TOLERANCE AND PLANT PROTECTION

Legumes are documented to be the earliest domesticated plants that have been grown for over 3000 years across China, Europe, Asia and America (Ahemad and Kibret, 2014). Leguminous plants that majorly include chickpea, soyabean, cowpea, lentils and groundnut have gained increased importance throughout the years due to their high nutritional content, besides which they have also been known for their ability to improve soil quality by fixing atmospheric nitrogen into the soil, with the help of nitrogen-fixing bacteria that colonise the roots of the plants (Dutta et al., 2022). Although legumes stand as a rich source of plant-based protein and offer significant ecological benefits, the production and yield of the crops are limited due to inadequate cultivation area as well as challenges from both abiotic stressors (drought, salinity, high and low temperature) and biotic threats (soil-borne pathogens). One distinctive approach for boosting legume productivity while fostering sustainable agriculture is to improve our understanding of the plant microbiome. The plant microbiome consists of different types of bacteria, fungi and other microorganisms in a plant ecosystem (the phyllosphere, endosphere, and rhizosphere) that play a beneficial role in mitigating environmental stress, aid in their resistance against pathogens by synthesising protective compounds and activating plant immunity and also promote plant growth and development by increasing nutrient uptake and regulating plant hormones. Prominent abiotic stresses arising as a result of climate change, that primarily affect agricultural fields and reduce crop productivity, include factors such as drought, high and low temperature, high salinity and nutrient deficiency in soil. Sustainable cultivation aims to promote crop cultivation in a sustainable way under challenging environmental conditions that demand the use of beneficial microorganisms. They have improved the antioxidant enzyme activity, enhanced nutrient uptake, regulated plant hormones synthesis and ACC deaminase activity, along with the production of siderophore. In this way, the microorganisms have helped to regulate plant biomass production and enhance crop yield. Along with this, bioengineering of beneficial microorganisms has provided a creative approach to sustain stress tolerance in plants (Figure 4.1).

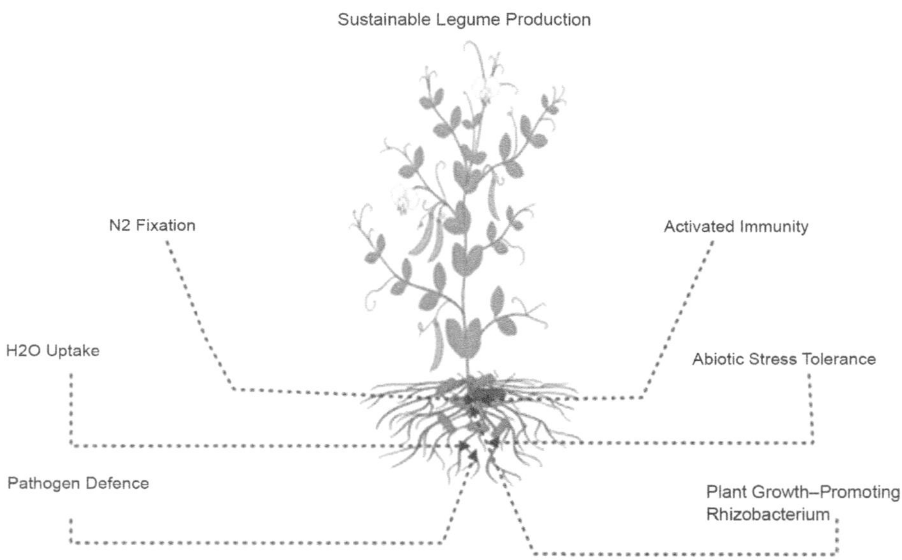

FIGURE 4.1 Diagrammatic representation of sustainable legume production.

Both biotic and abiotic stresses lead to major challenges for plants, as a result of which plants are bound to devise some tolerance mechanisms and the involvement of microorganisms is one of them. In a biotic stress environment, plants interact simultaneously with both antagonistic and beneficial factors, necessitating a delicate balance between activating defence mechanisms and preserving symbiotic relationships. Beneficial microorganisms play an important role by increasing nutrient uptake, strengthening plant immunity, and improving overall physiological parameters, resulting in a positive synergistic impact that boosts plant resistance to biotic stress (Omacini et al., 2001).

4.5.1 Amelioration of Abiotic Stressors

Abiotic stresses are the primary causes of the reduction of crop yield globally. Legumes cultivated in marginal or semi-arid environments need to tolerate factors like drought, salinity and high temperature. The associated microbiome, like plant growth–promoting rhizobacteria (PGPR) and plant growth–promoting fungi (PGPF), plays an essential role in helping leguminous plants cope with the adverse environment.

In limited water conditions, certain microbial isolates enhance water uptake and retention. Microorganisms such as *Bacillus* and *Pseudomonas* produce phytohormones like auxins (IAA) and gibberellins that promote root structure and increase the soil water uptake volume, along with nutrient uptake. The enzyme ACC deaminase reduces the concentration of the stress-ethylene precursor (ACC) in plants and effectively minimise the stress-ethylene production. High ethylene levels inhibit root growth during stress and thereby the microbial action sustains the hazardous effects of drought-induced premature senescence and root stunting (Glick, 2014).

Saline soils hinder the uptake of water through osmotic pressure and create ion toxicity (Na^+ and Cl^-accumulation). Some microbial strains can mitigate these effects by promoting the isolation of toxic ions by altering ion channel regulation in the epidermis of the root or by producing exopolysaccharides (EPS). They help in binding Na^+ ions in the rhizosphere region and, in turn, minimise uptake by the host (Han et al., 2015). Microbial communities can enhance the internal defence mechanism of plants through induced systemic tolerance (IST). Certain bacterial genera, like *Rhizobium* and non-rhizobium PGPR, are known to overcome heavy metals in the soil (Figure 4.2).

4.5.2 Biocontrol of Soil-Borne Pathogens

The use of microorganisms for tolerance against biotic stress has gained increasing importance because of its ecologically friendly alternative ways, which have helped to diminish the negative impact of dangerous chemical pesticides and fertilisers. They are now becoming an essential component of sustainable agriculture (Niu et al., 2020). Soil-borne pathogens are a significant biological stress to the growth and development of legumes, as they cause root rots, wilting and damping-off diseases (caused by *Fusarium, Pythium, Rhizoctonia,* and *Sclerotium* species). The

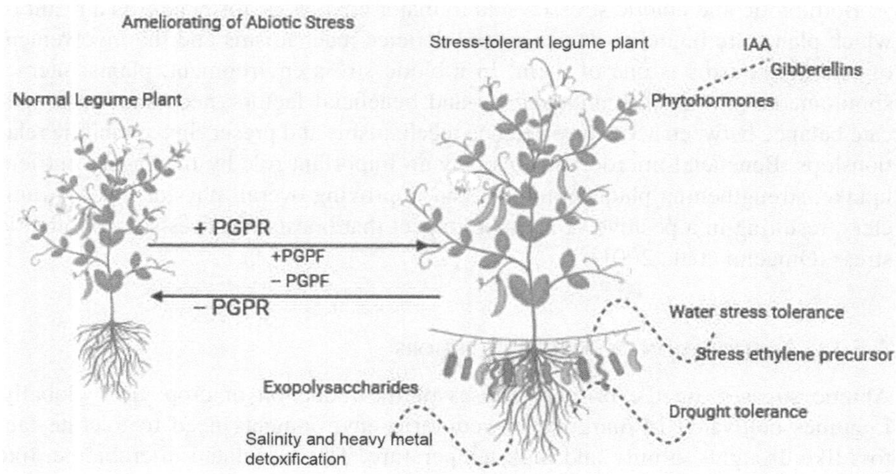

FIGURE 4.2 A diagrammatic representation about the involvement of PGPR in the amelioration of abiotic stressors.

indigenous and inoculated microbiome acts as a natural, sustainable way of defence. (Figure 4.3).

There are several mechanisms by which the microorganisms control soil-borne pathogens. (i) Direct antagonism and antibiosis – strains of *Trichoderma* (fungi), *Bacillus*, and *Pseudomonas* (bacteria) – directly antagonise pathogens. Antibiosis involves the secretion of secondary metabolites, which are toxic to the pathogen. *Pseudomonas* strains have been documented to produce hydrogen cyanide that can interfere with the respiration of pathogens. *Trichoderma* species show mycoparasitism; here, the biocontrol fungus directly attacks and enzymatically (by using chitinases and glucanases) degrades the cell wall of the pathogenic fungus (Harman et al., 2004). (ii) Competition for resources and niche exclusion – the rhizosphere is a highly competitive environment where nutrients and space are limited. Key symbiotic microbes such as PGPR and beneficial *Rhizobium* strains colonise the root surface. They form a biofilm that physically excludes pathogens from their infection sites. Furthermore, beneficial microbes compete with pathogens for root exudate nutrients, such as simple sugars and amino acids, which are essential for pathogen germination and proliferation (Weller, 2007). (iii) Induction of systemic resistance (ISR) – it is the activation of the immune system of the host plant. The recognition of microbe-associated molecular patterns (MAMPs) from beneficial microbes (e.g. flagellin, lipopolysaccharides (LPS), volatile organic compounds (VOCs)) by plants leads to the priming of defence pathways. This priming leads to a stronger immune response (Van Loon et al., 1998). ISR is mediated by signalling molecules (jasmonic acid and ethylene).

Microbiome-mediated stress tolerance in plants is necessary due to the presence of beneficial microorganisms within them. They improve the defence mechanisms of

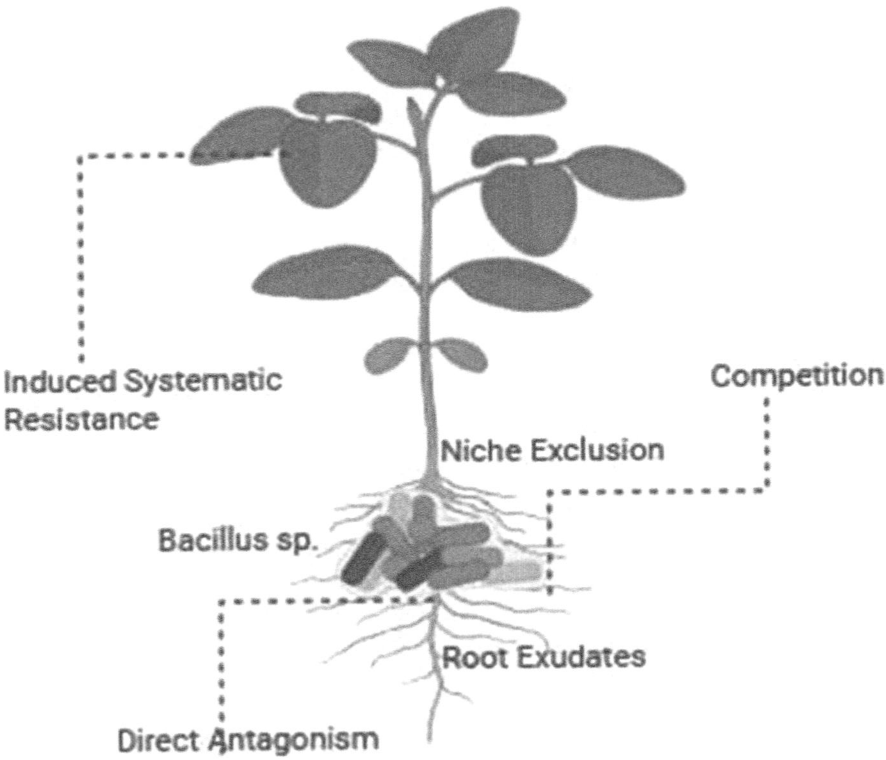

FIGURE 4.3 Pictorial representation indicating biocontrol of soil-borne pathogens.

plants to cope with the changing environment, as well as to sustain the abiotic factors that affect plant growth. These microorganisms are effective alternatives to chemical insecticides and pesticides because they harm neither the plant nor the ecosystem. They are an effective tool for food security.

4.6 FIELD EFFICACY AND IMPACT OF MICROBIAL INOCULANTS

The beneficial bacteria, fungi or their consortia, collectively known as microbial inoculants, have been demonstrated to be a valuable tool in sustainable agriculture; their field applications yet remain far from achieving the promised potential. Despite encouraging outcomes in controlled settings, trophic competition, antagonistic interactions with native microbiota, and abiotic factors (such as pH, temperature, and moisture) sometimes impair their performance in open-field settings (Khare and Arora, 2014). The primary problems are poor rhizosphere colonisation and a sharp drop in the inoculated microbial population, particularly when the physicochemical characteristics of the soil are poorly understood before application. The customised and standardised bioformulations, which are paired vis-à-vis with local agroecology and soil conditions, perform better and aid the plant in an efficient manner.

However, more research is needed in this direction to fully utilise the potential of microbial inoculants. Recently, liquid microbial inoculants have also become popular and have shown potential in field adaptability, stress tolerance, and shelf life, though the responses varied with different crops, genotypes, seasonality, and local microbial diversity. All-India Network Project on Soil Biodiversity & Biofertilizers (ICAR) demonstrated statistically significant production improvements in cereals and legumes utilising liquid formulations. The multi-omics approach could be very useful in determining the dependability and scalability of novel microbial strains and strain combinations and ultimately mitigating the problems of laboratory to field transitions (Díaz-Rodríguez et al., 2025).

4.7 SYSTEMIC CLASSIFICATION OF HIGH-IMPACT MICROBIAL SPECIES AND CONSORTIA

Integrative frameworks constituting the merging of phylogenetics, functional genomics, and ecological niche modelling are increasingly applied to classify microbial taxa and consortia that have substantial ecological and industrial impacts. In the context of plant–microbe symbioses, bioremediation, and bioenergy initiatives, notable taxa such as *Pseudomonas, Bacillus, Rhizobium, Azospirillum*, and *Trichoderma* are frequently identified as keystone species (Santamaría et al., 2025). The architecture of consortia, comprising either modular, hierarchical, or synergistic, plays a vital role in assessing field effectiveness and scalability. Furthermore, based on both origin and assembly methods, prominent microbial species and consortia have been systematically classified as natural (either undefined or defined), artificial, or synthetic (Ibrahim et al., 2021). Unlike defined consortia, which are specific groups pinpointed through metagenomics and certain culturing techniques, natural consortia are obtained from environments like soil or compost. Recent studies have highlighted the development and application of artificial and synthetic consortia involving selective microbial strains to perform targeted functions such as biological nitrogen fixation, phosphate solubilisation, disease management, and promoting plant growth (Minchev et al., 2021). Recent research highlights the use of synthetic assemblies for several leguminous crops that incorporate *Bacillus amyloliquefaciens, Pseudomonas chlororaphis,* and *Trichoderma harzianum* to improve plant health and manage both root and foliar diseases effectively (Cirillo et al., 2023). Combinations of *Trichoderma afroharzianum* and *Azotobacter chroococcum* have shown considerable enhancements in crop growth, resilience, and soil fertility for crops such as wheat and maize (Manjunath et al., 2023), even when facing challenging conditions. Furthermore, reports of yield increase between 15% and 30% across various agricultural systems demonstrate that the use of PGPR consortia is rapidly expanding to boost agricultural productivity (Joshi et al., 2025).

A robust, compatible microbial consortia formulation methodology and a multidisciplinary approach that could integrate biological, ecological, societal and

TABLE 4.2
Comparison between the Pros and Cons of Upscaling Microbial Consortia | Designed for Crop Enhancement.

Parameters	Pros	Cons
Plant growth	Enhance nutrient uptake, root development, and biomass yield	Variable efficacy across soil types and crop genotypes
Formulation and delivery	Liquid and encapsulated forms improve application ease	Shelf-life, carrier compatibility, and regulatory hurdles
Environmental impact	Reduces chemical fertiliser use, improves soil health	May disrupt native microbial communities if poorly matched
Scalability	Suitable for diverse agro-climatic zones with tailored consortia	Requires cold chain or stabilisers for shelf-life extension
Economic viability	Cost-effective in the long-term; potential for rural bio enterprise models	Initial formulation and QC costs; low farmer awareness
Policy and adoption	Aligns with sustainable agriculture and ESG goals	Needs stronger extension services and policy incentives

regulatory aspects will be very effective for scale-up production. Further, strategies like the use of machine learning techniques for strain selection, nanomaterial encapsulation for efficient delivery, and synthetic biology approaches may help to overcome upscaling and sustainability hurdles. There are instances where user-friendly regulator networks have been framed, such as the European Fertilizing Products Regulation (EU) 2019/1009 and India's FCO Amendment 2021. Further, the success of microbial consortia at the field level also depends upon the farmer education, extensive extension programs and cooperative research (Table 4.2).

4.8 SYNTHETIC MICROBIAL COMMUNITIES (SYNCOMS): DESIGNING AND IMPLEMENTATION

The usage of microbes to enhance crop productivity dates back to the 19th century and this has led to the formulation of microbial inoculants, which are effective in increasing crop productivity and sustainability (Singh and Trivedi, 2017). However, these practices did not show consistency in the field conditions, which can be due to insufficient colonisation of the inoculants on the roots, poor or less interaction of the inoculants with the native microorganisms and the soil micro-environment. The advent of SynComs, emerged as an exciting development in this area. SynCom systems are versatile and scalable, where microbial communities are assembled from well-defined and characterised microbial strains that helps in the understanding of microbial interactions among themselves, with the native microbiome as well as with the host (McCarty and Ledesma-Amaro, 2019). For example, SynComs developed from sugarcane and rhizosphere-associated microbes resulted in the production of greater biomass as well as drought tolerance in maize plants, improved yields and enhanced scab tolerance in potatoes (Wang et al., 2019), suggesting SynComs are functional at the inter-species level as well. Synthetic microbial systems have provoked interest in research surrounding the plant–microbe interaction, microbe-microbe interaction, etc., due to its capability in differentiating the role of secondary metabolites, signalling molecules, root exudate composition and exo-metabolites for the purpose of communications, colonisation and establishing microbial interactions in the rhizosphere (Dolinšek et al., 2016). SynComs can be designed mostly by either a top-down approach or a bottom-up approach. The bottom-up approach works by assembling a specific set of microorganisms of known characteristics. Microbial strains are chosen based on genotypic and phenotypic characterisation, isolation procedures, community functions as well as their functional attributes. Recently, studies have shown that the strains that are able to coexist in a complex yet stable habitat are unable to survive in pairwise culture systems. This indicates that the mutual interactions are important for the coexistence of multiple species. In contrast to the bottom-up approach, the top-down approach mostly relies on assembling a vast diversity of strains and gradually decreasing their complexity in a step-by-step fashion to understand the key functions as well as their specific taxonomic roles (Mehlferber et al., 2024). Other approaches include undergoing experimental evolution to enrich strains by multiple cycles of reinoculations or generating sub-communities by serial dilutions or random filtering. The bottom-up and top-down approaches for SynCom development are necessary and complementary, allowing scientists to create a bridge between the simple and complex systems and understand the structure and function of the microbiome (D'Souza et al., 2018). Successful implementations require a combination of advanced techniques like seed-biopriming, where microbial mixtures are coated on the seeds, encapsulation and nano formulations, designed to ensure successful colonisation on the roots and to protect the microbes before getting out-competed by the previously present native microbiome. Several factors are considered while implementing SynCom in natural and agricultural settings. Apart from the deployment of SynCom, ethical factors surrounding the research must also be considered, that is data should be retrievable, accessible, interoperable and reusable. Numerous data and strain repositories are available to achieve the goal, but effective reusability mostly depends on the availability of metadata (Huttenhower et al., 2023).

4.9 ADVANCED GENE EDITING (CRISPR/CAS9) AND ITS ROLE IN LEGUME SUSTAINABILITY

The recent molecular techniques, along with the multi-omics approaches, have facilitated translating the laboratory discoveries into large-scale practical disease resistance programs. CRISPR/Cas9 (clustered regularly interspaced short palindromic repeats and CRISPR-associated protein 9) is a relatively new technique that is being extensively used in functional genomics. It is a precise and efficient technique of gene manipulation used specifically for the knockout and knockdown of a specific gene of interest (Hendel et al., 2015). CRISPR/Cas9 technology is getting increased attention for its ability in managing plant diseases. Moreover, the CRISPR technology is used to genetically manipulate the microorganisms and enhance their effects in supporting overall crop health and productivity (Susan et al., 2019) as well as to generate new varieties that could be resistant to pathogens. Most large-seeded legumes are recalcitrant to genetic modification. Genetic modification is primarily achieved by *Agrobacterium*-mediated transformation (*Agrobacterium tumefaciens* C58 and *Agrobacterium rhizogenes* K599) as well as by the use of biolistic technologies. Almost 60% of successful CRISPR editing has been achieved by *Agrobacterium*-based transformation strategies. The transformation efficacy in lentils remained less than 1% in spite of numerous trials with a variety of explants. Definite optimisation and culture conditions are necessary for increasing the success rate of CRISPR technology. For example, soybean (*Glycine max*) fatty acyl-ACP thioesterase 1 (*GmFATA1*) and Glycine max fatty acyl-ACP thioesterase 2 (*GmFATA2*) are genes found in soybeans, involved in fatty acid biosynthesis. The use of CRISPR technology led to the knockout of the target genes, resulting in reduced fatty acid synthesis as well as growth defects. Overexpression of these acyl carrier protein (*ACP*) genes increased the fatty acid content, seed yield and vegetative growth (Liao et al., 2024). Knockout of mildew-resistant locus O (*MLO-7*), downy mildew-resistant genes (*DMR6*) and powdery mildew-resistant genes (*PMR4*) by CRISPR technology has led to the development of resistance against mildew, downy mildew and *Phytophthora* spp. respectively. Similarly, mutations in PUB17 (E3 ubiquitin ligase) increased resistance to *Botrytis cinerea* and *Alternaria solani* in tomato.

4.10 CONCLUSION

The plant-microbiome interaction has undoubtedly promoted sustainable agricultural systems, providing a high-throughput alternative to chemical fertilisers. The basis of this sustainability is synergistic interactions among the microbial communities, particularly between nitrogen-fixing rhizobium and the mycorrhizal fungi, leading to highly energy-driven processes. The function of the key microbiome lies beyond the biogeochemical cycles and extends to promoting growth, better nutrient sequestration, enhanced crop productivity, etc. The advent of various multi-omics approaches has led to a better understanding of the plant-microbiome interactions and the complex pathways facilitating those. Beneficial microbial inoculants are often integrated into the plant system to enhance crop yields. The incorporation of microbial technology into agricultural fields requires a lot of upgrades. Scientists have now shifted their focus more towards the active designing and construction of SynCom, based on their genotypic and functional attributes. Delivery systems are also getting improved, like nano-formulations and seed coating, to help the beneficial microbial inoculants survive under unfavourable climatic conditions. Once the microbes are well-associated with the plants and start to co-exist with the natural microbiome, they can function for a prolonged time period. The advent of multi-omic studies, as well as various genetic manipulation techniques like CRISPR technology, RNAi technology, has promoted effective plant–microbe interaction, thereby leading to enhanced crop productivity. Apart from the scientific advancements, policies must be updated between stakeholders and scientists for the deployment of the microbial systems as well as their safe and effective commercialisation to promote sustainable agriculture.

REFERENCES

Afridi, Muhammad Siddique, Muhammad Ammar Javed, Sher Ali, Flavio Henrique Vasconcelos De Medeiros, Baber Ali, Abdul Salam, Sumaira et al. "New opportunities in plant microbiome engineering for increasing agricultural sustainability under stressful conditions." *Frontiers in Plant Science* 13, (2022): 899464.

Ahemad, Munees, and Mulugeta Kibret. "Mechanisms and applications of plant growth promoting rhizobacteria: current perspective." *Journal of King Saud University-Science* 26, no. 1 (2014): 1–20.

Andrade-Domínguez, Andrés, Emmanuel Salazar, María del Carmen Vargas-Lagunas, Roberto Kolter, and Sergio Encarnación. "Eco-evolutionary feedbacks drive species interactions." *The ISME Journal* 8, no. 5 (2014): 1041–1054.

Ben Gaied, Roukaya, Clarisse Brígido, Imed Sbissi, and Mohamed Tarhouni. "Sustainable strategy to boost legumes growth under salinity and drought stress in semi-arid and arid regions." *Soil Systems* 8, no. 3 (2024): 84.

Cirillo, Valerio, Ida Romano, Sheridan L. Woo, Emilio Di Stasio, Nadia Lombardi, Ernesto Comite, Olimpia Pepe, Valeria Ventorino, and Albino Maggio. "Inoculation with a microbial consortium increases soil microbial diversity and improves agronomic traits of tomato under water and nitrogen deficiency." *Frontiers in Plant Science* 14, (2023): 1304627.

Díaz-Rodríguez, Alondra María, Fannie Isela Parra Cota, Luis Alberto Cira Chávez, Luis Fernando García Ortega, María Isabel Estrada Alvarado, Gustavo Santoyo, and Sergio de Los Santos-Villalobos. "Microbial inoculants in sustainable agriculture: advancements, challenges, and future directions." *Plants* 14, no. 2 (2025): 191.

Dolinšek, Jan, Felix Goldschmidt, and David R. Johnson. "Synthetic microbial ecology and the dynamic interplay between microbial genotypes." *FEMS Microbiology Reviews* 40, no. 6 (2016): 961–979.

D'Souza, Glen, Shraddha Shitut, Daniel Preussger, Ghada Yousif, Silvio Waschina, and Christian Kost. "Ecology and evolution of metabolic cross-feeding interactions in bacteria." *Natural Product Reports* 35, no. 5 (2018): 455–488.

Dutta, Asik, Ankita Trivedi, Chaitanya P. Nath, Debjyoti Sen Gupta, and Kali Krishna Hazra. "A comprehensive review on grain legumes as climate-smart crops: Challenges and prospects." *Environmental Challenges* 7, (2022): 100479.

Fan, Yinglun, Jinge Liu, Shanhua Lyu, Qi Wang, Shengming Yang, and Hongyan Zhu. "The soybean Rfg1 gene restricts nodulation by *Sinorhizobium fredii* USDA193." *Frontiers in Plant Science* 8, (2017): 1548.

Glick, Bernard R. "Bacteria with ACC deaminase can promote plant growth and help to feed the world." *Microbiological Research* 169, no. 1 (2014): 30–39.

Guo, Kaiyan, Jun Yang, Nan Yu, Li Luo, and Ertao Wang. "Biological nitrogen fixation in cereal crops: Progress, strategies, and perspectives." *Plant Communications* 4, no. 2 (2023).

Han, Qin, Fengli Wu, Xiaonan Wang, Hong Qi, Liang Shi, Ang Ren, Qinghai Liu, Mingwen Zhao, and Canming Tang. "The bacterial lipopeptide iturins induce *Verticillium dahliae* cell death by affecting fungal signalling pathways and mediate plant defence responses involved in pathogen-associated molecular pattern-triggered immunity." *Environmental Microbiology* 17, no. 4 (2015): 1166–1188.

Harman, Gary E., Charles R. Howell, Ada Viterbo, Ilan Chet, and Matteo Lorito. "*Trichoderma* species – opportunistic, avirulent plant symbionts." *Nature Reviews Microbiology* 2, no. 1 (2004): 43–56.

Hassani, M. Amine, Paloma Durán, and Stéphane Hacquard. "Microbial interactions within the plant holobiont." *Microbiome* 6, no. 1 (2018): 58.

Hendel, Ayal, Rasmus O. Bak, Joseph T. Clark, Andrew B. Kennedy, Daniel E. Ryan, Subhadeep Roy, Israel Steinfeld et al. "Chemically modified guide RNAs enhance CRISPR-Cas genome editing in human primary cells." *Nature Biotechnology* 33, no. 9 (2015): 985–989.

Huttenhower, Curtis, Robert D. Finn, and Alice Carolyn McHardy. "Challenges and opportunities in sharing microbiome data and analyses." *Nature Microbiology* 8, no. 11 (2023): 1960–1970.

Ibrahim, Maziya, Lavanya Raajaraam, and Karthik Raman. "Modelling microbial communities: Harnessing consortia for biotechnological applications." *Computational and Structural Biotechnology Journal* 19, (2021): 3892–3907.

Janati, Walid, Bouchra Benmrid, Wissal Elhaissoufi, Youssef Zeroual, Joshua Nasielski, and Adnane Bargaz. "Will phosphate bio-solubilization stimulate biological nitrogen fixation in grain legumes?" *Frontiers in Agronomy* 3, (2021): 637196.

Joshi, Diksha, Amit Kaushik, Reetesh Kumar, Aditi Arya, Gustavo Santoyo, Vipin Kumar Singh, Nikhil Kashyap et al. "Improving plant performance through microbiome manipulation: The potential role of current bioengineering approaches." *Bacteria* 4, no. 1 (2025): 12.

Khare, Ekta, and Naveen K. Arora. "Effects of soil environment on field efficacy of microbial inoculants." *Plant Microbes Symbiosis: Applied Facets*, (2014): 353–381.

Li, Jiayu, Juntao Wang, Hongwei Liu, Catriona A. Macdonald, and Brajesh K. Singh. "Application of microbial inoculants significantly enhances crop productivity: A meta-analysis of studies from 2010 to 2020." *Journal of Sustainable Agriculture and Environment* 1, no. 3 (2022): 216–225.

Liao, Wenying, Runze Guo, Kun Qian, Wanxuan Shi, James Whelan, and Huixia Shou. "The acyl–acyl carrier protein thioesterases *GmFATA1* and *GmFATA2* are essential for fatty acid accumulation and growth in soybean." *The Plant Journal* 118, no. 3 (2024): 823–838.

Manjunath, Mallappa, Anil Khokhar, Gajjala Ravindra Chary, Manmohanjit Singh, Sushil Kumar Yadav, Kodigal A. Gopinath, Narayana Jyothilakshmi, Karlapudi Srinivas, Mathyam Prabhakar, and Vinod Kumar Singh. "Microbial consortia enhance the yield of maize under sub-humid rainfed production system of India." *Frontiers in Sustainable Food Systems* 7, (2023): 1108492.

McCarty, Nicholas S., and Rodrigo Ledesma-Amaro. "Synthetic biology tools to engineer microbial communities for biotechnology." *Trends in Biotechnology* 37, no. 2 (2019): 181–197.

Mehlferber, Elijah C., Gontran Arnault, Bishnu Joshi, Laila P. Partida-Martinez, Kathryn A. Patras, Marie Simonin, and Britt Koskella. "A cross-systems primer for synthetic microbial communities." *Nature Microbiology* 9, no. 11 (2024): 2765–2773.

Minchev, Zhivko, Olga Kostenko, Roxina Soler, and María J. Pozo. "Microbial consortia for effective biocontrol of root and foliar diseases in tomato." *Frontiers in Plant Science* 12, (2021): 756368.

Niu, Ben, Weixiong Wang, Zhibo Yuan, Ronald R. Sederoff, Heike Sederoff, Vincent L. Chiang, and Rainer Borriss. "Microbial interactions within multiple-strain biological control agents impact soil-borne plant disease." *Frontiers in Microbiology* 11, (2020): 585404.

Omacini, Marina, Enrique J. Chaneton, Claudio M. Ghersa, and Christine B. Müller. "Symbiotic fungal endophytes control insect host–parasite interaction webs." *Nature* 409, no. 6816 (2001): 78–81.

Santamaría, Zaki, Pilar Navarro-Gómez, Juan A. Martínez-Mancebo, Maitane Juárez-Mugarza, Amando Flores, and Inés Canosa. "Genetic and species rearrangements in microbial consortia impact biodegradation potential." *The ISME Journal* 19, no. 1 (2025): wraf014.

Schaedel, Marie, Gabriela Hidrobo, and Julie Grossman. "From microns to meters: Exploring advances in legume microbiome diversity for agroecosystem benefits." *Frontiers in Sustainable Food Systems* 5, (2021): 668195.

Singh, Brajesh K., and Pankaj Trivedi. "Microbiome and the future for food and nutrient security." *Microbial Biotechnology* 10, no. 1 (2017): 50.

Suman, Jarupula, Amitava Rakshit, Siva Devika Ogireddy, Sonam Singh, Chinmay Gupta, and J. Chandrakala. "Microbiome as a key player in sustainable agriculture and human health." *Frontiers in Soil Science* 2, (2022): 821589.

Susan, Anannya, Manoj Kumar Yadav, Sonali Kar, S. Aravindan, U. Ngangkham, S. Raghu, S. R. Prabhukarthikeyan et al. "Molecular identification of blast resistance genes in rice landraces from northeastern India." *Plant Pathology* 68, no. 3 (2019): 537–546.

Timmusk, Salme, Lawrence Behers, Julia Muthoni, Anthony Muraya, and Anne-Charlotte Aronsson. "Perspectives and challenges of microbial application for crop improvement." *Frontiers in Plant Science* 8, (2017): 49.

Van, Loon, L. C., P. A. H. M. Bakker, and C. M. J. Pieterse. "Systemic resistance induced by rhizosphere bacteria." *Annual Review of Phytopathology* 36, no. 1 (1998): 453–483.

Wang, Zhenshuo, Yan Li, Lubo Zhuang, Yue Yu, Jia Liu, Lixia Zhang, Zhenjiang Gao et al. "A rhizosphere-derived consortium of *Bacillus subtilis* and *Trichoderma harzianum* suppresses common scab of potato and increases yield." *Computational and Structural Biotechnology Journal* 17, (2019): 645–653.

Weller, David M. "*Pseudomonas* biocontrol agents of soilborne pathogens: looking back over 30 years." *Phytopathology* 97, no. 2 (2007): 250–256.

5 Role of Symbiotic and Non-Symbiotic Microbes in Legume Rhizosphere

Ekta Narwal, Jairam Choudhary, Sujit Kumar Bishi, Madan Kumar, Seema Sangwan, and Prakash Chand Ghasal

5.1 INTRODUCTION

Leguminous plants occupy a central role in global agriculture due to their ecological, nutritional, and economic significance. Legumes, belonging to the family *Fabaceae*, are a cornerstone of global agriculture due to their nutritional, ecological, and economic significance. They are cultivated worldwide, ranging from soybeans in the Americas to pulses in Asia and Africa, and contribute directly to food security and environmental sustainability. With more than 22,000 recognised species distributed across various ecological niches, legumes represent one of the largest families of flowering plants. They are cultivated both as food crops (soybean, chickpea, lentil, pea, mung bean, cowpea, and groundnut), and as forage and green manure species (alfalfa, clover, stylosanthes, vetch). In many developing regions, legumes serve as the primary source of plant-based proteins, providing essential amino acids that complement cereal-based diets. Beyond human nutrition, they also contribute significantly to livestock feed and agroecosystem services such as soil fertility, erosion control, and carbon sequestration. Legumes are rich in protein (20–40%), dietary fibber, minerals (iron, zinc, magnesium), and vitamins, making them essential in combating malnutrition, particularly in developing countries where animal protein is scarce or expensive. Pulses such as chickpeas, lentils, and beans are staples in human diets, while soybean dominates global trade as both food and feed. With growing interest in plant-based diets, legumes have gained importance as sustainable protein alternatives (FAO, 2016; Ritchie & Roser, 2020). Legumes are unique in their ability to establish symbiotic associations with soil microbes, particularly nitrogen-fixing rhizobia, which reduces their reliance on external nitrogen inputs. One of the most significant contributions of legumes is their symbiotic association with *Rhizobium* and related bacteria, which fix atmospheric nitrogen into plant-available forms. This natural fertilisation reduces the need for synthetic nitrogen fertilisers, lowering greenhouse gas emissions and production costs while improving soil fertility for subsequent crops (Peoples et al., 2009). Legumes are widely used in crop rotations and intercropping systems to

DOI: 10.1201/9781003733324-6

enhance soil health and disrupt pest and disease cycles. Legumes play a vital role in climate-smart agriculture. Their lower carbon footprint compared to animal protein sources makes them important for climate mitigation strategies. Many species are drought-tolerant and adaptable to marginal environments, offering resilience in the face of climate change. Moreover, their diverse root exudates promote microbial biodiversity and soil ecological balance (Stagnari et al., 2017). Economically, soybean is the most cultivated legume, with production reaching ~390 million tonnes in 2023, concentrated in Brazil, the USA, Argentina, China, and India. Pulses account for ~95 million tonnes, with India, Canada, Myanmar, China, and Nigeria being major producers (FAO STAT, 2023). Legumes are traded extensively, providing income and livelihood opportunities for millions of smallholder farmers. Their role in livestock feed, processed foods, and plant-based protein industries further underlines their economic importance. Legumes directly contribute to several UN Sustainable Development Goals (SDGs), including Zero Hunger (SDG 2), Responsible Consumption and Production (SDG 12), Climate Action (SDG 13), and Life on Land (SDG 15). They improve nutritional security, reduce dependence on chemical fertilisers, enhance soil health, and promote sustainable farming practices (FAO, 2016).

5.2 THE RHIZOSPHERE AS A CRITICAL ECOLOGICAL INTERFACE

The rhizosphere, the narrow soil zone directly influenced by plant roots, is a dynamic environment where plants, soil, and microbes interact intensively. In legumes, the rhizosphere is of particular importance due to their ability to form symbiotic associations with nitrogen-fixing bacteria and other beneficial microbes. These interactions make legume rhizospheres hotspots of biological activity that influence soil fertility, plant health, and ecosystem sustainability. Legume roots release a variety of exudates, including sugars, amino acids, flavonoids, and organic acids, which shape microbial community structure in the rhizosphere. A defining feature is the symbiosis between legumes and *Rhizobium* or *Bradyrhizobium* species, leading to root nodule formation and biological nitrogen fixation. In addition to rhizobia, legume rhizospheres host arbuscular mycorrhizal fungi (AMF), plant growth–promoting rhizobacteria (PGPR), and diverse microbial consortia that enhance nutrient cycling and plant growth (Richardson et al., 2009). Legumes contribute substantially to soil nutrient enrichment. Nitrogen fixed in root nodules not only supports the host plant but also enriches the soil for subsequent crops in rotation. Root exudates mobilise phosphorus and other micronutrients, while interactions with AMF improve uptake efficiency. These processes enhance soil fertility and reduce the dependence on external chemical inputs (Peoples et al., 2009; Meena et al., 2018). The rhizosphere of legumes supports high microbial diversity, including beneficial bacteria (*Pseudomonas*, *Bacillus*, *Azospirillum*) and fungi, which suppress soil-borne pathogens through competition, antibiosis, and induced systemic resistance. The enrichment of soil organic matter and stimulation of enzymatic activities, such as dehydrogenase, phosphatase, and urease, further contribute to improved soil health (Stagnari et al., 2017).

5.3 MICROBIAL COMMUNITIES IN THE LEGUME RHIZOSPHERE

The rhizosphere of legumes represents a biologically active zone where root exudates interact with a wide spectrum of microorganisms. These microbial communities play crucial roles in nutrient cycling, plant growth promotion, and disease suppression. They can broadly be categorised into symbiotic microbes, which establish intimate associations with legume roots, and non-symbiotic microbes, which live freely in the rhizosphere but still influence plant health and soil fertility. Microbial communities in the legume rhizosphere form a dynamic and functionally diverse consortium that profoundly influences plant growth, nutrient cycling, and ecosystem stability. These communities include rhizobia, AMF, endophytes, actinobacteria, diazotrophs, and various non-symbiotic plant growth–promoting rhizobacteria, all of which interact synergistically with legumes to enhance nutrient acquisition, especially nitrogen and phosphorus (Mendes et al., 2013). Legume root exudates rich in sugars, amino acids, organic acids, and flavonoids act as chemical cues that shape the composition and functional potential of rhizosphere microbial assemblages, facilitating the recruitment of beneficial taxa involved in nodulation, nitrogen fixation, and mineral solubilisation (Sasse et al., 2018). These microbial communities also contribute to plant defence by producing antibiotics, siderophores, and elicitors that suppress soil-borne pathogens and prime plant immune responses (Berendsen et al., 2012). Furthermore, microbial networks in the rhizosphere improve legume resilience to abiotic stresses such as drought, salinity, and metal toxicity by enhancing osmoprotection, modulating phytohormone levels, and improving root system architecture (Compant et al., 2019). Through their collective metabolic activities and complex signalling interactions, rhizosphere microbial communities function as integral components of the legume holobiont, ensuring sustained productivity and ecological adaptability in diverse agroecosystems. The rhizosphere of legumes harbours both symbiotic and non-symbiotic microbial communities that play pivotal roles in plant growth, nutrient cycling, and stress tolerance. These microbes either establish intimate symbioses with the plant or exert beneficial effects as free-living associates, shaping the ecological dynamics of legume cultivation systems.

5.4 SYMBIOTIC MICROBES

5.4.1 Nitrogen-Fixing Rhizobia

Rhizobium, *Bradyrhizobium*, *Mesorhizobium*, and *Sinorhizobium* species establish nodules on legume roots, where they convert atmospheric nitrogen into ammonium via the nitrogenase enzyme complex. This process reduces the dependence on synthetic nitrogen fertilisers and enriches soil fertility (Peoples et al., 2009).

5.4.2 Arbuscular Mycorrhizal Fungi

AMF play a pivotal role in enhancing nutrient acquisition, stress tolerance, and overall symbiotic efficiency in the legume rhizosphere. By forming extensive extra radical hyphal networks, AMF significantly improve the uptake of phosphorus, micronutrients, and water, thereby supporting better root development and facilitating effective nodulation by rhizobia (Smith & Read, 2008). Their interaction with legumes also promotes the production of signalling molecules such as strigolactones, which enhance both mycorrhization and rhizobial infection processes, resulting in improved biological nitrogen fixation (Parniske, 2008). Moreover, AMF help alleviate abiotic stresses-including salinity, drought, and heavy-metal toxicity-by improving antioxidant capacity and maintaining ionic balance, which ultimately sustains legume growth and symbiotic functioning under adverse conditions (Ruiz-Lozano et al., 2016). Through these combined mechanisms, AMF act as key ecological engineers in the legume rhizosphere, reinforcing the synergistic roles of symbiotic and non-symbiotic microbes in promoting plant health and productivity.

5.4.3 Endophytes

Endophytic bacteria and fungi are essential contributors to legume health and productivity, functioning as internal symbionts that enhance nutrient acquisition, stress resilience, and nodulation efficiency. These microbes colonise internal tissues without causing disease and produce a variety of bioactive metabolites-such as phytohormones, siderophores, and antioxidant enzymes-that stimulate root growth, improve nitrogen fixation, and enhance rhizobial infection processes (Santoyo et al., 2016). Many endophytes also accelerate phosphorus and iron mobilisation through organic acid secretion and siderophore production, thereby supporting nutrient-limited legumes in the rhizosphere (Compant et al., 2010). In addition, endophytes help legumes withstand abiotic stresses, including drought, salinity, and heavy-metal toxicity, by modulating osmolyte production, enhancing ROS scavenging, and maintaining membrane stability (Cardoso et al., 2013). Through these mechanisms, endophytes act as key internal partners that complement rhizobial and mycorrhizal symbioses, strengthening the overall microbial network that sustains legume performance in diverse environments.

5.5 NON-SYMBIOTIC MICROBES: THE UNSUNG PARTNERS

While rhizobia are the hallmark symbionts of legumes, the rhizosphere is colonised by a much broader microbial community, including plant growth–promoting rhizobacteria (PGPR), free-living diazotrophs, phosphate-solubilising bacteria, siderophore-producing bacteria, actinobacteria, and fungi. Many of these microbes do not form nodules or establish direct intracellular symbioses with legumes, yet their presence in the rhizosphere is critical for plant nutrition, growth, and resilience. Non-symbiotic microbes interact with legumes through multiple mechanisms: enhancing nutrient availability, regulating plant hormones, suppressing pathogens, and mitigating abiotic stresses. Recent research emphasises that legume performance is not solely determined by rhizobia but by the collective function of diverse microbial consortia (Etesami & Santoyo, 2025). Both rhizobial and non-rhizobial bacteria are critical for legume survival, particularly under abiotic stresses such as salinity, drought, heavy metal toxicity, and nutrient imbalances. Although the ecological significance of non-rhizobial bacteria in legumes is not fully understood, their interactions with rhizobia inside nodules can influence plant fitness and behaviour

(Martínez-Hidalgo & Hirsch, 2017). While rhizobia have been the primary focus of research on legume–microbe interactions, increasing attention is being given to non-symbiotic microbes that colonise the rhizosphere or even inhabit nodules without directly fixing nitrogen. Non-rhizobial endophytes found within nodules are increasingly recognised for their synergistic role. They modulate plant hormonal balance, improve nutrient acquisition, and enhance tolerance to environmental stresses such as drought, salinity, heavy metal toxicity, and temperature extremes. The co-inoculation of legumes with rhizobia and non-rhizobial bacteria has often resulted in higher yields, improved nodulation, and enhanced resilience compared to inoculation with rhizobia alone (Etesami, 2022).

Non-rhizobial bacteria, many outside the Proteobacteria phylum, do not directly induce nodulation or nitrogen fixation but often exhibit multiple plant growth–promoting (PGP) traits These traits include:

- **Phytohormone production**: Indole-3-acetic acid (IAA), gibberellins, and cytokinins that stimulate root growth.
- **Enzymatic activity**: Hydrolytic enzymes such as lipase, cellulase, protease, and chitinase.
- **Nutrient mobilisation**: Phosphate solubilisation and siderophore production.
- **Stress tolerance**: 1-aminocyclopropane-1-carboxylate (ACC) deaminase activity reduces ethylene stress under salinity and drought.
- **Biocontrol effects**: Suppression of pathogens like *Fusarium oxysporum*, *Alternaria burnsii*, and *Rhizoctonia solani* through antibiosis, competition for iron (via siderophores), lytic enzyme production, degradation of virulence factors, and induced systemic resistance (Compant et al., 2005).

5.6 SYNERGISTIC INTERACTIONS IN RHIZO-MICROBIOME

Symbiotic and non-symbiotic microbes often interact synergistically. For example, AMF can enhance phosphorus availability, which improves rhizobial nodulation efficiency, while PGPR can stimulate root growth and facilitate microbial colonisation. Such cooperative networks promote nutrient cycling, improve stress resilience, and sustain ecosystem services in legume-based agroecosystems (Meena et al., 2018).

The mechanisms through which non-rhizobial bacteria enhance legume growth, survival, and symbiotic efficiency with rhizobia are diverse and multifaceted (Etesami et al., 2015). These microorganisms stimulate legume development by synthesising or modulating phytohormones; supplying essential nutrients such as phosphorus (via phosphate solubilisation), iron (through siderophore release), potassium, and even biologically fixed nitrogen; alleviating plant stress responses (Noori et al., 2018); and functioning as biocontrol agents against soil-borne pathogens.

Co-inoculation of legumes with non-rhizobial bacteria and rhizobia has also been reported to further improve plant performance by:

i. Stimulating the production of flavonoids that activate nod gene expression
ii. Increasing the number, biomass, and efficiency of nodules, thereby enhancing overall nodulation and plant growth

TABLE 5.1
Comparative Overview of Symbiotic and Non-symbiotic Microbial Communities in the Legume Rhizosphere

Category	Microbial Groups	Examples	Key Functions	Benefits to Legumes
Symbiotic Microbes	**Nitrogen-fixing bacteria (Rhizobia)**	*Rhizobium*, *Bradyrhizobium*, *Mesorhizobium*, *Sinorhizobium*	Form nodules; fix atmospheric N_2 into NH_4^+ via nitrogenase	Supply nitrogen to host; reduce fertiliser need; improve soil fertility
	Arbuscular Mycorrhizal Fungi (AMF)	*Glomus*, *Gigaspora*, *Acaulospora*	Form arbuscules in cortical cells; enhance nutrient uptake (P, Zn)	Improve P acquisition; enhance drought tolerance; soil aggregation; pathogen resistance
	Endophytes	*Burkholderia*, *Enterobacter*, *Piriformospora indica*	Colonise internal root/nodule tissues; some fix N_2, produce hormones	Promote growth, stress tolerance, and secondary metabolite production
Non-Symbiotic Microbes	**Free-living N-fixers**	*Azotobacter*, *Clostridium*	Fix nitrogen independently in rhizosphere	Supply limited N to soil; support microbial diversity
	PGPR (Plant Growth-Promoting Rhizobacteria)	*Pseudomonas*, *Bacillus*, *Azospirillum*	Produce phytohormones, solubilise P, induce systemic resistance	Stimulate root growth, nutrient uptake, and plant immunity
	Biocontrol agents	*Trichoderma*, *Streptomyces*	Produce antibiotics, lytic enzymes, and compete with pathogens	Suppress soil-borne diseases; reduce pesticide use
	Decomposers/ Nutrient cyclers	Saprophytic fungi (*Aspergillus*, *Penicillium*); bacteria	Decompose organic matter; release nutrients	Enhance soil fertility, organic matter recycling, and microbial diversity

iii. Boosting nitrogen assimilation and leghemoglobin content within nodule-
siv. Triggering defense pathways that strengthen resistance to pathogens
v. Improving mineral uptake and nutritional status of the host plant.

Collectively, these functions highlight the crucial role of non-rhizobial bacteria in strengthening legume-rhizobia symbiosis and promoting plant productivity under varying environmental conditions (Martínez-Hidalgo and Hirsch, 2017).

TABLE 5.2
Synergistic Interaction Microorganisms in the Rhizosphere to Benefit the Host Plant

Legume Crop	Interacting Microorganisms	Improved Role	Citation
Soybean	*Bradyrhizobium japonicum* + *Pseudomonas fluorescens*	Improved root growth and nutrient uptake	Egamberdieva et al. (2017a)
Soybean	*Bradyrhizobium japonicum*-SB1 *Bacillus thuringiensis*-KR1	Improved plant growth and nodulation	Mishra et al. (2009a)
Pea and Lentil	*Bacillus thuringiensis*-KR1 *Rhizobium leguminosarum*	Improved plant growth and nodulation	Mishra et al. (2009b)
Pea	*Rhizobium leguminosarum* + AMF (*Glomus intraradices*)	Improved phosphorus acquisition and nodulation	Smith & Read (2008)
Mung bean	*Pseudomonas aeruginosa* BHU B13–398 *Bacillus subtilis* BHU M	Improved biocontrol against soil pathogens and plant growth	Kumari et al. (2018)
Lupin	*Bradyrhizobium lupini* + *Penicillium bilaiae*	Improved P mobilisation and nodulation	Khan et al. (2010)
Faba bean	*Rhizobium TAL–1148*+ Bacilus subtilis	Reduced uptake of Ni	Elbagory et al. (2022)
Pongamia pinnata	*Rhizobium pisi* PZHK2+ *Ochrobacterium*	Enhanced activities of non-enzymatic antioxidants	Yu et al. (2022)
Soybean	*Bacilus japonicum E109* + *Azospirillum brasilense Az39*	Reduced As translocation to shoots	Armendariz et al. (2019)
Chickpea	*Mesorhizobium strain RC3* +	Reduced Cr uptake	Wani et al. (2008)
Horse gram	*Bacillus sp. AS03*	Reduced shoot translocation	
Horse gram	*Rhizobium sp. AS05*+ *Bacillus sp. AS03*	Reduced shoot translocation of Cr	Dhali et al. (2021)

5.6.1 Symbiotic Nitrogen Fixation: A Cornerstone of Legume Productivity

A wide range of microbes have the ability to fix the atmospheric nitrogen. Nearly about 2 genera of archaea (87 species), 38 bacterial genera and 20 cyanobacterial genera have been identified with the nitrogen fixation capability (Dixon et al., 1986; Sprent & Sprent, 1990). The hallmark of legumes is their symbiotic association with rhizobia, a group of α- and β-proteobacteria that colonise root hairs and induce the formation of nodules. Within these nodules, rhizobia differentiate into bacteroids, which house the enzyme nitrogenase. This enzyme catalyses the conversion of atmospheric nitrogen into ammonia, which can be assimilated by the plant into amino acids and other nitrogenous compounds. In return, the host plant provides carbohydrates and other organic compounds to sustain bacterial metabolism. The

TABLE 5.3
Plant Growth Promoting Microorganisms in Legume Rhizosphere.

Legume Crop	Microorganism	PGP Role	Reference
Soybean	*Bradyrhizobium japonicum*	Nitrogen fixation, nodulation	Hungria & Vargas (2000)
Common bean	*Rhizobium etli*	Enhanced nodulation and N fixation	Martínez-Romero, (2003)
Pea	*Rhizobium leguminosarum* bv. *viciae*	Improved nutrient uptake	Oldroyd et al. (2011)
Cowpea	*Bradyrhizobium yuanmingense*	Biological nitrogen fixation	Stepkowski et al. (2012)
Alfalfa	*Sinorhizobium meliloti*	Root growth promotion, N fixation	Jones et al. (2007)
Medicago sativa	*Sinorhizobium meliloti BL225C*	IAA production, Nitrogen-fixation and acdS positivity	Carelli et al. (2000)
Soybean	*Azospirillum brasilense*	Root elongation, hormone production	Bashan & de-Bashan (2010)
Soybean	*Sinorhizobium fredii strain SMH12*	Improved nodule number and water potentials	Kibido et al. (2020)
Faba bean	*Rhizobium leguminosarum*	Enhanced production of Osmoprotectants against drought stress	Amine-Khodja et al. (2022)
Phaseolus vulgaris	*Rhizobium etli*	Overexpressed Trehalose-6-Phosphate Synthase against drought stress	Suárez et al. (2008)
Chick Pea	*Mesorhizobium ciceri IC53*	Increased proline contents under salinity stress	Egamberdieva et al. (2017b)

global contribution of legume-rhizobia symbiosis to nitrogen inputs is estimated at least 70 million tonnes annually, a figure that underscores its ecological and agricultural importance (Brockwell et al., 1995). In cropping systems, legumes often improve the yield of subsequent cereal crops through nitrogen carryover effects. For example, maize yields following soybean or groundnut cultivation are significantly higher compared to continuous maize systems. Thus, rhizobial symbiosis not only benefits the legume host but also enhances the sustainability of mixed and rotational cropping systems.

5.6.2 Role of Nodule-Associated Non-Rhizobial Endophytes in Legume Rhizobium Symbiosis

Legumes play a crucial role in sustainable agriculture due to their capacity to fix atmospheric nitrogen through symbiosis with rhizobia. Growing evidence shows that non-rhizobial bacteria (NRB) residing within root nodules significantly enhance this symbiosis, especially under stress conditions. These NRB support nodule development and function by producing plant growth–promoting (PGP) metabolites, regulating phytohormone levels, and improving nutrient availability. They synthesise key hormones-including auxins (IAA), cytokinins, gibberellins, abscisic acid, jasmonic acid, and salicylic acid-as well as enzymes such as ACC deaminase, which reduce ethylene's negative impact on nodulation. NRB also aid nutrient cycling by solubilising essential minerals like phosphorus, potassium, silicates, zinc, and iron, all of which contribute to efficient nitrogen fixation. Co-inoculation of legumes with rhizobia and NRB carrying multiple PGP traits has demonstrated synergistic benefits by enhancing plant growth, yield, and tolerance to environmental stresses (Etesami & Santoyo, 2025).

5.7 ROLES BEYOND NITROGEN FIXATION

For many decades, nitrogen fixation was considered the sole contribution of rhizobia to legumes. However, research now indicates that rhizobia play additional roles in plant growth and stress adaptation.

5.7.1 Abiotic Stress Alleviation

Legumes, like all plants, are frequently exposed to abiotic stresses such as drought, salinity, acidity, alkalinity, heavy metal toxicity, and temperature extremes. These stresses reduce nodulation efficiency, impair nitrogenase activity, and limit productivity. Microbes mitigate these effects through multiple mechanisms.

5.7.2 Drought Stress

Drought stress reduces transpiration and water movement in legumes, which in turn limits the transport of nitrogen-fixation products and suppresses nitrogenase activity (McCulloch et al., 2021). It also leads to a decline in biomass and chlorophyll content, while inducing the accumulation of reactive oxygen species that can damage various biomolecules, including DNA. To cope with drought, legumes commonly enhance the production of antioxidants such as superoxide dismutase, catalase, ascorbic acid, and glutathione, along with osmoprotectants that help maintain cellular osmotic balance (Sarkar & Oba, 2018). Several rhizobial symbionts are known to improve drought tolerance in legumes, including *Bradyrhizobium diazoefficiens* SEMIA 5080 in soybean R01–581F, *Mesorhizobium huakuii* 7653R in *Astragalus sinicus* L., and *Sinorhizobium medicae* or *S. meliloti* in *Medicago truncatula* (Liu et al., 2022; Staudinger et al., 2016). Additionally, inoculation with *Rhizobium meliloti* has been shown to increase nodule formation and enhance photosynthesis in kidney bean, black bean, mung bean, and chickpea under water-limited conditions (Siddiqui et al., 2021). Rhizobia enhance antioxidant defence systems and osmolyte accumulation in host plants. Rhizobia-induced accumulation of Cu in *Medicago sativa* roots and increased expression of antioxidants have also been observed (Chen et al., 2018). Non-rhizobial microbes regulate hormonal pathways, mitigating the negative impacts of stress-induced hormones such as ethylene and abscisic acid.

5.7.3 Heavy Metal Stress

Soils naturally contain trace amounts of heavy metals needed for plant growth, but their elevated levels-due to human activities or natural deposits-can harm plant development. High metal concentrations generate reactive oxygen species that damage key cellular components and disrupt rhizobial symbiosis, reducing nodule formation and nitrogen fixation (Mondal et al., 2015; Goyal & Habtewold, 2023). However, rhizobia from metal-contaminated soils often evolve strong metal-tolerance mechanisms. These metal-tolerant strains can enhance legume growth, yield, and stress resilience under toxic levels of metals such as Cd, Ni, Hg, Cr, As, Al, Cu, and Pb. Their natural association with legumes helps plants establish and diversify even in metal-stressed environments. Both groups contribute to the immobilisation, detoxification, or sequestration of toxic heavy metals, thus allowing legumes to establish themselves in contaminated soils. A positive correlation between Rhizobium-induced differential methylation and expression of m6A RNA in soybean plants under Cd stress indicates a different mechanism of metal toxicity remediation (Han et al., 2022). The evolutionary success of legumes in diverse and often hostile environments is partly attributable to these microbial partnerships, which provide adaptive advantages that extend far beyond nitrogen nutrition (Goyal & Habtewold, 2023).

5.7.4 Soil PH

Soil pH strongly shapes soil properties, nutrient availability, and the overall functioning of microbial communities. Most soil microorganisms, including root-nodulating rhizobia, grow best near neutral pH, yet much of the world's arable land is either acidic or alkaline (Ferguson et al., 2013). Extreme pH conditions can hinder the establishment of legume–rhizobia symbiosis (Cooper, 1989), and delayed nodulation is commonly observed in legumes grown under acidic conditions. The ability of compounds such as genistein-an inducer of nod gene expression-to counteract acid-induced inhibition highlights that pH directly influences the regulation of symbiotic signalling (Hungria & Stacey, 1997). Soil pH also alters rhizosphere community composition, which in turn affects plant root responses (Han et al., 2020). Legumes adapted to acidic soils often produce modified Nod factors suitable for low-pH environments. Several rhizobial genes (actA, typA, atvA, lpiA, ubiF) are associated with improved acid tolerance and enhanced symbiotic performance (Tiwari et al., 2016)). For example, the acid-tolerant strain *Rhizobium tropici* CIAT899 synthesises more glutathione and produces ~1.8-fold more Nod factors under acidic conditions, many of which differ from those formed at neutral pH (Riccillo et al., 2000). Rhizobia overall show resilience across a wide pH range, with their high-pH stress responses involving elevated antioxidant activity, organic acid production, and protein profile adjustments.

5.7.5 Salinity Stress

Soil salinity-often intensified by the use of saline irrigation water and closely linked to drought-is a major factor reducing the effectiveness of legume–rhizobia symbiosis

(Chakraborty & Harris, 2022). High salt levels lead to toxic ion buildup and alter the quality of flavonoids released by legume roots, which disrupts nod factor production and early chemical communication between the partners. As a result, salt stress weakens rhizobial infection of root hairs in crops such as bean (Zahran & Sprent, 1986), soybean (Tu, 1981), pea (Borucki & Sujkowska, 2007), and chickpea (L'Taief et al., 2007), ultimately lowering nitrogen fixation (Aridhi et al., 2020). Some rhizobia, however, help legumes tolerate salinity by promoting indole-3-acetic acid synthesis, enhancing the accumulation of osmo-protectants, improving root osmotic water flow through changes in xylem potential and aquaporin levels, and modifying the plant's protein profile. Although nod factor production under high-salt conditions may be affected, certain rhizobia still maintain partially effective signalling (Goyal & Habtewold, 2023).

5.7.6 Biotic Stress Alleviation

Plants selectively recruit soil microbes that enhance their genetic and metabolic capacities. The resulting rhizosphere often acts as a microbially mediated barrier against soil pathogens through nutrient/space competition or direct antagonism (Dini-Andreote, 2020). Pathogens that bypass this barrier encounter the plant's innate defences-basal defences involving antimicrobial peptides, and inducible defences characterised by pathogen recognition, signalling, cell-wall reinforcement, and synthesis of toxic compounds (Balint-Kurti, 2019). Endophytes further strengthen plant disease resistance. For example, root-associated communities enriched in Chitinophagaceae and Flavobacteriaceae suppress fungal root diseases. Flavobacterium species possess gene clusters for non-ribosomal peptide synthetases and polyketide synthases, enabling strong antagonistic activity (Carrión et al., 2019). Although such evidence mostly comes from non-legume systems, it supports similar suppressive roles in other plants. Several studies highlight the biocontrol potential of rhizobia. Co-inoculation of *Mesorhizobium loti* with PGPR reduced white rot in Brassica by 99% (Chandra et al., 2007), and combining Rhizobium with *Trichoderma harzianum* protected groundnut from *Sclerotium rolfsii* (Ganesan et al., 2007). Cell-free rhizobial extracts have inhibited multiple fungal pathogens-*Macrophomina phaseolina*, *Rhizoctonia solani*, *Fusarium solani*, and *F. oxysporum*- and suppressed soybean root rot. *R. leguminosarum* bv. *phaseoli* (Parveen et al., 2019). Overall, the effectiveness of microbial biocontrol depends heavily on specific host genotypes and bacterial strains, underscoring the complexity of plant–microbe-pathogen interactions.

5.7.7 Nutrient Solubilisation

Microorganisms play a fundamental role in nutrient solubilisation in soil, acting as key drivers of biogeochemical cycles that determine plant nutrient availability. Many essential nutrients-such as phosphorus, potassium, sulphur, zinc, and iron-are often present in soils in insoluble or chemically bound forms that plants cannot directly absorb. Soil microbes, including bacteria, fungi, and actinomycetes, transform these unavailable nutrient pools into accessible forms through a variety of biochemical processes such as organic acid production, enzymatic mineral dissolution, siderophore release, and redox transformations (Narwal et al., 2023). By mobilising these nutrients, microbes not only enhance soil fertility and plant growth but also reduce the reliance on chemical fertilisers, making them central to sustainable and environmentally friendly agricultural systems. These nutrients are crucial in improving legume rhizobium symbiosis. Potassium (K) is essential for maintaining ionic balance, energy supply, and carbohydrate transport in legume nodules, thereby enhancing nodule formation and nitrogenase activity. It regulates key K transporters in roots and nodules, influencing ion homeostasis and signalling crucial for efficient symbiotic nitrogen fixation. Since most soil K is insoluble, improving its availability can significantly boost rhizobial performance, nodule development, and overall N_2-fixation efficiency. Silicon (Si) enhances symbiotic N_2 fixation in legumes by improving Si uptake through NIP transporters, strengthening nodule cell walls, and boosting nodule number and biomass. It optimises nitrogenase activity by supporting leghaemoglobin formation, improving gas exchange, and increasing nutrient uptake (P, Zn, K, Fe). Si also enhances photosynthesis and mitigates stresses like salinity, drought, and oxidative damage, allowing greater carbon allocation to rhizobia. Overall, Si supplementation significantly improves nodulation efficiency and N_2-fixing capacity in legumes. Iron (Fe) is essential for legume–rhizobium symbiosis because it forms key components of leghemoglobin, nitrogenase, and cytochromes that enable oxygen regulation and energy production for N_2 fixation. Efficient Fe transport-mediated by transporters like GmDMT1, NRAMP, YSL, and ZIP-is critical for delivering Fe to bacteroids through the symbiosome membrane. Fe deficiency severely reduces nodule initiation, development, and nitrogenase activity in many legumes. Since most soil Fe is insoluble, improving its availability is vital for maximising nodulation and N_2-fixing efficiency. Zinc (Zn) is essential for nodule development, bacteroid differentiation, and nitrogenase activity because it supports key Zn-dependent enzymes, antioxidant defences, and regulatory pathways that sustain N_2 fixation. Adequate Zn availability enables proper functioning of transporters like MtZIP6, MtYSL3, and ZnuABC, ensuring efficient Zn delivery to nodules. Zn deficiency disrupts metabolic activity, accelerates nodule senescence through regulators like FUN, and severely reduces N_2-fixation efficiency. Thus, maintaining sufficient Zn is crucial for optimal legume–rhizobium symbiosis and sustainable crop productivity. Nodule-associated bacteria and rhizospheric microbes are now recognised as key contributors to nutrient availability and the overall efficiency of legume–rhizobium symbiosis. Although rhizobia remain central to nitrogen fixation, numerous studies have shown that NRB within nodules enhance this process by improving the nutritional environment required for effective nodulation. These endophytic partners often possess strong nutrient-mobilising traits, including phosphate solubilisation, siderophore production, and mineral weathering, which increase the availability of essential elements such as phosphorus, iron, zinc, and potassium (Martínez-Hidalgo & Hirsch, 2017). By enhancing mineral acquisition, nodule associated bacteria support both rhizobial survival inside nodules and the metabolic demands of nitrogen fixation. Moreover, many bacteria synthesise phytohormones-such as indole-3-acetic acid, gibberellins, and cytokinins-that stimulate root growth and nodule formation, thereby optimising rhizosphere conditions for symbiotic establishment (Spaepen et al., 2007). Some strains also produce ACC deaminase, which reduces ethylene levels and prevents ethylene-induced inhibition of nodulation (Glick, 2014). In addition to improving nutrient cycling, nodule associated bacteria influence the internal redox environment of nodules by promoting antioxidant accumulation and mitigating oxidative stress, which is crucial under nutrient-deficient or stressful soil conditions. Nodule associated bacteria-rhizobium co-inoculation strategies show synergistic benefits, including higher nitrogen fixation rates, greater nutrient uptake, and improved plant growth under both optimal and stress conditions (Zgadzaj et al., 2016). Collectively, these findings highlight that nodule associated bacteria are not passive occupants of nodules but active contributors to nutrient mobilisation and symbiotic efficiency. Understanding their functional roles provides a promising avenue for developing next-generation microbial inoculants to improve legume productivity and sustainability in nutrient-limited agroecosystems (Etesami & Santoyo, 2025).

5.8 PRODUCTION OF GROWTH HORMONES

Auxin (IAA) plays a central concentration-dependent role in Rhizobium–legume symbiosis, promoting nodule initiation, cortical cell division, and nitrogenase activity at optimal levels (Etesami & Glick, 2024; Spaepen et al., 2007). Low to moderate IAA enhances N_2 fixation by 32–45% and upregulates key symbiotic genes such as nifA, fixK, and nifH (Ali et al., 2008; Defez et al., 2016). Rhizobial IAA production improves bacterial survival through enhanced EPS and trehalose biosynthesis (Bianco et al., 2009). However, excessive IAA triggers ethylene synthesis and suppresses nodulation, as seen in *Medicago truncatula* (Gamalero et al., 2023; Roy et al., 2017). During nodule organogenesis, rhizobia modulate polar auxin transport, creating localised auxin maxima essential for infection and primordium formation (Ng & Mathesius, 2018). Auxin interacts with cytokinin pathways and is fine-tuned by microRNAs such as miR393 and miR160, which modulate auxin receptor and ARF expression (Cai et al., 2017). Overall, auxin functions as both a promoter and inhibitor of nodulation, integrating hormonal and environmental cues to regulate optimal N_2 fixation. Ethylene is a key regulatory hormone that negatively influences legume-rhizobium symbiosis, particularly by inhibiting infection thread formation, suppressing nodule initiation, and accelerating nodule senescence, all of which reduce N_2-fixation efficiency (Guinel, 2015). Under stress conditions such as salinity, drought, or heavy metals, plants frequently overproduce ethylene, further restricting nodulation and nitrogenase activity. Ethylene-modulating microbes, especially those producing the enzyme ACC deaminase, play a crucial role in counteracting this inhibition. ACC deaminase degrades the ethylene precursor ACC into α-ketobutyrate and ammonia, thereby lowering ethylene levels and relieving its suppressive effects on nodulation (Glick, 2014). Co-inoculation of rhizobia with ACC deaminase-producing plant growth–promoting bacteria has been shown to enhance nodule formation, improve rhizobial competitiveness, and increase nitrogenase activity under stress conditions (Alemneh et al., 2020). These microbes also delay nodule senescence, maintaining N_2-fixation for longer periods (Nascimento et al., 2016). Furthermore, several stress-tolerant non-rhizobial nodule bacteria-such as Brachybacterium, Brevibacterium, Halomonas, Vibrio, Pseudomonas, and Paenibacillus-have demonstrated strong ACC deaminase activity and the ability to support plant performance under salinity and drought, enhancing conditions for effective symbiosis (Mukhtar et al., 2020). Together, these findings highlight that ethylene acts as a major brake on N_2 fixation, while ACC deaminase–producing microbes function as biological modulators capable of restoring optimal nodulation and sustaining symbiotic efficiency in challenging environments. Cytokinin is a key hormonal regulator of legume–rhizobium symbiosis, controlling early nodule organogenesis, bacteroid differentiation, and overall nitrogen-fixation efficiency. During infection, rhizobial signals activate cytokinin biosynthesis in host roots, triggering cortical cell division and nodule primordium formation through receptors such as CRE1/AHK4 (Frugier et al., 2008). Cytokinin also modulates the balance between nodule initiation and autoregulation, ensuring optimal investment in N_2-fixing structures (Gamas et al., 2017). Microbial partners-including Rhizobium, Mesorhizobium, Bacillus, Pseudomonas, and Azospirillum-synthesise active cytokinins such as trans-zeatin and isopentenyladenine via de novo pathways, directly enhancing nodulation and nitrogenase activity (Kejela, 2024; Uniyal et al., 2022). For example, *Mesorhizobium ciceri* ND-64 increases chickpea nitrogen fixation by elevating cytokinin levels in nodules (Lohosha et al., 2023), while *Pseudomonas fluorescens* G20–18 boosts plant stress tolerance through cytokinin mediated antioxidant activation (Großkinsky et al., 2016). Cytokinin-producing microbes also interact with other hormones: they counter ethylene's inhibitory effect on nodulation and improve nodule longevity by modulating cytokinin-ABA-ethylene crosstalk (Stefen et al., 2022). Despite their importance, the Cytokinin producing capabilities of many nodule non-rhizobial bacteria remain unexplored, even though these endophytes may significantly influence cytokinin-dependent processes such as nodule development, stress resilience, and nutrient mobilisation (Fahde et al., 2023). Understanding microbial cytokinin modulation is therefore central for optimising symbiotic nitrogen fixation and enhancing sustainable legume productivity. Gibberellic acid plays an important role in legume growth, nodulation, and nitrogen fixation by promoting cell division, root elongation, and nodule expansion, thereby supporting the metabolic demands of symbiotic N_2 fixation (Magome & Kamiya, 2016). In legumes, gibberellic acid influences nodule size and function, with recent evidence showing that soybean converts rhizobia-derived gibberellic acid precursors (GA9) into active GA4 through host GA3-oxidases, enhancing nodule development and bacteroid fitness (Nett et al., 2022). Microbial partners-including Serratia, Bacillus, Azospirillum, and Pseudomonas-synthesise gibberellic acid through terpenoid pathways and modulate endogenous plant GA levels to improve shoot growth, biomass accumulation, and stress resilience (Devi et al., 2016; Egamberdieva et al., 2017). Despite these benefits, the GA-producing capacity of many nodule non-rhizobial bacteria remains uncharacterised, representing a key gap in understanding hormone-mediated optimisation of nitrogen fixation.

5.9 MYCORRHIZAL FUNGI AND LEGUMES

AMF, belonging to the phylum Glomeromycota, form symbiotic associations with most legumes. Unlike rhizobia, AMF colonise root cortical cells and form arbuscules-branched structures that facilitate nutrient exchange.

The benefits of AMF to legumes include:

- Enhanced phosphorus uptake, especially in P-deficient soils
- Improved drought tolerance through increased root water absorption and accumulation of osmolytes
- Suppression of root pathogens through induced systemic resistance
- Synergistic interactions with rhizobia: Co-inoculation of legumes with rhizobia and AMF often results in additive or synergistic effects on growth and yield. For example, soybean co-inoculated with *Glomus intraradices* and *Bradyrhizobium japonicum* shows improved nodulation and nitrogen fixation compared to single inoculations.

AMF also influence soil structure through glomalin production, enhancing soil aggregation and long-term fertility (Narwal et al., 2018; Singh et al., 2017).

REFERENCES

Alemneh, A., Zhou, Y., Ryder, M., & Denton, M. (2020). Mechanisms in plant growth-promoting rhizobacteria that enhance legume–rhizobial symbioses. Journal of Applied Microbiology, 129(3), 1133–1156. https://doi.org/10.1111/jam.14754

Ali, B., Hayat, S., Hasan, S. A., & Ahmad, A. (2008). A comparative effect of IAA and 4-Cl-IAA on growth, nodulation and nitrogen fixation in *Vigna radiata* (L.) Wilczek. Acta Physiologiae Plantarum, 30(1), 35–41. https://doi.org/10.1007/s11738-007-0088-4

Amine-Khodja, I. R., Boscari, A., Riah, N., Kechid, M., Maougal, R. T., Belbekri, N., & Djekoun, A. (2022). Impact of two strains of Rhizobium leguminosarum on the adaptation to terminal water deficit of two cultivars Vicia faba. Plants, 11(4), 515.

Aridhi, F., Sghaier, H., Gaitanaros, A., Khadri, A., Aschi-Smiti, S., & Brouquisse, R. (2020). Nitric oxide production is involved in maintaining energy state in alfalfa (Medicago sativa L.) nodulated roots under both salinity and flooding. Planta, 252, 22.

Armendariz, A. L., Talano, M. A., Nicotra, M. F. O., Escudero, L., Breser, M. L., Porporatto, C., & Agostini, E. (2019). Impact of double inoculation with Bradyrhizobium japonicum E109 and Azospirillum brasilense Az39 on soybean plants grown under arsenic stress. Plant Physiology and Biochemistry, 138, 26–35.

Balint-Kurti, P. (2019). The plant hypersensitive response: Concepts, control and consequences. Molecular Plant Pathology, 20, 1163–1178.

Bashan, Y., & de-Bashan, L. E. (2010). How the plant growth-promoting bacterium Azospirillum promotes plant growth – a critical assessment. Advances in Agronomy, 108, 77–136.

Berendsen, R. L., Pieterse, C. M. J., & Bakker, P. A. H. M. (2012). The rhizosphere microbiome and plant health. Trends in Plant Science, 17(8), 478–486. https://doi.org/10.1016/j.tplants.2012.04.001

Bianco, C., Imperlini, E., & Defez, R. (2009). Legumes like more IAA. Plant Signaling & Behavior, 4, 763–765. https://doi.org/10.4161/psb.4.8.9166

Borucki, W., & Sujkowska, M. (2007). The effects of sodium chloride-salinity upon growth, nodulation, and root nodule structure of pea (Pisum sativum L.) plants. Acta Physiologiae Plantarum, 30, 293–301.

Brockwell, J., Bottomley, P. J., & Thies, J. E. (1995). Manipulation of rhizobia microflora for improving legume productivity and soil fertility: A critical assessment. Plant and Soil, 174(1), 143–180.

Cai, Z., Wang, Y., Zhu, L., Tian, Y., Chen, L., Sun, Z., Ullah, I., & Li, X. (2017). GmTIR1/GmAFB3-based auxin perception regulated by miR393 modulates soybean nodulation. New Phytologist, 215, 672–686.

Cardoso, E. J. B. N., Vasconcellos, R. L. F., Bini, D., Miyauchi, M. Y. H., Santos, C. A. D., Alves, P. R. L., . . . Nogueira, M. A. (2013). Soil health: Looking for suitable indicators. What should be considered to assess the effects of use and management on soil health? Scientia Agricola, 70, 274–289.

Carelli, M., Gnocchi, S., Fancelli, S., Mengoni, A., Paffetti, D., Scotti, C., & Bazzicalupo, M. (2000). Genetic diversity and dynamics of Sinorhizobium meliloti populations nodulating different alfalfa cultivars in Italian soils. Applied and Environmental Microbiology, 66(11), 4785–4789.

Carrión, V. J., Perez-Jaramillo, J., Cordovez, V., Tracanna, V., de Hollander, M., Ruiz-Buck, D., . . . Mendes, L. W. (2019). Pathogen-induced activation of disease-suppressive functions in the endophytic root microbiome. Science, 366, 606–612. 10.1126/science.aaw9285

Chakraborty, S., & Harris, J. M. (2022). At the crossroads of salinity and Rhizobium–legume symbiosis. Molecular Plant–Microbe Interactions, 35, 540–553.

Chandra, S., Choure, K., Dubey, R. C., & Maheshwari, D. (2007). Rhizosphere-competent Mesorhizobium loti MP6 induces root hair curling, inhibits Sclerotinia sclerotiorum and enhances growth of Indian mustard. Brazilian Journal of Microbiology, 38, 124–130.

Chen, J., Liu, Y. Q., Yan, X. W., Wei, G. H., Zhang, J. H., & Fang, L. C. (2018). Rhizobium inoculation enhances copper tolerance by affecting copper uptake and regulating the ascorbate–glutathione cycle in Medicago sativa. Ecotoxicology and Environmental Safety, 162, 312–323.

Compant, S., Clément, C., & Sessitsch, A. (2010). Plant growth-promoting bacteria in the rhizo- and endosphere of plants: Their role, colonization, mechanisms involved and prospects for utilization. Soil Biology & Biochemistry, 42(5), 669–678. https://doi.org/10.1016/j.soilbio.2009.11.024

Compant, S., Duffy, B., Nowak, J., Clément, C., & Barka, E. A. (2005). Use of plant growth-promoting bacteria for biocontrol of plant diseases: Principles, mechanisms of action, and future prospects. Applied and Environmental Microbiology, 71(9), 4951–4959.

Compant, S., Samad, A., Faist, H., & Sessitsch, A. (2019). A review on the plant microbiome: Ecology, functions, and emerging trends in microbial application. Journal of Advanced Research, 19, 29–37. https://doi.org/10.1016/j.jare.2019.03.004

Cooper, J. E. (1989). Nodulation of legumes by rhizobia in acid soils. In V. Vančura & F. Kunc (Eds.), Developments in Soil Science (Vol. 18, pp. 57–61). Elsevier.

Defez, R., Esposito, R., Angelini, C., & Bianco, C. (2016). Overproduction of indole-3-acetic acid in free-living rhizobia induces transcriptional changes resembling those occurring in nodule bacteroids. Molecular Plant-Microbe Interactions, 29, 484–495.

Devi, K. A., Pandey, P., & Sharma, G. D. (2016). Plant growth-promoting endophyte Serratia marcescens AL2–16 enhances the growth of Achyranthes aspera L. HAYATI Journal of Biosciences, 23, 173–180.

Dhali, S., Pradhan, M., Sahoo, R. K., Mohanty, S., & Pradhan, C. (2021). Alleviating Cr (VI) stress in horse gram (Macrotyloma uniflorum Var. Madhu) by native Cr-tolerant nodule endophytes isolated from contaminated site of Sukinda. Environmental Science and Pollution Research, 28(24), 31717–31730.

Dini-Andreote, F. (2020). Endophytes: The second layer of plant defence. Trends in Plant Science, 25, 319–322.

Dixon, R. O. D., & Wheeler, C. T. (1986). Nitrogen Fixation in Plants. Glasgow, United Kingdom: Blackie.

Egamberdieva, D., Wirth, S. J., Alqarawi, A. A., Abd_Allah, E. F., & Hashem, A. (2017a). Phytohormones and beneficial microbes: Essential components for plants to balance stress and fitness. Frontiers in Microbiology, 8, 2104.

Egamberdieva, D., Wirth, S. J., Shurigin, V. V., Hashem, A., & Abd_Allah, E. F. (2017b). Endophytic bacteria improve plant growth, symbiotic performance of chickpea (Cicer arietinum L.) and induce suppression of root rot caused by Fusarium solani under salt stress. Frontiers in Microbiology, 8, 1887.

Egamberdieva, D., Wirth, S. J., Jabborova, D., Räsänen, L. A., and Liao, H. (2017c). Coordination between Bradyrhizobium and Pseudomonas alleviates salt stress in soybean through altering root system architecture. Journal of Plant Interactions, 12, 100–107. https://doi.org/10.1080/17429145.2017.1294212.

Elbagory, M., El-Nahrawy, S., & Omara, A. E. D. (2022). Synergistic interaction between symbiotic N2 fixing bacteria and Bacillus strains to improve growth, physiological parameters, antioxidant enzymes and Ni accumulation in Faba bean plants (Vicia faba) under nickel stress. Plants, 11(14), 1812.

Etesami, H., Alikhani, H. A., & Hosseini, H. M. (2015). Indole-3-acetic acid (IAA) production trait, a useful screening to select endophytic and rhizosphere competent bacteria for rice growth promoting agents. *MethodsX*, 2, 72–78.

Etesami, H. (2022). Root nodules of legumes: A suitable ecological niche for isolating non-rhizobial bacteria with biotechnological potential in agriculture. Current Research in Biotechnology, 4, 78–86.

Etesami, H., & Glick, B. R. (2024). Bacterial indole-3-acetic acid: A key regulator for plant growth, plant–microbe interactions, and agricultural resilience. Microbiological Research, 281, 127602.

Etesami, H., & Santoyo, G. (2025). Boosting Rhizobium–legume symbiosis: The role of nodule non-rhizobial bacteria in hormonal and nutritional regulation under stress. Microbiological Research, 128192.

Fahde, S., Boughribil, S., Sijilmassi, B., & Amri, A. (2023). Rhizobia: A promising source of plant growth-promoting molecules and their non-legume interactions. Agriculture, 13, 1279.

FAO. (2016). Pulses: Nutritious Seeds for a Sustainable Future. Rome: Food and Agriculture Organization of the United Nations.

FAOSTAT. (2023). Food and Agriculture Organization of the United Nations. Available at: https://www.fao.org/faostat

Ferguson, B. J., Lin, M. H., & Gresshoff, P. M. (2013). Regulation of legume nodulation by acidic growth conditions. Plant Signaling & Behavior, 8, e23426.

Frugier, F., Kosuta, S., Murray, J. D., Crespi, M., & Szczyglowski, K. (2008). Cytokinin: Secret agent of symbiosis. Trends in Plant Science, 13, 115–120.

Gamalero, E., Lingua, G., & Glick, B. R. (2023). Ethylene, ACC, and the plant growth-promoting enzyme ACC deaminase. Biology, 12, 1043.

Gamas, P., Brault, M., Jardinaud, M.-F., & Frugier, F. (2017). Cytokinins in symbiotic nodulation: When, where, what for? Trends in Plant Science, 22, 792–802.

Ganesan, S., Kuppusamy, R., & Sekar, R. (2007). Integrated management of stem rot disease using Rhizobium and Trichoderma harzianum. Turkish Journal of Agriculture and Forestry, 31, 103–108.

Glick, B. R. (2014). Bacteria with ACC deaminase can promote plant growth and help to feed the world. Microbiological Research, 169, 30–39. https://doi.org/10.1016/j.micres.2013.09.009

Goyal, R. K., & Habtewold, J. Z. (2023). Evaluation of legume–rhizobial symbiotic interactions beyond nitrogen fixation. Microorganisms, 11(6), 1454.

Großkinsky, D. K., Tafner, R., Moreno, M. V., Stenglein, S. A., García de Salamone, I. E., Nelson, L. M., . . . Roitsch, T. (2016). Cytokinin production by Pseudomonas fluorescens G20–18 determines biocontrol activity. Scientific Reports, 6, 23310.

Guinel, F. C. (2015). Ethylene, a hormone at the center stage of nodulation. Frontiers in Plant Science, 6, 1121.

Han, Q., Ma, Q., Chen, Y., Tian, B., Xu, L., Bai, Y., . . . Li, X. (2020). Variation in rhizosphere microbial communities and its association with the symbiotic efficiency of rhizobia in soybean. ISME Journal, 14, 1915–1928.

Han, X., Wang, J., Zhang, Y., Kong, Y., Dong, H., Feng, X., . . . Xin, D. (2022). Changes in the m6A RNA methylome accompany rhizobia-promoted root growth under cadmium stress. Journal of Hazardous Materials, 441, 129843.

Hungria, M., & Stacey, G. (1997). Molecular signals exchanged between host plants and rhizobia. Soil Biology & Biochemistry, 29, 819–830.

Hungria, M., & Vargas, M. A. T. (2000). Environmental factors affecting N2 fixation in grain legumes in the tropics, with an emphasis on Brazil. Field Crops Research, 65(2–3), 151–170.

Jones, K. M., Kobayashi, H., Davies, B. W., Taga, M. E., & Walker, G. C. (2007). How rhizobial symbionts invade plants: The Sinorhizobium–Medicago model. Nature Reviews Microbiology, 5(8), 619–633.

Kejela, T. (2024). Phytohormone-producing rhizobacteria and their role in plant growth. In New Insights into Phytohormones. IntechOpen. https://doi.org/10.5772/intechopen.1002823

Khan, M. S., Zaidi, A., Ahemad, M., Oves, M., & Wani, P. A. (2010). Plant growth promotion by phosphate solubilizing fungi–current perspective. Archives of Agronomy and Soil Science, 56(1), 73–98.

Kibido, T., Kunert, K., Makgopa, M., Greve, M., & Vorster, J. (2020). Improvement of rhizobium-soybean symbiosis and nitrogen fixation under drought. Food and Energy Security, 9(1), e177.

Kumari, P., Meena, M., Gupta, P., Dubey, M. K., Nath, G., & Upadhyay, R. S. (2018). Plant growth promoting rhizobacteria and their biopriming for growth promotion in mung bean (Vigna radiata (L.) R. Wilczek). Biocatalysis and Agricultural Biotechnology, 16, 163–171.

Liu, Y., Guo, Z., & Shi, H. (2022). Rhizobium symbiosis leads to increased drought tolerance in Chinese milk vetch. Agronomy, 12, 725.

Lohosha, O., Vorobei, Y. O., & Leonova, N. (2023). Symbiotic efficiency and cytokinin activity of new Mesorhizobium ciceri strains. Mikrobiolohichnyi Zhurnal, 85, 3–11.

L'Taief, B., Sifi, B., Zaman-Allah, M., Drevon, J. J., & Lachaâl, M. (2007). Effect of salinity on root-nodule conductance to oxygen diffusion in the Cicer arietinum–Mesorhizobium ciceri symbiosis. Journal of Plant Physiology, 164(8), 1028–1036.

Magome, H., & Kamiya, Y. (2016). Inactivation processes. In Annual Plant Reviews (Vol. 49: Gibberellins, pp. 73–94). Wiley.

Martínez-Hidalgo, P., & Hirsch, A. M. (2017). The nodule microbiome: N2-fixing rhizobia do not live alone. Phytobiomes Journal, 1, 70–82.

Martínez-Romero, E. (2003). Diversity of Rhizobium–Phaseolus vulgaris symbiosis: Overview and perspectives. Plant and Soil, 252(1), 11–23.

McCulloch, L. A., Piotto, D., & Porder, S. (2021). Drought and soil nutrient effects on symbiotic nitrogen fixation in Neotropical legumes. Biotropica, 53, 703–713.

Meena, R. S., Das, A., Yadav, G. S., & Lal, R. (Eds.). (2018). Legumes for soil health and sustainable management (p. 541). Singapore: Springer.

Mendes, R., Garbeva, P., & Raaijmakers, J. M. (2013). The rhizosphere microbiome: Significance of plant–microbe interactions. FEMS Microbiology Reviews, 37(5), 634–663. https://doi.org/10.1111/1574-6976.12028

Mishra, P. K., Mishra, S., Selvakumar, G., Bisht, J. K., Kundu, S., Gupta, H. S. (2009a). Coinoculation of Bacillus thuringiensis-KR1 with Rhizobium leguminosarum enhances plant growth and nodulation of pea (Pisum sativum L.) and lentil (Lens culinaris L.). World Journal of Microbiology and Biotechnology, 25, 753–761.

Mishra, P. K., Mishra, S., Selvakumar, G., Kundu, S., Shankar Gupta, H. (2009b). Enhanced soybean (Glycine max L.) plant growth and nodulation by Bradyrhizobium japonicum-SB1 in presence of Bacillus thuringiensis-KR1. Acta Agriculturae Scandinavica, Section B – Soil & Plant Science, 59, 189–196.

Mondal, N. K., Das, C., & Datta, J. K. (2015). Effect of mercury on seedling growth, nodulation, and ultrastructure of Vigna radiata. Environmental Monitoring and Assessment, 187, 241.

Mukhtar, S., Hirsch, A. M., Khan, N., Malik, K. A., Humm, E. A., Pellegrini, M., . . . Clum, A. (2020). Impact of soil salinity on the cowpea nodule microbiome. Phytobiomes Journal, 4, 364–374.

Narwal, E., Annapurna, K., Choudhary, J., & Sangwan, S. (2018). Effect of arbuscular mycorrhizal fungal colonization on nutrient uptake in rice aerobic conditions. International Journal of Current Microbiology and Applied Sciences, 7(04), 1072–1093.

Narwal, E., Choudhary, J., Jat, N. K., Meena, A. L., Ghasal, P. C., Dutta, D., . . . Choudhary, M. (2023). Plant growth-promoting microbes: Key players in organic agriculture. In Plant Growth Promoting Microorganisms of Arid Region (pp. 139–160). Singapore: Springer Nature Singapore.

Nascimento, F. X., Rossi, M. J., & Glick, B. R. (2016). Role of ACC deaminase in stress control of leguminous plants. In Plant Growth-Promoting Actinobacteria (pp. 179–192). Springer.

Nett, R. S., Bender, K. S., & Peters, R. J. (2022). Production of the plant hormone gibberellin by rhizobia increases host nodule size. ISME Journal, 16, 1809–1817.

Ng, J. L., & Mathesius, U. (2018). Acropetal auxin transport inhibition in indeterminate nodule formation. Frontiers in Plant Science, 9, 169.

Noori, F., Etesami, H., Zarini, H. N., Khoshkholgh-Sima, N. A., Salekdeh, G. H., & Alishahi, F. (2018). Mining alfalfa (Medicago sativa L.) nodules for salinity tolerant non-rhizobial bacteria to improve growth of alfalfa under salinity stress. Ecotoxicology and Environmental Safety, 162, 129–138.

Oldroyd, G. E. D., Murray, J. D., Poole, P. S., & Downie, J. A. (2011). The rules of engagement in the legume–rhizobial symbiosis. Annual Review of Genetics, 45, 119–144.

Parniske, M. (2008). Arbuscular mycorrhiza: The mother of plant root endosymbioses. Nature Reviews Microbiology, 6(10), 763–775. https://doi.org/10.1038/nrmicro1987

Parveen, G., Noreen, R., Shafique, H. A., Sultana, V., Ehteshamul-Haque, S., & Athar, M. (2019). Role of rhizobia in suppressing soybean root diseases under soil amendment. Planta Daninha, 37, e019172336.

Peoples, M. Á., Brockwell, J., Herridge, D. F., Rochester, I. J., Alves, B. J. R., Urquiaga, S., . . . Jensen, E. S. (2009). The contributions of nitrogen-fixing crop legumes to the productivity of agricultural systems. Symbiosis, 48(1), 1–17.

Riccillo, P. M., Muglia, C. I., de Bruijn, F. J., Roe, A. J., Booth, I. R., & Aguilar, O. M. (2000). Glutathione is involved in environmental stress responses in Rhizobium tropici. Journal of Bacteriology, 182, 1748–1753.

Richardson, A. E., Barea, J. M., McNeill, A. M., & Prigent-Combaret, C. (2009). Acquisition of phosphorus and nitrogen in the rhizosphere and plant growth promotion by microorganisms. Plant Soil 321:305–339.

Ritchie, H., & Roser, M. (2020). Diet compositions. *Our World in Data.* Available at: https://ourworldindata.org/diet-compositions

Roy, S., Robson, F., Lilley, J., Liu, C.-W., Cheng, X., Wen, J., . . . Bone, C. (2017). MtLAX2, an auxin influx transporter, is required for nodule organogenesis. Plant Physiology, 174, 326–338.

Ruiz-Lozano, J. M., Aroca, R., Zamarreño, Á. M., Molina, S., Andreo-Jiménez, B., Porcel, R., García-Mina, J. M., Ruyter-Spira, C., & López-Ráez, J. A. (2016). Arbuscular mycorrhizal symbiosis induces strigolactone biosynthesis under drought and improves drought tolerance in lettuce and tomato. Plant, Cell & Environment, 39(2), 441–452. https://doi.org/10.1111/pce.12631

Santoyo, G., Moreno-Hagelsieb, G., del Carmen Orozco-Mosqueda, M., & Glick, B. R. (2016). Plant growth-promoting bacterial endophytes. Microbiological Research, 183, 92–99.

Sarker, U., & Oba, S. (2018). Drought stress effects on growth, ROS markers, compatible solutes and antioxidants in Amaranthus tricolor. Applied Biochemistry and Biotechnology, 186, 999–1016.

Sasse, J., Martinoia, E., & Northen, T. (2018). Feed your friends: Do plant exudates shape the root microbiome? Trends in Plant Science, 23(1), 25–41. https://doi.org/10.1016/j.tplants.2017.09.003

Siddiqui, Z. S., Ali, F., & Uddin, Z. (2021). Sustainable effect of Sinorhizobium meliloti on nodulation and photosynthetic traits under low moisture stress. Letters in Applied Microbiology, 72, 714–724.

Singh, G., Sangwan, S., Narwal, E., & Annapurna, K. (2017). Microbial proteins and soil carbon sequestration. In Adaptive Soil Management: From Theory to Practices (pp. 131–139). Singapore: Springer Singapore.

Smith, S. E. & Read, D. J. (2008). Mycorrhizal Symbiosis. 3rd Edition, Academic Press, London.

Spaepen, S., Vanderleyden, J., & Remans, R. (2007). Indole-3-acetic acid in microbial signaling. FEMS Microbiology Reviews, 31, 425–448.

Sprent, J. I., & Sprent, P. (1990). Nitrogen Fixing Organisms: Pure and Applied Aspects (Vol. 256). London: Chapman and Hall.

Stagnari, F., Maggio, A., Galieni, A., & Pisante, M. (2017). Multiple benefits of legumes for agriculture sustainability: An overview. Chemical and Biological Technologies in Agriculture, 4(1), 2.

Staudinger, C., Mehmeti-Tershani, V., Gil-Quintana, E., Gonzalez, E. M., Hofhansl, F., Bachmann, G., & Wienkoop, S. (2016). Evidence for a rhizobia-induced drought stress response strategy in Medicago truncatula. Journal of Proteomics, 136, 202–213.

Stefen, D. L. V., Nunes, F. R., Rodolfo, G. R., Segatto, C., Anastácio, T. C., & Lajus, C. R. (2022). How phytohormones synthesized by PGPR affect plant growth?. In Secondary metabolites and volatiles of PGPR in plant-growth promotion (pp. 119–131). Cham: Springer International Publishing.

Stepkowski, T., Moulin, L., Krzyżańska, A., McInnes, A., Law, I. J., & Howieson, J. (2012). European origin of Bradyrhizobium populations infecting cowpea and other legumes in Africa. ISME Journal, 6, 1250–1261.

Suárez, R., Wong, A., Ramírez, M., Barraza, A., Orozco, M. D. C., Cevallos, M. A., . . . Iturriaga, G. (2008). Improvement of drought tolerance and grain yield in common bean by overexpressing trehalose-6-phosphate synthase in rhizobia. Molecular Plant-Microbe Interactions, 21(7), 958–966.

Tiwari, S., Sarangi, B. K., & Thul, S. T. (2016). Identification of arsenic-resistant endophytes for remediation. Journal of Environmental Management, 180, 359–365.

Tu, J. C. (1981). Effect of salinity on rhizobium–root hair interaction in soybean. Canadian Journal of Plant Science, 61, 231–239.

Uniyal, S., Bhandari, M., Singh, P., Singh, R. K., & Tiwari, S. P. (2022). Cytokinin biosynthesis in cyanobacteria for crop improvement. Frontiers in Genetics, 13, 933226.

Wani, P. A., Khan, M. S., & Zaidi, A. (2008). Chromium-reducing and plant growth-promoting Mesorhizobium improves chickpea growth in chromium-amended soil. Biotechnology Letters, 30(1), 159–163.

Yu, X., Shoaib, M., Cheng, X., Cui, Y., Hussain, S., Yan, J., . . . & Zou, T. (2022). Role of rhizobia in promoting non-enzymatic antioxidants to mitigate nitrogen-deficiency and nickel stresses in Pongamia pinnata. Ecotoxicology and environmental safety, 241, 113789.

Zahran, H. H., & Sprent, J. I. (1986). Effects of sodium chloride and polyethylene glycol on root-hair infection and nodulation of Vicia faba L. plants by Rhizobium leguminosarum. Planta, 167(3), 303–309.

Zgadzaj, R., Garrido-Oter, R., Jensen, D. B., Koprivova, A., Schulze-Lefert, P., & Radutoiu, S. (2016). Root nodule symbiosis in Lotus japonicus drives the establishment of distinctive rhizosphere, root, and nodule bacterial communities. Proceedings of the National Academy of Sciences, 113(49), E7996-E8005.

6 Nutrient Cycling Microbes in Legume Rhizosphere

Jyotsana Tilgam, Kumar M., Sandeep Jaiswal, Jyoti Prakash Singh, Adarsh Kumar, and Shobit Thapa

6.1 INTRODUCTION

The rhizosphere, the narrow zone of soil surrounding and influenced by plant roots, is one of the most dynamic and biologically active ecosystems on Earth (Hinsinger et al. 2009; Raaijmakers et al. 2009). It harbours an immense diversity of organisms, including bacteria, fungi, oomycetes, nematodes, protozoa, algae, viruses, archaea, and arthropods, all interconnected through complex trophic networks fuelled by root-derived carbon inputs such as exudates and mucilage. These rhizodeposits profoundly influence microbial diversity, activity, and functional potential, enabling plants to selectively enrich beneficial microbes that promote growth, enhance nutrient acquisition, and suppress pathogens (Ashajyothi et al. 2022). Leguminous plants hold a pivotal role in sustainable agriculture due to their ability to establish symbiotic associations with nitrogen-fixing rhizobia and to engage in beneficial interactions with mineral-solubilising microorganisms (Kishore et al. 2015; Naorem et al. 2022; Nabati et al. 2025). These associations enable legumes to fix atmospheric nitrogen, mobilise insoluble phosphorus and micronutrients, and improve soil structure and organic matter content. Consequently, legumes reduce reliance on synthetic fertilisers, enhance soil fertility, and promote agroecosystem stability, making them essential components of crop rotation and intercropping systems (Paungfoo-Lonhienne et al. 2017; Punyalue et al. 2018; Zhao et al. 2022). It is estimated that agriculturally significant legume species contribute approximately 40 to 90 teragrams of nitrogen annually through biological nitrogen fixation (Galloway et al. 2008; Taylor et al. 2020). Nutrient cycling in soils is governed by intricate biological processes largely mediated by rhizosphere microorganisms. These microbes regulate the transformation, mobilisation, and recycling of essential nutrients such as nitrogen (N), phosphorus (P), potassium (K), sulphur (S), and trace elements, thereby sustaining soil fertility and plant productivity. They also produce phytohormones, siderophores, and volatile organic compounds (VOCs) that stimulate root development and enhance plant resilience to abiotic and biotic stresses (Raaijmakers et al. 2009). For instance, a split-root model study combined with molecular profiling revealed that *Trifolium* and other legumes undergo systemic microbial shifts under polycyclic aromatic hydrocarbon (PAH)

DOI: 10.1201/9781003733324-7

contamination, characterised by enrichment of *Verrucomicrobia*, *Actinobacteria*, and the betaproteobacterium *Denitratisoma*, highlighting their potential role in rhizoremediation (Kawasaki et al. 2012). Soil microorganisms also play a crucial role in nutrient cycling through the decomposition of plant litter and utilisation of root exudates, key processes linking plant and soil mineral nutrition. Diverse plant communities enhance microbial diversity by producing varied litter and exudates, which influence soil organic carbon and microbial composition. Microbial groups such as arbuscular and ectomycorrhizal fungi facilitate plant phosphorus uptake, while saprotrophic fungi contribute to litter decomposition and organic P mobilisation (Battini et al. 2017). Furthermore, microbes can solubilise inorganic P or immobilise it within their biomass. The exchange of plant-derived carbon for mineral nutrients such as phosphorus underscores the complexity of quantifying these interactions under natural conditions. Thus, the legume rhizosphere serves as a biological hotspot where nutrient cycling, plant–microbe communication, and soil health converge. Although extensive research has elucidated the mechanisms of rhizobial nitrogen fixation, transferring this legume-specific symbiosis to non-legume crops remains a challenge. Understanding the distinct cellular responses induced by rhizobia and mycorrhizal fungi is essential for achieving this long-standing goal and realising nitrogen-fixing capabilities in cereals (Geurts et al. 2012). A deeper understanding of these interactions is essential for developing microbial-based strategies that support sustainable intensification, soil restoration, and climate-smart agriculture. In this context, legumes offer an ideal model for exploring the ecological and functional dynamics of the rhizosphere microbiome, providing valuable insights into how beneficial microbial networks can be harnessed to optimise nutrient cycling and crop productivity.

6.2 THE LEGUME RHIZOSPHERE: A UNIQUE MICROBIAL NICHE

The legume rhizosphere represents a highly dynamic and selective microenvironment shaped by complex interactions among plants, microbes, and soil physicochemical properties. Root exudates, comprising organic acids, amino acids, sugars, phenolics, and volatile organic compounds, play a central role in recruiting and regulating rhizosphere microorganisms. These exudates not only attract beneficial microbes but also facilitate symbiotic interactions with rhizobia and mycorrhizal fungi while serving as a first line of biochemical defence against soilborne pathogens (Hawkins et al. 2000; Miransari 2011). Through these chemical cues, plants actively shape the structure of their rhizosphere microbiome, while microbial communities, in turn, modify soil composition via organic matter degradation and secondary metabolite production. Legume–rhizobia symbioses are characterised by remarkable specificity – no single rhizobial strain can nodulate all legume species, and vice versa. For example, Rhizobium leguminosarum bv. trifolii forms symbiosis exclusively with clover species (Trifolium spp.) (Van der Drift et al. 1996), whereas Sinorhizobium fredii NGR234 displays a broad host range, establishing associations with up to 112 legume genera (Pueppke et al. 1999). This specificity operates at both species and genotype levels and may vary across different stages of interaction. In some cases, rhizobia may infect roots but fail to induce nodulation, or nodules may form but exhibit variable nitrogen-fixation efficiency. Such outcomes depend on finely tuned molecular recognition mechanisms that have co-evolved between host plants and their symbionts. Understanding the genetic and molecular basis of this specificity is vital for engineering improved plant and bacterial strains with enhanced biological nitrogen fixation efficiency. In addition to Rhizobium and Bradyrhizobium, several non-traditional nitrogen-fixing bacteria such as Burkholderia and Achromobacter have been reported to effectively nodulate legumes like cowpea (Vigna unguiculata) (Azarias Guimarães et al. 2012).

Microorganisms also improve the bioavailability of trace elements such as iron, zinc, and manganese. For instance, siderophore-producing bacteria chelate iron from insoluble forms, enhancing its plant uptake, while certain rhizobacteria (e.g. *Bacillus subtilis* GB03) can activate plant genes involved in iron acquisition (Zhang et al. 2009). Many rhizobacteria also contribute to mineral weathering, releasing nutrients in deficient soils and promoting plant growth. Genetic studies in *Lotus japonicus* mutants have shown that core symbiosis genes not only regulate nodulation but also influence the composition of root-associated microbial communities. Legume nodules, being nutrient-rich and low-oxygen microhabitats, harbour specialised microbial communities distinct from those in roots but consistent across legume species, providing key insights into the evolution and ecology of plant–microbe symbioses. Intercropping systems further illustrate the ecological significance of legume–microbe interactions. Legume–cereal intercropping enhances arbuscular mycorrhizal fungal diversity and abundance, thereby improving phosphorus acquisition and overall plant growth (Zhao et al. 2022). Similarly, intercropping *Arachis pintoi* (Pinto peanut) with litchi trees improved soil potassium availability, bacterial community structure, and microbial diversity, contributing to enhanced soil fertility and resilience. Maize–legume intercropping systems have been shown to increase AMF spore density, enrich soil macrofauna, improve nitrogen accumulation, and enhance soil biodiversity and productivity (Punyalue et al. 2018). Collectively, these interactions underscore the central role of legumes and their rhizosphere microbiota in driving sustainable agricultural systems. Through biological nitrogen fixation, efficient nutrient cycling, and improved soil structure, legumes not only sustain their own productivity but also enhance the fertility and ecological resilience of entire cropping systems.

6.3 NUTRIENT CYCLE IN LEGUME RHIZOSPHERE ELABORATED BY MICROORGANISMS

Major transformation process of mineral nutrients occurring in soil are mineralisation, assimilation, immobilisation, and oxido-reductive processes. Based on mineral nutrients the processes mediated by microorganisms in rhizosphere vary. This section of the chapter deals with nutrient cycling operating in legume rhizosphere mediated by microorganisms. The cycling of nitrogen, phosphorus and sulphur is discussed.

6.3.1 Nitrogen Cycle

In terrestrial ecosystems, legumes increase soil nitrogen cycling and N-availability thereby contributing to overall productivity of the ecosystem (Gou et al. 2023).

Nitrogen cycling in legume rhizosphere mediated by microorganisms involves transformations like biological nitrogen fixation, mineralisation, nitrification and denitrification. Biological nitrogen fixation (BNF) is mediated by both symbiotic nitrogen fixers and free-living nitrogen fixers. Mineralisation involves decomposition of complex organic compounds containing nitrogen leading to the release of ammonium, which are either taken up by plants or further oxidised to nitrate by a process called nitrification. The nitrates released through the process of nitrification are again taken up by the plants. Under the condition of reduced oxygen levels, nitrates become the source for a process called denitrification through which the nitrogen is lost to atmosphere as greenhouse gas, nitrous oxide.

6.3.1.1 Biological Nitrogen Fixation

BNF is an important component of nitrogen cycle as it circumvents the energy requirement to convert gaseous nitrogen to ammonia. Prokaryotes are the only group of organisms that enable this process, and they mediate this energy intensive process with the help of nitrogenase enzymes, energy in the form of ATPs and some cofactors like Manganese (Naorem et al. 2022). Nitrogen is different from other elements in the sense that they are not completely recycled. The nitrogen entering into the biosphere is lost continuously through various processes like denitrification, ammonia volatilisation, nitrate leaching to groundwater, surface runoff and erosion. These losses therefore make BNF, the most important component of nitrogen cycle for replenishing the lost nitrogen in biosphere (Ladha et al. 2022). BNF in rhizosphere of leguminous plants operates through two processes: symbiotic nitrogen fixation (SNF) and free-living nitrogen fixation (FLNF). SNF in leguminous plants is mediated by a specific group of bacteria called rhizobia, which form specialised structures in legume roots called nodules. This symbiotic process of nitrogen fixation is termed as legume-rhizobia symbiosis. Of the total nitrogen fixed annually (1.75×10^{11} Kg), more than 45 % is contributed by legume – rhizobia symbiosis. The amount of nitrogen fixed by legumes is a factor of legume species, management strategies, soil physicochemical properties, and climate (Kebede 2021). It has been reported that BNF by various legumes ranges from as low as 16.3–71.9 Kg/ha by common bean in Canada to as high as 118.6–311 Kg/ha by fava bean in Greece and Italy (Vasconcelos et al. 2020). The legume-rhizobium association is very specific, that is specific leguminous plants are nodulated by specific rhizobium. The specificity of rhizobium to specific leguminous plants is due to the interaction of flavonoids produced by legumes and NodD protein produced by rhizobia. The NodD proteins in different rhizobia respond to specific flavonoids released by specific legumes and this NodD-flavonoid interaction determines the legume-rhizobium compatibility for nodule formation. For instance, 4,4'-dihydroxy-2'methoxychalone of Alfalfa recognises NodD protein of compatible rhizobium Sinorhizobium meliloti but could not recognise NodD of Bradyrhizobium japonicum which is a compatible rhizobium for Glycine max (Yu and Zhu 2025). The taxonomy of rhizobium has shown overwhelming development since the early classification based on cross-inoculation groups. Currently, rhizobia-forming nodules are distributed over 24 genera and more than 230 species (Rajkumari et al. 2022). Some of the important leguminous crops and their nodulating bacteria are enlisted in Table 6.1. FLNF in rhizosphere is mediated by heterotrophic bacteria and some methanogenic archaea where there is no obvious symbiosis between plants and the nitrogen fixers. It is a significant process that adds nitrogen in the terrestrial ecosystems and the magnitude of nitrogen fixed by FLNF out volumes the SNF. A report suggests that FLNF contributes 6 Kg Nitrogen per hectare per year against the 4.5 Kg per hectare per year by SNF in tropical forest systems. Free-living nitrogen fixers are distributed across different bacterial phyla like Cyanobacteria, Firmicutes, and Proteobacteria (Smercina et al. 2019). Some of the well-known free living nitrogen fixers are Azotobacter, Azospirillum, Azoarcus, Burholderia, Clostridium, Gluconacetobacter, Herbaspirillum, Klebsiella, Anaebena, Nostoc, Pseudomonas, Phyllobacterium and archaea like Methanosarcina and Methanococcus (Khan et al. 2021). Interestingly free-living nitrogen fixers of Bradyrhizobium spp. have been

TABLE 6.1
Common Leguminous Plants and Their Symbiotic Partners

Leguminous plants	Valid scientific names	Nodulating rhizobia	Reference
Chickpea	*Cicer arietinum*	*Mesorhizobium ciceri, M. silanurunense, M. plurifarium, M. helmanticense*	Kumar et al. (2024)
Pea	*Pisum sativum*	*Rhizobium leguminosarum* bv. *viciae, Sinorhizobium meliloti*	Laguerre et al. (2007); Ilahi et al. (2021)
Pigeonpea	*Cajanus cajan*	*Bradyrhizobium yuanmingense, B. elkanii, B. pachyrhizi, B. liaoningense, B. japonicum, B. ivorense*	Kumar et al. (2023)
Lentil	*Lens culinaris*	*R. lentis*	Pathak et al. (2024)
Soybean	*Glycine max*	*B. japonicum*	Yu and Zhu (2025)
Greengram	*Vigna radiata*	*B. yuanmingense, B. subterraneum*	Kumar et al. (2023)
Cowpea	*V. unguiculata*	*B. yuanmingense, B. subterraneum*	Kumar et al. (2023)
Blackgram	*V. mungo*	*Sinorhizobium terangae, S. mexicanum*	Kumar et al. (2023)
Lucerne	*Medicago sativa*	*S. meliloti*	Yu and Zhu (2025)
Clovers	*Trifolium* spp.	*R. leguminosarum* bv. *trifolii*	Reeve et al. (2013)
Groundnut	*Arachis hypogaea*	*B. arachidis*	Wang et al. (2013)
Touch-me-not/ Sleeping grass	*Mimosa pudica*	*Burkholderia* sp., *Cupriavidus* sp.	Barret and Parker (2006)
Common bean	*Phaseolus vulgaris*	*R. leguminosarum* bv. *phaseoli, R. elti, R. gallicum*	Silva et al. (2003)
Dhaincha	*Sesbania* spp.	*R. huautlense, Azorhizobium caulinodans, A. doebereinereae*	Cummings et al. (2009)
Common Kidney-Vetch	*Anthyllis vulneria*	*M. loti*	Ampomah and Huss-Danell
Lotus	*Lotus corniculatus*	*M. loti*	Ampomah and Huss-Danell

isolated from soybean rhizosphere in China. Bradyrhizobium and Azorhizobium are the only members of rhizobial group that can fix nitrogen non-symbiotically (Tao et al. 2021). In general, the diversity of diazotrophs is lower in legume rhizosphere as compared to grasses, however, the legume rhizosphere exhibits strong dominance of key diazotrophs like Klebsiella, and Pseudomonas (Shen et al. 2025).

6.3.1.2 Mineralisation, Nitrification, and Other Oxidoreductive Processes

Nitrogen cycling in legume systems is strongly driven by microbial processes in the rhizosphere. Organic nitrogen is converted to plant-available ammonium through mineralisation, which is enhanced by rhizosphere priming and microbial activity stimulated by legume root exudates (Moreau et al. 2019; Henneron et al. 2020). A large proportion (40–75%) of biologically fixed nitrogen in legumes becomes available to subsequent crops through mineralisation, contributing 30–110 kg N ha^{-1} depending on the conditions (Kumar et al. 2020). Legume rhizodeposition is a major nitrogen source for soil microbes, contributing up to 70% of total soil N inputs (Wang et al. 2021). Microbial immobilisation temporarily retains mineral N under high C:N conditions and can favour biological nitrogen fixation by reducing mineral N availability (Saiz et al. 2021). Nitrification converts ammonium to nitrate, increasing N losses, though plants can suppress this via biological nitrification inhibition (BNI). While uncommon in legumes, evidence from peanut- and soybean-based rotations suggests legumes may influence nitrifier populations (Subbarao et al. 2007; Paungfoo-Lonhienne et al. 2017; Zhou et al. 2020). Denitrification by free-living and symbiotic diazotrophs releases N_2O and N_2, occurring in both rhizosphere and nodules (Sánchez and Minamisawa 2019). Legume residues with low C:N ratios further enhance denitrification. In contrast, DNRA conserves nitrogen by reducing nitrate to ammonium and is favoured in legume rhizospheres rich in carbon and low in nitrate (Pandey et al. 2020).

6.3.2 Phosphorus Cycle

Phosphorus (P) is a key macronutrient required for nucleic acids, ATP, membranes, and stress tolerance, with higher demand in legumes due to energy-intensive nitrogen fixation. Unlike nitrogen, the P cycle is closed, as P becomes immobilised by metal ions and is mainly supplied through mineral weathering (Wang et al. 2024). Only a small fraction of soil P is plant-available as orthophosphate, while over 90% exists in unavailable organic or mineral forms (Shen et al. 2011; Ducousso-Détrez et al. 2022). Microbes regulate P availability through mineralisation of organic P via phosphatases and solubilisation of inorganic P using acids and chelators (Kishore et al. 2015). Although legumes enhance rhizosphere phosphatase activity, their overall impact on soil P cycling is modest compared to cereals (Maltais-Landry et al. 2014; Koukoyi et al. 2025).

6.3.3 Sulphur Cycling

Sulphur (S) in soil cycles through assimilation, mineralisation, and redox reactions, with plants and microbes mainly taking it up as sulphate (SO_4^{2-}). Sulphur is especially critical for legumes, as deficiency restricts nodulation and symbiotic nitrogen fixation. Most soil sulphur (75–90%) occurs in organic forms, while only a small fraction is inorganic; microbial mineralisation by bacteria and fungi releases plant-available sulphur (Chaudhary et al. 2023). The released sulphur undergoes microbial oxidation or reduction depending on soil conditions. Sulphur oxidation is mediated by bacteria such as Thiobacillus, Acidithiobacillus, Pseudomonas, and Alcaligens using Sox and SOR enzyme systems, and is favoured in the legume rhizosphere by root exudates (Nabati et al. 2025). In contrast, sulphur reduction occurs in anaerobic micro-sites and is driven by sulphate-reducing bacteria like Desulfovibrio and Desulfotomaculum.

6.4 PLANT–MICROBIAL INTERACTIONS AND FUNCTIONAL NETWORKS WITH SPECIAL EMPHASIS ON LEGUMES AND THEIR RHIZOSPHERE

6.4.1 Structure and Communication within the Rhizosphere Microbiome

The rhizosphere constitutes a dynamic interface where plant roots interact with a diverse microbial community that governs nutrient cycling, plant growth, and ecosystem stability. The abundance and functionality of these microbial populations are primarily shaped by host genotype, soil type, and environmental factors such as pH, temperature, and nutrient availability (Hassani et al. 2018; Xiong et al. 2021). Dominated by Proteobacteria, Actinobacteria, Bacteroidetes, Firmicutes, and Acidobacteria, rhizosphere consortia represent the most functionally diverse microbial habitat in terrestrial ecosystems (Berg et al. 2020). Healthy plants exert a strong selective influence on microbial recruitment through root exudates – a complex mixture of organic acids, amino acids, sugars, and secondary metabolites – that attract beneficial microorganisms while suppressing pathogens (Dennis et al. 2010). This plant-driven selection establishes beneficial microbial communities that confer enhanced nutrient acquisition, abiotic stress tolerance, and immunity (Hassani et al. 2018; Jacoby et al. 2021). Microorganisms, in turn, communicate through quorum sensing (QS) systems that rely on *N*-acyl homoserine lactones (AHLs), as well as volatile organic and inorganic compounds (VOCs, VICs) that modulate community behaviours across short and long distances (Fuqua et al. 2001). Alongside QS, small molecules such as trehalose, thiamine, and oxalic acid act as signalling agents, shaping community dynamics and cooperative metabolism (Scherlach and Hertweck 2020). The process of microbial attachment and colonisation follows a biphasic model, beginning with reversible adhesion and culminating in biofilm formation (Wheatley and Poole 2018). Microbes secrete extracellular polymers and cyclic-di-AMP to strengthen attachment and coordinate cellular responses. Specificity in host colonisation is mediated by lectins and arabinogalactan proteins (AGPs) on plant surfaces that recognise compatible microbial surface carbohydrates, ensuring symbiont selectivity (Imberty and Varrot 2008). This interaction marks the onset of a complex exchange of molecular signals that define compatibility and symbiotic potential.

6.4.2 Molecular Dialogue and Symbiotic Signalling in Legume Rhizosphere

Legumes are powerful models for studying plant–microbe communication. Root exudates rich in flavonoids activate rhizobial NodD, inducing Nod factor production that is perceived by LysM receptors to trigger calcium signalling, root hair deformation, and nodule formation (Cooper 2004; Hassan and Mathesius 2012; Shumilina et al. 2023). At the same time, plants activate innate immunity via PRRs, MAPK pathways, and ROS, which rhizobia suppress using T3SS/T4SS effector proteins to enable symbiosis (Yang et al. 2010; Teulet et al. 2022). Effectors such as NopL fine-tune host signalling for efficient nodulation (Ge et al. 2016). Nodule number and function are systemically regulated by autoregulation of nodulation involving CLE/CEP peptides, NARK/SUNN receptors, miR2111, and TML in response to nitrogen status (Ferguson et al. 2019; Tsikou et al. 2018). Recent work shows zinc acts as a second messenger controlling nitrogen fixation and nodule senescence (Lin et al. 2024, 2025), while transcription factors maintain redox balance (Deng et al. 2019). Some legumes use Nod factor–independent infection pathways yet retain core symbiotic genes, highlighting evolutionary streamlining of efficient symbiosis (Ibañez and Fabra 2011; Quilbé et al. 2022; Raul et al. 2022).

6.4.3 Functional Roles of Microbes in Nutrient Cycling and Ecosystem Dynamics

Rhizosphere microorganisms play central roles in nutrient cycling, converting atmospheric and mineral-bound nutrients into plant-available forms. *Bradyrhizobium* and *Rhizobium* spp. fix atmospheric N_2 through nitrogenase complexes encoded by *nifHDK*, while phosphate-solubilising microbes (PSB) such as *Bacillus*, *Pseudomonas*, and *Aspergillus* release organic acids (gluconate, citrate, oxalate) that dissolve Ca–P, Fe–P, and Al–P complexes (Awasthi et al. 2011; Thepbandit and Athinuwat, 2024). Potassium-solubilising bacteria (KSB) like *Paenibacillus* and *Pantoea* release exopolysaccharides and siderophores, enhancing K and Fe mobilisation (Etesami et al. 2017; Hider and Kong 2011). These microbes also regulate hormonal balance by producing indole acetic acid (IAA), cytokinins, and gibberellins, thereby linking nutrient metabolism with plant developmental signalling (Egamberdieva et al. 2017). Under phosphorus-limited conditions, legumes exhibit increased acid phosphatase activity and exude carboxylates such as malate, oxalate, and tartrate, which stimulate PSB proliferation and enhance P bioavailability (Dai et al. 2025). Integration of legumes into rice-based cropping systems has been shown to improve soil pH buffering capacity, cation exchange, and the abundance of N-cycling genes (*nifH, nirK/S, narGH*), thereby reducing denitrification losses and improving nitrogen retention (Cai and Zhao 2024). Diversified intercropping, such as peanut with maize or oilseed rape, stimulates flavonoid and coumarin biosynthesis through activation of the phenylpropanoid pathway, enhancing *Bradyrhizobium* nodulation and nitrogen fixation (Qiao et al. 2024). In natural ecosystems, legumes profoundly influence biogeochemical cycling. *Epihov et al. (2021)* demonstrated that N_2-fixing trees double silicate weathering rates in tropical soils, leading to acidification and release of Fe-,

Al-, and Mo-bound nutrients essential for nitrogenase function. Their microbiomes, enriched with Fe(III)-reducing Acidobacteria, particularly *Candidatus Acidoferrum*, drive redox cycling and mineral nutrient liberation, coupling nitrogen fixation with phosphorus mobilisation. These processes create positive feedback loops that sustain both symbiotic partners and surrounding vegetation, enhancing nutrient turnover and carbon sequestration.

6.4.4 Multi-Symbiont Functional Networks and Ecological Stability

Legumes maintain multi-symbiont associations involving rhizobia, arbuscular mycorrhizal fungi (AMF), and endophytic bacteria that synergistically regulate nitrogen, phosphorus, and carbon fluxes in the rhizosphere. These associations are coordinated through the Common Symbiosis Pathway (CSP) – a signalling network mediated by *SYMRK*, *CCaMK*, and *CYCLOPS* – that facilitates dual colonisation and mutual benefit (Oldroyd and Downie, 2004; Kistner et al. 2005). In *Lotus japonicus*, more than 3,600 AMF-responsive genes have been identified, encoding transporters for phosphate, nitrate, ammonium, and potassium, thereby linking microbial symbiosis with nutrient flux and stress adaptation (Handa et al. 2015). Endophytic cohabitants such as *Bacillus*, *Pseudomonas*, and *Micromonospora* act in concert with rhizobia within nodules to enhance N_2 fixation, nutrient solubilisation, and hormone regulation (Mishra et al. 2009). Functional redundancy and buffering within these networks ensure ecosystem resilience. In *Trifolium pratense*, the Rhizobium leguminosarum-dominated root microbiome interacts with *Pantoea*, *Sphingomonas*, and *Flavobacterium*, forming a functionally balanced system where co-inoculation mitigates deleterious single-strain effects (Hartman et al. 2017). Likewise, in *Pisum sativum*, root-associated Proteobacteria – notably *Rhizobium*, *Pseudomonas*, and *Nitrobacter* – enrich for genes associated with nitrogen fixation (*nifQ*), phosphate solubilisation (*pqqC*), siderophore synthesis (*enterobactin*), and IAA biosynthesis, enhancing growth under acidic and metal-stressed soils (Chaudhari et al. 2020).

Collectively, these interactions illustrate that legumes act as ecosystem engineers, structuring their rhizosphere microbiomes through chemical signalling, metabolic cooperation, and feedback regulation. Through symbiotic nitrogen fixation, nutrient mobilisation, and microbial redox cycling, legumes sustain soil fertility, improve nutrient-use efficiency, and foster ecological resilience. Understanding these multi-layered networks offers transformative potential for designing microbiome-assisted crop improvement and sustainable agricultural systems that rely on natural symbiotic capacities rather than external inputs.

6.5 ENVIRONMENTAL AND MANAGEMENT EFFECTS ON NUTRIENT-CYCLING MICROBES IN LEGUME RHIZOSPHERE

Leguminous crop rhizospheres are highly dynamic micro-ecosystems with complex interactions between soil particles, plant roots, and various microbial communities. These microorganisms play critical roles in the cycling of essential nutrients such as nitrogen (N), phosphorus (P), sulphur (S), and other micronutrients through processes such as biological nitrogen fixation, mineralisation, solubilisation, and nutrient mobilisation. The composition, functionality, and efficiency of these nutrient-cycling microbial communities are significantly affected by environmental factors and agronomic management approaches. A thorough understanding of these interactions is required to maximise microbial potential for improving soil fertility, increasing crop yield, and maintaining the long-term viability of legume-based cropping systems (Du et al. 2025).

6.5.1 Environmental Determinants of Microbial Function

Environmental circumstances function as significant ecological filters, determining which microbial populations survive in the legume rhizosphere and how effectively they carry out nutrient transformation activities. Abiotic stresses such as soil acidity, temperature extremes, and moisture fluctuations have a direct impact on cellular integrity, regulate the expression of genes involved in nutrient cycling, and modulate the synthesis of critical signalling molecules such as flavonoids and Nod factors that facilitate plant–microbe communication (Rasheela et al. 2024).

6.5.1.1 Soil Acidity, Alkalinity, and Salinity

The distribution, diversity, and metabolic activity of rhizosphere bacteria are most strongly influenced by soil pH out of all the environmental factors. The majority of rhizobial species, particularly those in the Rhizobium and Mesorhizobium families, are sensitive to acidic environments and exhibit decreased nodulation effectiveness and survival in soils with pH values below 5.5 (Zhang et al. 2020). Additionally, acidic environments limit biological nitrogen fixation by limiting the availability of important micronutrients like molybdenum, which is a crucial part of the nitrogenase enzyme complex (Kafeel et al. 2023). However, certain Bradyrhizobium strains associated with tropical legumes have evolved particularly high acid tolerance, demonstrating the genetic and physiological adaptability of the legume-rhizobia symbiosis (Goyal et al. 2023). Alkaline and saline soils pose additional constraints to microbial functioning. High pH leads to the precipitation of phosphorus and micronutrients, reducing their availability to phosphate-solubilising bacteria (PSB) and other nutrient-mobilising microbes. Salinity imposes osmotic stress and ion toxicity, which disrupt cellular processes and reduce microbial proliferation (Dey et al. 2021). In rhizobia, salinity stress not only affects free-living stages but also impairs infection thread formation and nodule development. To survive under such stress, salt-tolerant microbes accumulate compatible solutes such as glycine betaine and trehalose to maintain osmotic balance and stabilise proteins and membranes (Kumawat et al. 2022a).

6.5.1.2 Temperature and Moisture Dynamics

Soil microbes' growth kinetics, enzymatic activity, and metabolic rate are all influenced by temperature. For development and nutrient transformation, every microbial strain has an ideal temperature range (Gonzalez et al. 2023). Enzymes essential for phosphorus solubilisation and nitrogen fixation may become denatured if exposed to temperatures higher than this range. Temperatures above 35 °C disrupt nodulation and root hair infection in certain rhizobial strains, postponing or preventing successful symbiosis (Serova et al. 2023). Temperature and soil moisture work together to control microbial motility, substrate transport, and gas diffusion. Reduced soil water potential during drought reduces nutrient transport and prevents flagellated bacteria from moving towards root exudates. Additionally, drought reduces the flow of photosynthates from shoots to nodules, which lowers nitrogen-fixation rates by limiting the energy supply to bacteroids (Dollete et al. 2024). Waterlogged or flooded soils, on the other hand, produce anaerobic conditions that are unsuitable for aerobic phosphorus solubilisers but promote denitrifying and methanogenic bacteria. Despite rhizobia's microaerophilic nature, prolonged oxygen deprivation during floods inhibits nitrogen fixation and nodule formation (Das et al. 2025). Therefore, both drought and excessive wetness play important roles in controlling the microbial activity and composition of the rhizosphere.

6.5.1.3 Soil Texture and Organic Matter Content

Aeration, water retention, and the physical microhabitats available to microorganisms are all influenced by the texture of the soil. Clayey soils, due to their finer particles and higher surface area, provide physical protection from desiccation and predation, thus sustaining microbial biomass. On the other hand, an excessive amount of clay can hinder root elongation and oxygen diffusion. Despite being well-aerated, sandy soils are more prone to quick drying and nutrient leaching, which frequently leads to decreased enzyme activity and microbial population (Naorem et al. 2023). For heterotrophic nutrient-cycling bacteria, such as PSB and phosphate-mineralising microorganisms, soil organic matter (SOM) is the primary source of energy and nutrients (Sinduja et al. 2024). High levels of SOM ensure a consistent and balanced supply of carbon and nutrients, promoting a more diversified and functionally active microbial population (Kumar et al. 2021). Labile carbon molecules in root exudates, such as organic acids, amino acids, and sugars, cause PSB and other beneficial microorganisms to release organic acids and enzymes that mobilise phosphorus from mineral and organic complexes (Rawat et al. 2021). Conversely, continual conventional tillage frequently depletes SOM and disturbs soil structure, resulting in decreased microbial biomass and enzyme activity. Such degradation has a direct impact on biological nitrogen fixation and other important microbial processes, lowering the sustainability of legume-based agricultural systems (Fan et al. 2022).

6.5.2 Management Practices and Microbial Ecology

Agricultural management practices act as significant ecological determinants that shape the structure, abundance, and activity of nutrient-cycling microbial communities in the legume rhizosphere. These interventions alter soil physicochemical and biological parameters, resulting in either favourable or detrimental settings for beneficial microbes. As a result, the long-term health and fertility of legume-based systems are significantly influenced by how management strategies interact with microbial ecology (Waoo et al. 2025).

6.5.2.1 Tillage Systems and Soil Disturbance

Tillage regulates the physical and biological environment of the soil, impacting microbial habitat, residue distribution, and nutrient dynamics. Conventional tillage

(CT), which includes deep plowing and frequent soil inversion, disturbs fungal hyphal networks, particularly arbuscular mycorrhizal fungi (AMF), which help legumes absorb phosphorus. The increased aeration caused by CT stimulates quick decomposition of soil organic matter (SOM), resulting in a transitory boost in microbial respiration followed by a long-term drop in biomass and enzyme activity (Han et al. 2025). In contrast, conservation tillage (reduced or zero tillage) maintains soil structure, enhances residue cover, and supports the stratification of microbial communities near the surface. No-till systems sustain higher microbial biomass, enzyme activity, and nutrient-cycling efficiency in the rhizosphere, particularly in the upper 5–10 cm soil layer (Taskin et al. 2021). Conservation tillage favours the persistence of inoculated rhizobia and phosphate-solubilising bacteria (PSB) by maintaining a more stable microenvironment (Kumawat et al. 2022b). Moreover, the continuous supply of carbon from root residues in no-till fields supports a resilient microbial food web, strengthening the linkage between soil structure and nutrient cycling efficiency.

6.5.2.2 Application of Fertilisers and Agrochemicals

The application of chemical fertilisers and agrochemicals, while improving short-term crop productivity, can have mixed or adverse effects on soil microbial ecology. High nitrogen fertiliser input often suppresses biological nitrogen fixation (BNF) by legumes. The presence of mineral N in the rhizosphere reduces the plant's need to form nodules and synthesise nitrogenase, resulting in less effective nodules (Xu et al. 2025). This physiological feedback reduces the long-term symbiotic efficiency and can lead to decreased reliance on microbial N cycling. Similarly, broad-spectrum pesticides and fungicides, though effective against target pathogens, frequently exhibit non-target effects on beneficial microorganisms. Certain fungicides inhibit AMF colonisation, disrupting P uptake, while some herbicides modify root exudate chemistry, altering microbial composition in the rhizosphere. The degree of such disruption varies with the chemical nature of the compound, its persistence, and the adaptive capacity of the native microbial community (Ruuskanen et al. 2023).

6.5.2.3 Crop Rotation and Intercropping Systems

Cropping system design plays a vital role in determining the diversity and activity of nutrient-cycling microbes. Crop rotation, particularly legume–cereal sequences, enhances soil fertility and microbial resilience by diversifying carbon inputs and interrupting pest and disease cycles (Watt et al. 2006). Different plant species release distinct root exudates like organic acids, amino acids, and sugars that shape microbial community composition, stimulating diverse nutrient-cycling guilds (Chen et al. 2024). Intercropping, especially cereal–legume combinations, further enhances nutrient-use efficiency by spatially and temporally coupling N and P cycles. For instance, in legume–cereal intercrops, cereals utilise available soil N, thereby reducing N feedback inhibition on legume nodulation and N-fixation (Picone et al. 2025). The intensified rhizosphere interactions promote higher microbial biomass, enzyme activity, and nutrient solubilisation. Microbial consortia in such systems often exhibit synergistic behaviour, where PSB and mycorrhizal fungi mobilise phosphorus to meet the higher demand created by increased N-fixation (Kebede 2021). Consequently, crop diversification serves as a low-cost, ecologically sound strategy to stimulate beneficial rhizosphere microbes and maintain long-term soil fertility.

6.5.2.4 Inoculation and Biofertiliser Application

Microbial inoculation represents a direct means to manipulate rhizosphere microbiota for enhanced nutrient cycling. In legume production, inoculation with effective rhizobial strains remains a cornerstone of sustainable nutrient management. However, inoculation success depends on strain competitiveness, adaptability to local edaphic conditions, and compatibility with host plants. Recent advances have focused on multi-functional biofertilisers that combine nitrogen-fixing rhizobia with other beneficial microbes, such as PSB, potassium-mobilising bacteria, and AMF. These multi-strain inoculants address multiple nutrient constraints simultaneously. For instance, soybean co-inoculated with Bradyrhizobium japonicum and Pseudomonas sp. capable of phosphate solubilisation exhibited significantly higher N and P uptake than single inoculations (Monica et al. 2025). Such synergistic formulations enhance root growth, improve nodulation, and strengthen plant nutrient status under both optimal and stress conditions. The effectiveness of microbial inoculants, however, is highly dependent on appropriate field management reduced tillage, adequate moisture, and limited exposure to toxic chemicals enhance inoculant survival and colonisation (Kumar et al. 2025). Adoption of quality-assured biofertilisers, along with residue retention and organic amendments, creates a conducive microenvironment that favours microbial persistence and functional expression.

6.6 BIOTECHNOLOGICAL AND ECOLOGICAL APPLICATIONS

Legume rhizosphere harbours a diverse microbial taxon that actively participate in various nutrient transformation activities. The application of beneficial microbial strains in nutrient cycling is becoming crucial and of paramount importance in the paradigm shift towards sustainable agricultural intensification. The transformation of basic microbial ecology into economically viable product specifically biostimulant and biofertilisers, offers dual advantage: fulfilling the need of global production while simultaneously mitigating severe environmental pollution caused by conventional agrochemicals. The legume rhizosphere forms a central platform for sustainable agriculture, where the nutrient-cycling microbes are exploited and manipulated using biotechnology. This integration aims to create resilient, nutrient-efficient, and climate-smart agroecosystems.

6.6.1 Advanced Microbial Biotechnology and Product Synthesis

Rhizosphere microbes may be harnessed for deciphering key metabolic pathways that can be enhanced for high-value product generation. Optimising nutrient mobilisation and harnessing microbial bioproduct synthesis represent synergistic strategies for sustainable agriculture. The molecular biology may assist in creating a super strain of Rhizobia with high nitrogen-fixing ability, better tolerance to stresses, and high competitiveness against native soil microbes. Important compounds like organic acid or enzymes may be produced from microbes, with the help of molecular biology, to release macro nutrient or micronutrient bound to the soil. The production of biostimulant or microbial byproduct may also be altered with the help of such techniques. The functional effectiveness and survival of strains in diverse environmental conditions, biotechnological interventions may be employed to convert them into more stable inoculant. The PGPRs are grown in bioreactors to produce important byproducts such as phytohormones and siderophore. Additionally, certain rhizobacteria, like Bacillus species, are used to synthesise industrial biopolymers, specifically polyhydroxyalkanoates (PHAs) and biosurfactants, which offer a sustainable supply of biodegradable plastics and environmentally friendly materials for environmental remediation.

6.6.2 The Molecular Frontier: Genetic Engineering and Synthetic Symbioses

Innovative methods like genome editing and adaptive laboratory evolution are being employed to enhance microbial attributes including rhizosphere competence, thermotolerance, and osmotic adjustment and ensure their survival and usefulness under increasingly varied climatic circumstances. An especially ambitious objective is to introduce the complex nif gene cluster into non-leguminous staple crops like rice and maize to expand biological nitrogen fixation (BNF) to these hosts using the conserved common symbiosis signalling mechanism shared across plant kingdoms. This will minimise the need for synthetic nitrogen fertilisers worldwide. Smart nutrient management, which combines precision agriculture with engineered microbial inoculants and uses biosensors to detect enzymatic activities like nitrogenase, alkaline phosphatase to enable real-time monitoring of microbial performance, improves agroecosystem sustainability, input efficiency, and adaptive nutrient management (Li et al. 2023). This supports adaptive nutrient management and maximises input use efficiency while minimising environmental losses.

6.6.3 Precision Ecology and System Integration

The engineering of the soil microbial community and the incorporation of new microbial technology into precision farming systems are gaining importance in modern agricultural. Advancement in high-throughput sequencing and multi-omics approaches, keystone microbial taxa are identified for effective utilisation in agriculture and allied sector (Ranjan et al. 2025). The information generated through modern techniques helps in generating synthetic microbial communities (SynComs); a combination of customised strains that offers synergistic benefits that outweigh those of conventional single-strain inoculants. Furthermore, functional gene markers (nifH, phoD) serve as bioindicators to monitor soil health and guide focused nutrient management. Legume–microbe associations enhance carbon sequestration through root exudation, microbial necromass, and stabilisation of organic matter in soil aggregates. Legume–cereal intercropping increases beneficial microbial networks, decreases nutrient losses, and enhances nitrogen use efficiency. Additionally, legumes can be used to remediate contaminated soils through engineered bioremediation, which uses metal-tolerant or genetically enhanced PGPR. This allows plants to extract and tolerate heavy metals and pesticide residues, supporting environmental restoration and crop productivity.

6.7 CHALLENGES AND FUTURE PERSPECTIVES

Legume rhizosphere microbiomes have enormous ecological and economic potential, but a number of interrelated barriers hinder their widespread and dependable use. The effectiveness of microbial inoculants is still highly context-dependent, impacted by crop genotype, soil characteristics, climate, and interactions with native microbial communities. Abiotic stress or competition may prevent introduced strains from establishing in the field, and ecological complexity makes outcome prediction much more difficult. Although it stabilises ecosystem processes, functional redundancy within native microbiota frequently compromises the consistency of single-strain formulations (Louca et al. 2018). This emphasises the necessity of creating artificial microbial consortia that are resilient in a variety of settings. The mechanistic interpretation and validation of microbial activity constitute another significant constraint. In order to measure in situ activity, integrative methods like metatranscriptomics, metaproteomics, and stable isotope probing are required because gene presence alone does not ensure functional expression (Prosser 2020). Furthermore, regulatory and biosafety concerns are raised by problems with inoculant stability, persistence, and ecological safety, particularly with regard to the displacement of native species (Mitter et al. 2019). Particularly for smallholder agricultural systems, technological and financial obstacles such as formulation, shelf-life stability, and cost-effective distribution provide additional challenges (Schreiter et al. 2018). Strong regulatory frameworks and long-term, multi-site monitoring are necessary to guarantee effectiveness, safety, and farmer acceptance at scale.

Precision microbiome engineering is a viable avenue for the development of bio-inoculants of the future. Elite strains with exceptional stress tolerance and rhizosphere competency may be produced using omics-driven strain selection, synthetic community design, and genetic improvement (Prosser 2020). While combining microbial techniques with sustainable agronomic practices can produce synergistic advantages, advanced delivery technologies such encapsulation, biochar-based carriers, and controlled-release hydrogels can improve establishment and shelf life (Mitter et al. 2019). Legume–microbe interactions under climatic stress should be given top priority in future studies to guarantee functional resilience. In the end, the scalability and effect of microbial technologies in sustainable agriculture might be transformed by the thoughtful creation of microbial consortia, bolstered by strong legislative frameworks, farmer involvement, and climate adaption tactics.

6.8 CONCLUSION

The legume rhizosphere serves as a dynamic hub of plant–microbe interactions that regulate nutrient cycling, plant growth, and ecosystem stability. Rhizosphere microbes, including *Rhizobium, Bradyrhizobium, Bacillus,* and *Pseudomonas*, facilitate nitrogen fixation, phosphorus and potassium solubilisation, and hormonal regulation, linking nutrient metabolism with plant development. These multi-symbiont networks, involving arbuscular mycorrhizal fungi and endophytes, enhance soil fertility and stabilise nutrient fluxes, positioning legumes as ecosystem engineers that shape their rhizosphere microbiomes through sophisticated signalling and cooperative metabolism. Despite their potential, legume-associated microbiomes face challenges including inconsistent inoculant performance influenced by soil, climate, and native microbes. Abiotic stress, competition, and functional redundancy limit consistency, emphasising the need for resilient microbial consortia. Advances in omics tools and precision microbiome engineering can help identify and design effective strains. Improved delivery systems and supportive policies will be crucial for stable, safe, and scalable use of microbial technologies to enhance nutrient cycling, soil health, and climate-resilient farming.

REFERENCES

Ashajyothi, M., K. Charishma, A. Patel, S. Paul, Y. N. Venkatesh, I. Prakash, and J. Tilgam. 2022. "Rhizosphere Microbiome: Significance in Sustainable Crop Protection." In *Rhizosphere Microbes: Biotic Stress Management*, 283–309. Singapore: Springer Nature Singapore.

Awasthi, R., R. Tewari, and H. Nayyar. 2011. "Synergy between Plants and P-Solubilizing Microbes in Soils: Effects on Growth and Physiology of Crops." *International Research Journal of Microbiology* 2: 484–503.

Azarias Guimarães, A., P. M. Duque Jaramillo, R. S. A. Nóbrega, L. A. Florentino, K. B. Silva, and F. M. de Souza Moreira. 2012. "Genetic and Symbiotic Diversity of Nitrogen-Fixing Bacteria Isolated from Agricultural Soils in the Western Amazon by Using Cowpea as the Trap Plant." *Applied and Environmental Microbiology* 78(18): 6726–6733.

Barrett, C. F., and M. A. Parker. 2006. "Coexistence of *Burkholderia*, *Cupriavidus*, and *Rhizobium* sp. Nodule Bacteria on Two *Mimosa* spp. in Costa Rica." *Applied and Environmental Microbiology* 72(2): 1198–1206. https://doi.org/10.1128/AEM.72.2.1198-1206.2006.

Battini, F., M. Grønlund, M. Agnolucci, M. Giovannetti, and I. Jakobsen. 2017. "Facilitation of Phosphorus Uptake in Maize Plants by Mycorrhizosphere Bacteria." *Scientific Reports* 7: 4686. https://doi.org/10.1038/s41598-017-04959-0.

Berg, G., D. Rybakova, D. Fischer, T. Cernava, M. C. C. Vergès, T. Charles, et al. 2020. "Microbiome Definition Re-visited: Old Concepts and New Challenges." *Microbiome* 8: 1–22.

Cai, S., and X. Zhao. 2024. "Responses of Bacterial Communities and Nitrogen-Cycling Microbiomes to the Conversion from Cereals to Legumes in the Rice-Based System." *Applied Soil Ecology* 195: 105241.

Chaudhari, D., K. Rangappa, A. Das, J. Layek, S. Basavaraj, B. K. Kandpal, Y. Shouche, and P. Rahi. 2020. "Pea (*Pisum sativum* L.) Plant Shapes Its Rhizosphere Microbiome for Nutrient Uptake and Stress Amelioration in Acidic Soils of the North-East Region of India." *Frontiers in Microbiology* 11: 968.

Chaudhary, S., S. S. Sindhu, R. Dhanker, and A. Kumari. 2023. "Microbes-mediated Sulphur Cycling in Soil: Impact on Soil Fertility, Crop Production and Environmental Sustainability." *Microbiological Research* 271: 127340.

Cooper, J. E. 2004. "Multiple Responses of Rhizobia to Flavonoids during Legume Root Infection." In *Advances in Botanical Research*, vol. 41, 1–62. Cambridge: Academic Press.

Cummings, S. P., P. Gyaneshwar, P. Vinuesa, F. T. Farruggia, M. Andrews, D. Humphry, G. N. Elliott, A. Nelson, C. Orr, D. Pettitt, G. R. Shah, S. R. Santos, H. B. Krishnan, D. Odee, F. M. S. Moreira, J. I. Sprent, J. P. W. Young, and E. K. James. 2009. "Nodulation of *Sesbania* Species by *Rhizobium* (*Agrobacterium*) Strain IRBG74 and Other Rhizobia." *Environmental Microbiology* 11(10): 2510–2525. https://doi.org/10.1111/j.1462-2920.2009.01975.x.

Dai, Y., H. W. Tian, Q. Zhu, C. X. Yi, Y. M. Wang, J. Pang, et al. 2025. "Adaptation Strategies of Eight Crop Legume Species to Low Phosphorus Availability: Root Morphology, Exudates and Microbiome Interactions." *Environmental and Experimental Botany* 238: 106252.

Das, A. K., D. S. Lee, Y. J. Woo, S. Sultana, A. Mahmud, and B. W. Yun. 2025. "The Impact of Flooding on Soil Microbial Communities and Their Functions: A Review." *Stresses* 5(2): 30.

Deng, J., F. Zhu, J. Liu, Y. Zhao, J. Wen, T. Wang, and J. Dong. 2019. "Transcription Factor bHLH2 Represses *CYSTEINE PROTEASE77* to Negatively Regulate Nodule Senescence." *Plant Physiology* 181(4): 1683–1703.

Dennis, P. G., A. J. Miller, and P. R. Hirsch. 2010. "Are Root Exudates More Important than Other Sources of Rhizodeposits in Structuring Rhizosphere Bacterial Communities?" *FEMS Microbiology Ecology* 72(3): 313–327.

Dey, G., P. Banerjee, R. K. Sharma, J. P. Maity, H. Etesami, A. K. Shaw, et al. 2021. "Management of Phosphorus in Salinity-Stressed Agriculture for Sustainable Crop Production by Salt-Tolerant Phosphate-Solubilizing Bacteria – A Review." *Agronomy* 11(8): 1552.

Dollete, D., R. A. Lumactud, C. N. Carlyle, K. Szczyglowski, B. Hill, and M. S. Thilakarathna. 2024. "Effect of Drought Stress on Symbiotic Nitrogen Fixation, Soil Nitrogen Availability, and Soil Microbial Diversity in Forage Legumes." *Plant and Soil* 495(1): 445–467.

Du, J., Q. Gao, F. Sun, B. Liu, Y. Jiao, and Q. Liu. 2025. "Agricultural Soil Microbiomes at the Climate Frontier: Nutrient-Mediated Adaptation Strategies for Sustainable Farming." *Ecotoxicology and Environmental Safety* 295: 118161.

Ducousso-Détrez, A., J. Fontaine, A. Lounès-Hadj Sahraoui, and M. Hijri. 2022. "Diversity of Phosphate Chemical Forms in Soils and Their Contributions on Soil Microbial Community Structure Changes." *Microorganisms* 10(3): 609. https://doi.org/10.3390/microorganisms10030609.

Egamberdieva, D., S. J. Wirth, A. A. Alqarawi, E. F. Abd_Allah, and A. Hashem. 2017. "Phytohormones and Beneficial Microbes: Essential Components for Plants to Balance Stress and Fitness." *Frontiers in Microbiology* 8: 2104.

Epihov, D. Z., K. Saltonstall, S. A. Batterman, L. O. Hedin, J. S. Hall, M. van Breugel, et al. 2021. "Legume–Microbiome Interactions Unlock Mineral Nutrients in Regrowing Tropical Forests." *Proceedings of the National Academy of Sciences* 118(11): e2022241118.

Etesami, H., S. Emami, and H. A. Alikhani. 2017. "Potassium Solubilizing Bacteria (KSB): Mechanisms, Promotion of Plant Growth, and Future Prospects – A Review." *Journal of Soil Science and Plant Nutrition* 17(4): 897–911.

Fan, H., S. Jia, M. Yu, X. Chen, A. Shen, and Y. Su. 2022. "Long-Term Straw Return Increases Biological Nitrogen Fixation by Increasing Soil Organic Carbon and Decreasing Available Nitrogen in Rice–Rape Rotation." *Plant and Soil* 479(1): 267–279.

Ferguson, B. J., C. Mens, A. H. Hastwell, M. Zhang, H. Su, C. H. Jones, X. Chu, and P. M. Gresshoff. 2019. "Legume Nodulation: The Host Controls the Party." *Plant, Cell & Environment* 42(1): 41–51.

Fuqua, C., M. R. Parsek, and E. P. Greenberg. 2001. "Regulation of Gene Expression by Cell-to-Cell Communication: Acyl-Homoserine Lactone Quorum Sensing." *Annual Review of Genetics* 35: 439–468.

Galloway, J. N., A. R. Townsend, J. W. Erisman, M. Bekunda, Z. Cai, J. R. Freney, et al. 2008. "Transformation of the Nitrogen Cycle: Recent Trends, Questions, and Potential Solutions." *Science* 320: 889–892. https://doi.org/10.1126/science.1136674.

Ge, Y. Y., Q. W. Xiang, C. Wagner, D. Zhang, Z. P. Xie, and C. Staehelin. 2016. "The Type III Effector NopL of *Sinorhizobium* sp. Strain NGR234 Is a Mitogen-Activated Protein Kinase Substrate." *Journal of Experimental Botany* 67(8): 2483–2494.

Geurts, R., A. Lillo, and T. Bisseling. 2012. "Exploiting an Ancient Signalling Machinery to Enjoy a Nitrogen Fixing Symbiosis." *Current Opinion in Plant Biology* 15: 438–443.

Gonzalez, J. M., and B. Aranda. 2023. "Microbial Growth under Limiting Conditions: Future Perspectives." *Microorganisms* 11(7): 1641.

Gou, X., P. B. Reich, L. Qiu, M. Shao, G. Wei, J. Wang, and X. Wei. 2023. "Leguminous Plants Significantly Increase Soil Nitrogen Cycling across Global Climates and Ecosystem Types." *Global Change Biology*. https://doi.org/10.1111/gcb.16742.

Goyal, R. K., and J. Z. Habtewold. 2023. "Evaluation of Legume–Rhizobial Symbiotic Interactions beyond Nitrogen Fixation That Help Host Survival and Diversification in Hostile Environments." *Microorganisms* 11(6): 1454.

Han, Z., et al. 2025. "Long-Term Conservation Tillage Increases Soil Organic Carbon Stability by Modulating Microbial Nutrient Limitations and Aggregate Protection." *Agronomy* 15(7): 1571.

Handa, Y., H. Nishide, N. Takeda, Y. Suzuki, M. Kawaguchi, and K. Saito. 2015. "RNA-Seq Transcriptional Profiling of an Arbuscular Mycorrhiza Provides Insights into Regulated and Coordinated Gene Expression in *Lotus japonicus* and *Rhizophagus irregularis*." *Plant and Cell Physiology* 56(8): 1490–1511.

Hartman, K., M. G. van der Heijden, V. Roussely-Provent, J. C. Walser, and K. Schlaeppi. 2017. "Deciphering Composition and Function of the Root Microbiome of a Legume Plant." *Microbiome* 5(1): 2.

Hassan, S., and U. Mathesius. 2012. "The Role of Flavonoids in Root–Rhizosphere Signalling: Opportunities and Challenges for Improving Plant–Microbe Interactions." *Journal of Experimental Botany* 63(9): 3429–3444.

Hassani, M. A., P. Durán, and S. Hacquard. 2018. "Microbial Interactions within the Plant Holobiont." *Microbiome* 6: 1–17.

Hawkins, H. J., A. Johansen, and E. George. 2000. "Uptake and Transport of Organic and Inorganic Nitrogen by Arbuscular Mycorrhizal Fungi." *Plant and Soil* 226: 275–285.

Henneron, L., P. Kardol, D. A. Wardle, C. Cros, and S. Fontaine. 2020. "Rhizosphere Control of Soil Nitrogen Cycling: A Key Component of Plant Economic Strategies." *New Phytologist* 228: 1269–1282. https://doi.org/10.1111/nph.16760.

Hider, R. C., and X. L. Kong. 2011. "Glutathione: A Key Component of the Cytoplasmic Labile Iron Pool." *Biometals* 24(6): 1179–1187.

Hinsinger, P., A. G. Bengough, D. Vetterlein, and I. M. Young. 2009. "Rhizosphere: Biophysics, Biogeochemistry and Ecological Relevance." *Plant and Soil* 321: 117–152.

Ibañez, F., and A. Fabra. 2011. "Rhizobial Nod Factors Are Required for Cortical Cell Division in the Nodule Morphogenetic Programme of the Aeschynomeneae Legume *Arachis*." *Plant Biology* 13(5): 794–800.

Ilahi, H., J. Hsouna, W. Ellouze, T. Gritli, S. Chihaoui, F. Barhoumi, M. N. Elfeddy, S. Bachkouel, L. Ouahmane, J. T. Tambong, and B. Mnasri. 2021. "Phylogenetic Study of Rhizobia Nodulating Pea (*Pisum sativum*) Isolated from Different Geographic Locations in Tunisia." *Systematic and Applied Microbiology* 44(2): 126221.

Imberty, A., and A. Varrot. 2008. "Microbial Recognition of Human Cell Surface Glycoconjugates." *Current Opinion in Structural Biology* 18: 567–576.

Jacoby, R. P., A. Koprivova, and S. Kopriva. 2021. "Pinpointing Secondary Metabolites That Shape the Composition and Function of the Plant Microbiome." *Journal of Experimental Botany* 72(1): 57–69.

Kafeel, U., U. Jahan, and F. A. Khan. 2023. "Role of Mineral Nutrients in Biological Nitrogen Fixation." In *Sustainable Plant Nutrition*, 87–106. Cambridge, MA: Academic Press.

Kawasaki, A., E. R. Watson, and M. A. Kertesz. 2012. "Indirect Effects of Polycyclic Aromatic Hydrocarbon Contamination on Microbial Communities in Legume and Grass Rhizospheres." *Plant and Soil* 358: 169–182.

Kebede, E. 2021. "Contribution, Utilization, and Improvement of Legumes-Driven Biological Nitrogen Fixation in Agricultural Systems." *Frontiers in Sustainable Food Systems* 5: 767998. https://doi.org/10.3389/fsufs.2021.767998.

Khan, S., S. Nadir, S. Iqbal, J. Xu, H. Gui, A. Khan, and L. Ye. 2021. "Towards a Comprehensive Understanding of Free-Living Nitrogen Fixation." *Circular Agricultural Systems* 1(1): 13. https://doi.org/10.48130/CAS-2021-0013.

Kishore, N., P. K. Pindi, and S. R. Reddy. 2015. "Phosphate-Solubilizing Microorganisms: A Critical Review." In *Plant Biology and Biotechnology*, 307–333. Springer.

Kistner, C., T. Winzer, A. Pitzschke, L. Mulder, S. Sato, T. Kaneko, et al. 2005. "Seven *Lotus japonicus* Genes Required for Transcriptional Reprogramming of the Root during Fungal and Bacterial Symbiosis." *The Plant Cell* 17(8): 2217–2229.

Koukoyi, G., G. Degla, and A. Alla. 2025. "A Mathematical Model of Phosphorus Cycle in Rhizosphere." *CEUR Workshop Proceedings* 4036 (Paper 3).

Kumar, R., B. Farda, A. Mignini, R. Djebaili, L. Koolman, A. Paul, et al. 2025. "Microbial Solutions in Agriculture: Enhancing Soil Health and Resilience through Bio-Inoculants and Bioremediation." *Bacteria* 4(3): 28.

Kumar, R., K. Saurabh, N. Kumawat, P. K. Sundaram, J. S. Mishra, D. K. Singh, et al. 2021. "Sustaining Productivity through Integrated Use of Microbes in Agriculture." In *Role of Microbial Communities for Sustainability*, 109–145. Singapore: Springer Singapore.

Kumar, S., R. S. Meena, R. Datta, S. K. Verma, G. S. Yadav, G. Pradhan, A. Molaei, G. K. M. Rahman, and H. A. Mashuk. 2020. "Legumes for Carbon and Nitrogen Cycling: An Organic Approach." In *Sustainable Agriculture Reviews 52*, edited by Eric Lichtfouse, 1–25. Springer.

Kumar, S. C., M. Kumar, R. Singh, and A. K. Saxena. 2024. "Population and Genetic Diversity of Rhizobia Nodulating Chickpea in Indo-Gangetic Plains of India." *Brazilian Journal of Microbiology* 55(4): 4057–4075.

Kumar, S. C., P. Singh, M. Kumar, M. V. Singh Rajawat, W. A. Ansari, D. L. N. Rao, and A. K. Saxena. 2023. "Population and Diversity of Pigeonpea Rhizobia from the Indo-Gangetic Plains of India." *Symbiosis* 90(2): 213–230.

Kumawat, C., A. Kumar, J. Parshad, S. S. Sharma, A. Patra, P. Dogra, et al. 2022a. "Microbial Diversity and Adaptation under Salt-Affected Soils: A Review." *Sustainability* 14(15): 9280.

Kumawat, K. C., et al. 2022b. "Co-Inoculation of Indigenous *Pseudomonas oryzihabitans* and *Bradyrhizobium* sp. Modulates the Growth, Symbiotic Efficacy, Nutrient Acquisition, and Grain Yield of Soybean." *Pedosphere* 32(3): 438–451.

Ladha, J. K., M. B. Peoples, P. M. Reddy, J. C. Biswas, A. Bennett, M. L. Jat, and T. J. Krupnik. 2022. "Biological Nitrogen Fixation and Prospects for Ecological Intensification in Cereal-Based Cropping Systems." *Field Crops Research* 283: 108541.

Laguerre, G., G. Depret, V. Bourion, and G. Duc. 2007. "*Rhizobium leguminosarum* bv. *viciae* Genotypes Interact with Pea Plants in Developmental Responses of Nodules, Roots and Shoots." *New Phytologist* 176(3): 564–571.

Li, L., Hu, Z., Tan, G., Fan, J., Chen, Y., Xiao, Y., Wu, S., Zhi, Q., Liu, T., Yin, H. and Tang, Q. 2023. "Enhancing Plant Growth in Biofertilizer-amended Soil through Nitrogen-transforming Microbial Communities." *Frontiers in Plant Science* 14: 1259853.

Li, X., R. Xiao, and Y. Tao. 2025. "Enhancing Plant Stress Resilience and Agricultural Sustainability through Rhizosphere Microbiome Optimization." *Plant and Soil* 513(2): 1711–1723.

Lin, P., J. Wang, P. Chen, Z. Fu, K. Luo, Y. Li, et al. 2024. "Relay Intercropped Soybean Promotes Nodule Development and Nitrogen Fixation by Root Exudate Deposition." *Frontiers in Plant Science* 15: 1447447.

Louca, S., M. F. Polz, F. Mazel, M. B. N. Albright, J. A. Huber, M. I. O'Connor, M. Ackermann, A. S. Hahn, D. S. Srivastava, S. A. Crowe, M. Doebeli, and L. W. Parfrey. 2018. "Function and Functional Redundancy in Microbial Systems." *Nature Ecology & Evolution* 2(6): 936–43. https://doi.org/10.1038/s41559-018-0519-1.

Maltais-Landry, G., K. Scow, and E. Brennan. 2014. "Soil Phosphorus Mobilization in the Rhizosphere of Cover Crops Has Little Effect on Phosphorus Cycling in California Agricultural Soils." *Soil Biology and Biochemistry* 78: 255–262.

Miransari, M. 2011. "Arbuscular Mycorrhizal Fungi and Nitrogen Uptake." *Archives of Microbiology* 193: 77–81.

Mishra, P. K., S. Mishra, G. Selvakumar, J. K. Bisht, S. Kundu, and H. S. Gupta. 2009. "Co-Inoculation of *Bacillus thuringiensis*-KR1 with *Rhizobium leguminosarum* Enhances Plant Growth and Nodulation of Pea (*Pisum sativum* L.) and Lentil (*Lens culinaris* L.)." *World Journal of Microbiology and Biotechnology* 25(5): 753–761.

Mitter, E. K., M. Tosi, D. Obregón, K. E. Dunfield, and J. J. Germida. 2019. "Rethinking Crop Nutrition in Times of Modern Microbiology: Innovative Biofertilizer Technologies." *Frontiers in Sustainable Food Systems* 3: 76. https://doi.org/10.3389/fsufs.2019.00076

Monica, S., S. Panneerselvam, R. Rajasekaran, S. Nalliappan, A. Kailappan, and A. Rangasamy. 2025. "Recent Advances in Bioinoculant Formulations and Their Shelf-Life: A Review." *Current Microbiology* 82(11): 1–21.

Moreau, D., R. D. Bardgett, R. D. Finlay, D. L. Jones, and L. Philippot. 2019. "A Plant Perspective on Nitrogen Cycling in the Rhizosphere." *Functional Ecology* 33(4): 540–552.

Nabati, J., A. Yousefi, A. Hasanfard, Z. Nemati, N. Kahrom, and A. M. Kurdestani. 2025. "Enhancing Chickpea Yield through the Application of Sulfur and Sulfur-Oxidizing Bacteria." *Scientific Reports* 15(1): 1169. https://doi.org/10.1038/s41598-024-84971-3.

Naorem, A., S. Jayaraman, Y. P. Dang, R. C. Dalal, N. K. Sinha, C. S. Rao, and A. K. Patra. 2023. "Soil Constraints in an Arid Environment – Challenges, Prospects, and Implications." *Agronomy* 13(1): 220.

Naorem, A., J. Tilgam, P. Priyadarshini, Y. Tak, A. Bharati, and A. Patel. 2022. "Nitrogenase Enzyme Complex: Functions, Regulation, and Biotechnological Applications." In *Advances in Plant Nitrogen Metabolism*, 142–154. Boca Raton, FL: CRC Press.

Oldroyd, G. E., and J. A. Downie. 2004. "Calcium, Kinases and Nodulation Signalling in Legumes." *Nature Reviews Molecular Cell Biology* 5(7): 566–576.

Pandey, C. B., U. Kumar, M. Kaviraj, K. J. Minick, A. K. Mishra, and J. S. Singh. 2020. "DNRA: A Short-Circuit in Biological N-Cycling to Conserve Nitrogen in Terrestrial Ecosystems." *Science of the Total Environment* 738: 139710.

Pathak, D., D. Singh, M. N. Jha, and K. B. Choudhary. 2024. "Harnessing the Efficacy of a *Rhizobium* Strain in Individual and Consortium Mode to Promote Sustainable Lentil Production and Biofortification under Diverse Soil Conditions." *Journal of Soil Science and Plant Nutrition* 24(3): 5175–5194. https://doi.org/10.1007/s42729-024-01783-x.

Paungfoo-Lonhienne, C., W. Wang, Y. K. Yeoh, and N. Halpin. 2017. "Legume Crop Rotation Suppressed Nitrifying Microbial Community in a Sugarcane Cropping Soil." *Scientific Reports* 7(1): 16707. https://doi.org/10.1038/s41598-017-16790-2.

Picone, R., G. Pietramellara, G. Guggenberger, S. I. Pathan, and N. Gentsch. 2025. "The Contemporary Plant–Soil Feedback in Legume–Cereal Intercropping Systems: A Review of Carbon, Nutrient, and Microbial Dynamics." *Plant and Soil*: 1–21.

Prosser, J. I. 2020. "Putting Science Back into Microbial Ecology: A Question of Approach." *Philosophical Transactions of the Royal Society B: Biological Sciences* 375(1798): 20190240. https://doi.org/10.1098/rstb.2019.0240.

Pueppke, S. G., and W. J. Broughton. 1999. "Rhizobium sp. NGR234 and R. fredii USDA257 Share Exceptionally Broad, Nested Host Ranges." *Molecular Plant–Microbe Interactions* 12: 293–318.

Punyalue, A., S. Jamjod, and B. Rerkasem. 2018. "Intercropping Maize with Legumes for Sustainable Highland Maize Production." *Mountain Research and Development* 38(1): 35–44.

Qiao, M., R. Sun, Z. Wang, K. Dumack, X. Xie, C. Dai, et al. 2024. "Legume Rhizodeposition Promotes Nitrogen Fixation by Soil Microbiota under Crop Diversification." *Nature Communications* 15(1): 2924.

Quilbé, J., N. Nouwen, M. Pervent, R. Guyonnet, J. Cullimore, F. Gressent, et al. 2022. "A Mutant-Based Analysis of the Establishment of Nod-Independent Symbiosis in the Legume *Aeschynomene evenia*." *Plant Physiology* 190(2): 1400–1417.

Raaijmakers, J. M., T. C. Paulitz, C. Steinberg, C. Alabouvette, and Y. Moënne-Loccoz. 2009. "The Rhizosphere: A Playground and Battlefield for Soilborne Pathogens and Beneficial Microorganisms." *Plant and Soil* 321: 341–361.

Rajkumari, J., P. Katiyar, S. Dheeman, P. Pandey, and D. K. Maheshwari. 2022. "The Changing Paradigm of Rhizobial Taxonomy and Its Systematic Growth Upto Postgenomic Technologies." *World Journal of Microbiology and Biotechnology* 38(11): 206. https://doi.org/10.1007/s11274-022-03417-7.

Ranjan, A., S. Kaushik, and J. Tilgam. 2025. "Omics Techniques to Decipher Plant Root–Microbe Interactions." In *Plant–Microbiome Interactions for Climate-Resilient Agriculture*, 213–227. Singapore: Springer Nature Singapore.

Rasheela, A. R. P., M. F. Khalid, D. A. Abumaali, J. M. Alatalo, and T. Ahmed. 2024. "Impact of Abiotic Stressors on Soil Microbial Communities: A Focus on Antibiotics and Their Interactions with Emerging Pollutants." *Soil Systems* 9(1): 2.

Raul, B., O. Bhattacharjee, A. Ghosh, P. Upadhyay, K. Tembhare, A. Singh, et al. 2022. "Microscopic and Transcriptomic Analyses of Dalbergoid Legume Peanut Reveal a Divergent Evolution Leading to Nod-Factor-Dependent Epidermal Crack-Entry and Terminal Bacteroid Differentiation." *Molecular Plant-Microbe Interactions* 35(2): 131–145.

Rawat, P., S. Das, D. Shankhdhar, and S. C. Shankhdhar. 2021. "Phosphate-Solubilizing Microorganisms: Mechanism and Their Role in Phosphate Solubilization and Uptake." *Journal of Soil Science and Plant Nutrition* 21(1): 49–68.

Reeve, W., R. Tian, S. De Meyer, V. Melino, J. Terpolilli, J. Ardley, and N. Kyrpides. 2013. "Genome Sequence of the Clover-Nodulating *Rhizobium leguminosarum* bv. *trifolii* Strain TA1." *Standards in Genomic Sciences* 9(2): 243–253. https://doi.org/10.4056/sigs.4156641.

Ruuskanen, S., B. Fuchs, R. Nissinen, P. Puigbò, M. Rainio, K. Saikkonen, and M. Helander. 2023. "Ecosystem Consequences of Herbicides: The Role of Microbiome." *Trends in Ecology & Evolution* 38(1): 35–43.

Saiz, E., F. Sgouridis, F. P. Drijfhout, M. Peichl, M. B. Nilsson, and S. Ullah. 2021. "Chronic Atmospheric Reactive Nitrogen Deposition Suppresses Biological Nitrogen Fixation in Peatlands." *Environmental Science and Technology* 55(2): 1310–1318. https://doi.org/10.1021/acs.est.0c06180.

Sánchez, C., and K. Minamisawa. 2019. "Nitrogen Cycling in Soybean Rhizosphere: Sources and Sinks of Nitrous Oxide (N_2O)." *Frontiers in Microbiology* 10: 1943.

Scherlach, K., and C. Hertweck. 2020. "Chemical Mediators at the Bacterial-Fungal Interface." *Annual Review of Microbiology* 74: 267–290.

Schreiter, S., G. C. Ding, H. Heuer, G. Neumann, M. Sandmann, R. Grosch, and K. Smalla. 2018. "Effect of the Soil Type on the Microbiome in the Rhizosphere of Field-Grown Lettuce." *Frontiers in Microbiology* 9: 473. https://doi.org/10.3389/fmicb.2018.00473.

Serova, T. A., P. G. Kusakin, A. B. Kitaeva, E. V. Seliverstova, A. P. Gorshkov, D. A. Romanyuk, et al. 2023. "Effects of Elevated Temperature on *Pisum sativum* Nodule Development: I – Detailed Characteristic of Unusual Apical Senescence." *International Journal of Molecular Sciences* 24(24): 17144.

Shen, J., L. Yuan, J. Zhang, H. Li, Z. Bai, X. Chen, and F. Zhang. 2011. "Phosphorus Dynamics: From Soil to Plant." *Plant Physiology* 156(3): 997–1005. https://doi.org/10.1104/pp.111.175271.

Shen, W., C. Yan, H. Yan, X. Zhou, Y. Wang, X. Yang, and Y. Deng. 2025. "Organic Fertilization Shapes Diazotrophic Microbiomes in Legume and Grass Rhizospheres of the Qinghai-Tibet Plateau." *Journal of Environmental Sciences*. https://doi.org/10.1016/j.jes.2024.08.019.

Shumilina, J., A. Soboleva, E. Abakumov, O. Y. Shtark, V. A. Zhukov, and A. Frolov. 2023. "Signaling in Legume–Rhizobia Symbiosis." *International Journal of Molecular Sciences* 24(24): 17397.

Silva, C., P. Vinuesa, L. E. Eguiarte, E. Martínez-Romero, and V. Souza. 2003. "*Rhizobium etli* and *Rhizobium gallicum* Nodulate Common Bean (*Phaseolus vulgaris*) in a Traditionally Managed Milpa Plot in Mexico: Population Genetics and Biogeographic Implications." *Applied and Environmental Microbiology* 69(2): 884–893. https://doi.org/10.1128/AEM.69.2.884-893.2003.

Sinduja, M., P. M. Brindhavani, G. K. Dinesh, J. J. John, K. Mathiyarasi, S. Velusamy, et al. 2024. "Role of Microbes and Microbiomes in Natural Resource Management and the Regulation of Biogeochemical Processes and Nutrient Cycling." In *Microbes as Agents of Change for Sustainable Development*, 1–28. Sharjah: Bentham Science Publishers.

Smercina, D. N., S. E. Evans, M. L. Friesen, and L. K. Tiemann. 2019. "To Fix or Not to Fix: Controls on Free-Living Nitrogen Fixation in the Rhizosphere." *Applied and Environmental Microbiology* 85(6): e02546-18. https://doi.org/10.1128/AEM.02546-18.

Subbarao, G. V., Rondon, M., Ito, O., Ishikawa, T., Rao, I. M., Nakahara, K., Lascano, C. and Berry, W. L. 2007. "Biological Nitrification Inhibition (BNI)—Is It a Widespread Phenomenon? *Plant and Soil* 294(1): 5–18.

Tao, Jinjin, Sishuo Wang, Tianhua Liao, and Haiwei Luo. 2021. "Evolutionary Origin and Ecological Implication of a Unique *nif* Island in Free-Living *Bradyrhizobium* Lineages." *The ISME Journal* 15(11): 3195–3206.

Taskin, E., R. Boselli, A. Fiorini, C. Misci, F. Ardenti, F. Bandini, L. Guzzetti, D. Panzeri, N. Tommasi, A. Galimberti, and M. Labra. 2021. "Combined Impact of No-Till and Cover Crops with or without Short-Term Water Stress as Revealed by Physicochemical and Microbiological Indicators." *Biology* 10(1): 23.

Taylor, B. N., E. L. Simms, and K. J. Komatsu. 2020. "More than a Functional Group: Diversity within the Legume–Rhizobia Mutualism and Its Relationship with Ecosystem Function." *Diversity* 12: 50. https://doi.org/10.3390/d12020050.

Teulet, A., A. Camuel, X. Perret, and E. Giraud. 2022. "The Versatile Roles of Type III Secretion Systems in Rhizobium-Legume Symbioses." *Annual Review of Microbiology* 76: 45–65.

Thepbandit, W., and D. Athinuwat. 2024. "Rhizosphere Microorganisms Supply Availability of Soil Nutrients and Induce Plant Defense." *Microorganisms* 12 (3): 558.

Tsikou, D., Z. Yan, D. B. Holt, N. B. Abel, D. E. Reid, L. H. Madsen, et al. 2018. "Systemic Control of Legume Susceptibility to Rhizobial Infection by a Mobile MicroRNA." *Science* 362 (6411): 233–236.

Van der Drift, K. M., H. P. Spaink, G. V. Bloemberg, A. A. van Brussel, B. J. Lugtenberg, J. Haverkamp, et al. 1996. “Rhizobium leguminosarum bv. trifolii Produces Lipo-Chitin Oligosaccharides with nodE-Dependent Highly Unsaturated Fatty Acyl Moieties.” *Journal of Biological Chemistry* 271: 22563–22569.

Vasconcelos, M. W., M. A. Grusak, E. Pinto, A. Gomes, H. Ferreira, B. Balázs, . . . P. Iannetta. 2020. “The Biology of Legumes and their Agronomic, Economic, and Social Impact.” In *The Plant Family Fabaceae: Biology and Physiological Responses to Environmental Stresses*, 3–25. Singapore: Springer Singapore.

Wang, Rui, Yue Li Chang, Wen Tao Zheng, Dan Zhang, Xiao Xia Zhang, Xin Hua Sui, En Tao Wang, Jia Qi Hu, Li Ya Zhang, and Wen Xin Chen. 2013. “Bradyrhizobium arachidis sp. nov., Isolated from Effective Nodules of *Arachis hypogaea* Grown in China.” *Systematic and Applied Microbiology* 36(2): 101–105.

Wang, Shuzhen, Jie Yan, Kate V. Heal, Haiyang Li, Yuanchun Yu, and Chuifan Zhou. 2024. “Rhizosphere Microbial Roles in Phosphorus Cycling during Successive Plantings of Chinese Fir Plantations.” *Forest Ecology and Management* 570: 122227.

Wang, X., Y. Yang, K. Pei, J. Zhou, L. Peixoto, A. Gunina, . . . Y. Kuzyakov. 2021. “Nitrogen Rhizodeposition by Legumes and Its Fate in Agroecosystems: A Field Study and Literature Review.” *Land Degradation & Development* 32(1): 410–419.

Waoo, A. A., C. Vyas, and B. A. Chopade. 2025. “Soil Bacteria: An Invisible Army for Climate Resilience and Mitigation.” In *Climate Change and Soil Microorganisms for Environmental Sustainability*, 429–449. Singapore: Springer Nature Singapore.

Watt, M., J. A. Kirkegaard, and J. B. Passioura. 2006. “Rhizosphere Biology and Crop Productivity – A Review.” *Soil Research* 44(4): 299–317.

Wheatley, R. M., and P. S. Poole. 2018. “Mechanisms of Bacterial Attachment to Roots.” *FEMS Microbiology Reviews* 42(4): 448–461.

Xiong, C., Y. G. Zhu, J. T. Wang, B. Singh, L. L. Han, and J. P. Shen. 2021. “Host Selection Shapes Crop Microbiome Assembly and Network Complexity.” *New Phytologist* 229(2): 1091–1104.

Xu, Y., Q. Gao, L. Xue, J. Zhang, and C. Wang. 2025. “Optimized Nitrogen Fertilizer Management Enhances Soybean (*Glycine max* (L.) Merril.) Yield and Nitrogen Use Efficiency by Promoting Symbiotic Nitrogen Fixation Capacity.” *Frontiers in Plant Science* 16: 1604251.

Yang, S., F. Tang, M. Gao, H. B. Krishnan, and H. Zhu. 2010. “R Gene-Controlled Host Specificity in the Legume–Rhizobia Symbiosis.” *Proceedings of the National Academy of Sciences* 107 (43): 18735–18740.

Yu, Xiaocheng, and Hongyan Zhu. 2025. “Enacting Partner Specificity in Legume–Rhizobia Symbioses.” *Abiotech* 6(2): 311–327.

Zhang, H., Y. Sun, X. Xie, M.-S. Kim, S. E. Dowd, and P. W. Paré. 2009. “A Soil Bacterium Regulates Plant Acquisition of Iron via Deficiency-Inducible Mechanisms.” *The Plant Journal* 58(4): 568–577.

Zhang, J., D. Singh, C. Guo, Y. Shang, and S. Peng. 2020. “Rhizobia at Extremes of Acidity, Alkalinity, Salinity, and Temperature.” In *Microbial Versatility in Varied Environments: Microbes in Sensitive Environments*, 51–65. Singapore: Springer Singapore.

Zhao, Y., C. Yan, F. Hu, Z. Luo, S. Zhang, M. Xiao, . . . H. Fan. 2022. “Intercropping Pinto Peanut in Litchi Orchard Effectively Improved Soil Available Potassium Content, Optimized Soil Bacterial Community Structure, and Advanced Bacterial Community Diversity.” *Frontiers in Microbiology* 13: 868312.

Zhou, Yi, Christopher J. Lambrides, Jishun Li, Qili Xu, Ruey Toh, Shenzhong Tian, Peizhi Yang, Hetong Yang, Maarten Ryder, and Matthew D. Denton. 2020. “Nitrifying Microbes in the Rhizosphere of Perennial Grasses Are Modified by Biological Nitrification Inhibition.” *Microorganisms* 8(11): 1687.

7 Integrating the Role of Beneficial Microbes in Biotic and Abiotic Stress Tolerance of Legumes

Neeraj Kumar Dubey, Sohan Lal, Gunja Gond, Ran Vijay Singh, Satendra Pal Singh, Alok Ranjan, Pradeep Kumar Yadav, Neeshma Jaiswal, Rashmi Srivastava, and Rakesh Srivastava

7.1 INTRODUCTION

The legume family, with ~750 genera and 20,000 species, is the third-largest angiosperm group and a key food source after cereals (Ejigui et al., 2007). Legumes provide proteins, amino acids, vitamins, minerals, and carbohydrates, making them essential in regions where animal proteins are scarce (Graham and Vance, 2003). They fix atmospheric nitrogen via rhizobia, improving soil fertility, reducing synthetic fertiliser use, and supporting sustainable agriculture (Rubiales and Mikic, 2015). Globally, legumes serve multiple roles: food (e.g. *Glycine max, Cicer arietinum, Lens culinaris*), oil (*Arachis hypogaea, Pongamia pinnata*), fibber (*Crotalaria juncea*), fodder (*Trifolium alexandrinum, Medicago sativa*), timber (*Dalbergia sissoo, Pterocarpus marsupium*), dyes (*Indigofera tinctoria, Butea monosperma*), gums (*Astragalus gummifer, Cyamopsis tetragonoloba*), ornamentals (*Lathyrus odoratus, Clitoria ternatea*), and medicinal plants (*Glycyrrhiza glabra, Psoralea corylifolia*) (Singh et al., 2022). Despite their importance, legume productivity is limited by biotic and abiotic stresses. Associations with rhizospheric microbes enhance resilience, growth, and soil fertility, benefiting subsequent crops This multifunctionality nutritional, ecological, and economic underscores legumes' critical role in global agriculture, particularly in South Asia.

7.2 NUTRIENT CYCLING MICROBES

Legumes play a central role in sustainable agriculture, not only as a major source of food and fodder but also through their unique rhizospheric associations with microbes that improve soil fertility and plant health (Figure 7.1).

DOI: 10.1201/9781003733324-8

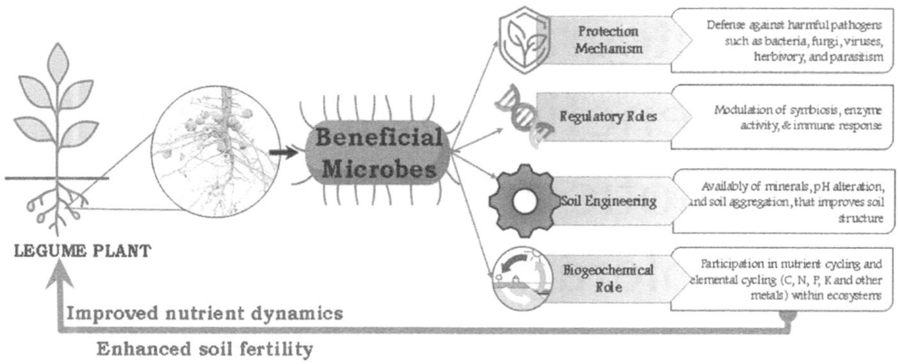

FIGURE 7.1 Mechanisms by which beneficial bacteria support healthy plant growth and overall fitness.

Nutrient-cycling microbes are essential for maintaining ecosystem health and productivity, as they break down organic matter and transform nutrients into mineral forms that plants can readily absorb. In leguminous plants, this process is largely driven by nitrogen-fixing bacteria (rhizobia), which form symbiotic associations with the roots. These rhizobia convert atmospheric nitrogen into a usable form for plants (Wobeng et al., 2020). Consequently, legumes enhance soil fertility by improving nutrient dynamics, increasing nitrogen input, enhancing phosphorus availability, and modifying physicochemical soil properties through rhizobia-nodulated intensive root systems (Wobeng et al., 2020). Legume roots face constant biotic and abiotic stresses (Sindhu et al., 2010) and respond by forming diverse microbial interactions associative, endophytic, symbiotic, pathogenic, and parasitic (Sindhu et al., 2010). Root exudates foster beneficial rhizosphere microbes that enhance growth and yield via secondary metabolites and nutrient provision, including fixed nitrogen, phosphorus, iron, potassium, and sulphur (Sindhu et al., 2010). Key bacterial associates include *Ensifer fredii, Rhizobium tropici, Rhizobium etli,* and *Azospirillum brasilense* in soybean and bean, whose root colonisation relies on motility, aggregation, surface proteins, pili, lectins, flavonoids, and bacterial polysaccharides, progressing from reversible to irreversible attachment via cellulose fibrils (Rodríguez-Navarro et al., 2007). Other rhizobia Rhizobium, *Bradyrhizobium, Microvirga, Phyllobacterium, Burkholderia, Cupriavidus*, and *Bosea* infect diverse legumes, supporting nitrogen fixation, soil fertility, nutrient cycling, and plant resilience (Radl et al., 2014). Understanding these symbioses is vital for enhancing legume productivity and sustainable agriculture.

7.3 IMPACT OF BIOTIC STRESS ON LEGUMINOUS CROPS PRODUCTION

Biotic stresses severely limit legume yield and quality worldwide, being caused by pathogenic bacteria, fungi, viruses, nematodes, parasitic weeds, and herbivorous insects (Kharayat et al., 2017; Srivastava et al., 2022). These pathogens not only reduce productivity but also impair seed nutritional composition (Grau et al., 2004). Examples include bacterial blight in pea causing up to 20% yield loss (Hollaway et al., 2007) and fungal diseases such as Botrytis grey mould and Fusarium wilt in chickpea causing up to 40% loss (Pande et al., 2006). Numerous viral infections (over 44 viruses) threaten legumes including pea, lentil, faba bean, and chickpea (Grigoras et al., 2010). Fungal diseases such as rust, wilt, root rot, blight, and mildew cause severe yield declines in major legumes including chickpea, faba bean, pea, bean, soybean, mung bean, and lentil (Pande et al., 2000). Nematodes cause about 12% productivity loss and predispose plants to secondary infections (Pande et al., 2000; Correia et al., 2023). Parasitic weeds such as Orobanche and Alectra also cause significant regional damage (Ohlson and Timko, 2022). Insects, including aphids, pod borers, beetles, stem borers, and leaf miners, inflict severe field and storage losses, with several crops experiencing 30–40% yield reduction (Pande et al., 2000; Kanobe et al., 2015; Chen et al., 2018; Mishra et al., 2022). Collectively, these stresses pose major threats to global legume production, underscoring the need for resistant cultivars and integrated pest management strategies.

7.4 IMPACT OF ABIOTIC STRESS ON LEGUMINOUS CROPS PRODUCTION

Abiotic stresses severely constrain legume growth and yield, with drought, heat, chilling, salinity, and frost acting individually or together to reduce productivity (Srivastava et al., 2014; Dubey et al., 2025). Drought causes the greatest losses, often up to ~50% in pea, chickpea, soybean, and groundnut (Bakhoum et al., 2023). Heat stress, particularly during reproductive stages, similarly reduces yield (Sher et al., 2024), and combined heat–drought can lower productivity by 40–57% (Kumari et al., 2021). Chilling stress impairs reproductive success (Kaur et al., 2011), though seed priming improves tolerance (Abdela et al., 2020). Salinity may reduce yields up to 50% (Parmar et al., 2024), while frost also causes notable losses (Alharbi et al., 2021). Collectively, these stresses necessitate resilient cultivars and adaptive management.

7.5 BENEFICIAL MICROBES FOR BIOTIC AND ABIOTIC STRESS ENHANCE TOLERANCE IN LEGUMES

The use of plant growth–promoting rhizobacteria (PGPR) has emerged as a promising strategy to enhance tolerance of host plants against both biotic and abiotic stresses (Ahmad et al., 2022). For instance, *Pseudomonas fluorescens* and *Mesorhizobium ciceri* treatments in chickpea (*Cicer arietinum* L.) improve drought resistance in treated plants (Abdela et al., 2020). Similarly, *Bradyrhizobium japonicum* and *Bacillus thuringiensis* applications in soybean (*Glycine max*) alter root structure and length, thereby conferring drought tolerance. In mungbean (*Vigna radiata*), *Ochrobactrum pseudogrignonense* enhances root and shoot length and biomass under drought conditions (Saikia et al., 2018). Likewise, *Pseudomonas putida* stimulates gibberellin secretion in soybean during drought stress (Kang et al., 2014), while *Pseudomonas aeruginosa* induces proline accumulation in mungbean under similar stress (Sarma and Saikia, 2014). In addition, *M. ciceri* and *P. fluorescens* enhance proline, water, and carotenoid levels in chickpea during water stress, thus promoting drought tolerance (Abdela et al., 2020). Other PGPR strains such as *Acinetobacter calcoaceticus* AC06 and *Bacillus amyloliquefaciens* BA01 improve drought tolerance in groundnut (*Arachis hypogaea*) (Eswaran et al., 2024). Chemical compounds like proline also help plants tolerate heat and chilling stresses by improving pollen viability, ovule health, pollen germination, and seed setting in chickpea (*C. arietinum*) (Kaur et al., 2011). Proline further enhances tolerance to drought stress. Salinity stress poses a severe threat as well, capable of reducing legume yield by up to 100% due to ion toxicity and impaired photosynthesis. PGPR such as *Pseudomonas chlororaphis* TSAU13, *P. extremorientalis* TSAU20, *P. trivialis* 3Re27, *P. fluorescens*, *P. putida*, *Bacillus subtilis*, *Mesorhizobium*, *Rhizobium* spp., and arbuscular mycorrhizal fungi have been reported to improve yield in common bean (*Phaseolus vulgaris*), soybean (*G. max*), pea (*Pisum sativum*), faba bean (*Vicia faba*), chickpea (*C. arietinum*), and mungbean (*V. radiata*) under saline stress (Nadeem et al., 2019). These microbes enhance photosynthetic efficiency, nutrient uptake, and root-shoot growth in stressed conditions. Additionally, bio-priming pea seeds with aqueous extracts of *Typha angustifolia* L. leaves induces salinity tolerance by stimulating proline, sugar, flavonoid, and alkaloid accumulation (Ghezal et al., 2016). Microbial treatments also bring about nutritional and physiological changes in legumes. For example, treating chickpea seeds with two efficient endophytic strains (*Enterobacter* sp. BHUJPCS-2 and BHUJPCS-8) enhances the nutritional value of foliage and seeds (Mukherjee et al., 2024). Salinity remains a serious challenge, affecting nearly 20–50% of agricultural land (Kumar and Sharma, 2020). In mungbean, *Pantoea* sp. and *Enterococcus* increase salinity tolerance through phytohormonal activity (Panwar et al., 2016). Similarly, peanut plants treated with *Brachybacterium saurashtrense*, *Brevibacterium casei*, *etc.* show improved biomass and salt resistance. Phytohormones such as ACC (1-aminocyclopropane-1-carboxylic acid) also contribute to salt tolerance in mungbean when mediated by *Pseudomonas syringae*, *P. fluorescens*, and *Rhizobium phaseoli* (Ahmad et al., 2011). In soybean and pea, treatments with *P. putida* and *Arthrobacter protophormiae* increase ACC deaminase activity, chlorophyll content, biomass, and salinity tolerance (Barnawal et al., 2014). Other mechanisms of stress tolerance involve hormonal regulation (Agarwal et al., 2020; Verma et al., 2016). For example, *P. putida*-treated soybean induces salicylic acid (SA), gibberellins (GA), and abscisic acid (ABA), conferring drought and salinity resistance. *P. aeruginosa* in groundnut induces indole-3-acetic acid (IAA), ammonia, and phenols, promoting salt resistance. Similarly, *P. pseudoalcaligenes* in chickpea facilitates phosphate solubilisation and IAA production, contributing to salt tolerance (Patel et al., 2012). In mungbean, *Rhizobium* and *Pseudomonas* enhance chlorophyll levels and photosynthesis, conferring resistance (Ahmad et al., 2013). *Rhizobium leguminosarum* in lentil (*Lens culinaris*) increases IAA production and salinity tolerance (Jida and Assefa, 2011). Pea treated with *A. protophormiae*, *R. leguminosarum*, and *Glomus mosseae* influences ethylene (ET) production, thereby enhancing salt tolerance (Barnawal et al., 2014). Likewise, pigeon pea (*Cajanus cajan*), black gram (*Vigna mungo*), and faba bean

(*V. faba*) treated with *Bradyrhizobium*, *P. fluorescens*, and *P. putida* show improved water absorption, photosynthetic activity, and salinity resistance (Bhat et al., 2022; Metwali et al., 2015). Interestingly, tolerance against abiotic stresses can also provide cross-protection against biotic challenges. For instance, similarly, microbial consortia of *P. aeruginosa*, *Trichoderma harzianum*, and *B. subtilis* protect plants against *Sclerotinia sclerotiorum* (white mould disease) by activating defence responses, including proteomic changes linked to metabolism, phenylpropanoid pathways, and stress regulation. Moreover, *Pseudomonas guariconensis* induces systemic resistance in groundnut seedlings (Patel et al., 2015), while late leaf spot (LLS) caused by *Phaeoisariopsis personata* is reduced by chitinolytic *Bacillus circulans* and *Serratia marcescens* (Kishore et al., 2005). In addition to microbial interventions, cultural practices such as soil preparation, residue destruction, drainage, optimal fertilisation, crop rotation, and pathogen-free seeds help reduce pathogen pressure (Grau et al., 2004). Rhizobacteria also decrease root-borne diseases and induce systemic resistance to future pathogens. Furthermore, several legume cultivars – including chickpea, urd bean (*Vigna mungo*), pigeon pea (*C. cajan*), cowpea (*Vigna unguiculata*), and mungbean (*V. radiata*) – exhibit natural resistance against pathogens such as *Fusarium*, yellow mosaic virus, *Phytophthora drechsleri*, and *Meloidogyne* spp. (Sandhu et al., 2025). In conclusion, microbial treatments not only mitigate the harmful impacts of abiotic stresses like drought, heat, salinity, and frost but also provide resilience against biotic challenges such as fungi, viruses, and nematodes. This dual role highlights the immense potential of PGPR and other beneficial microbes in sustainable legume cultivation.

7.6 CONCLUSION

Leguminous plants provide numerous benefits to humankind, serving as sources of food, fodder, fibber, and oil. In addition, their symbiotic association with *Rhizobium* bacteria enhances nitrogen fixation, which improves soil fertility and reduces the need for synthetic nitrogen inputs. As a result, legume-mediated cropping systems, whether intercropping or alternate cropping, represent an effective strategy to promote sustainable agriculture and maintain a healthy ecosystem. However, legumes are exposed to more than 100 different types of stresses, including abiotic factors and herbivorous insects, which together can cause 15–100% yield losses. This book chapter summarises the role of diverse microbial flora in enhancing both biotic and abiotic stress tolerance in legume crops, offering valuable insights into reducing pesticide dependence and inorganic fertiliser use, which otherwise degrade soil fertility. Importantly, such beneficial microbes not only induce desirable traits in host plants but also contribute to overall soil health. Pathogen attacks in legumes are often interlinked; for example, infection by nematodes or parasitic weeds can predispose plants to additional pathogen invasions. Root-borne diseases are particularly difficult to control since pesticides are less effective in the underground environment. Moreover, excessive use of pesticides and inorganic fertilisers has resulted in pest resistance and loss of soil fertility, further complicating crop protection. Consequently, alternative approaches that rely on beneficial soil microflora are essential to mitigate both current biotic stresses and future abiotic challenges.

Certain chemicals, such as proline, play a key role in enhancing plant tolerance to heat and chilling stresses. Interestingly, beneficial microbes can induce the accumulation of proline in treated plants, thereby improving their capacity to withstand multiple abiotic stresses. Thus, integrating microbial interventions with conventional legume management practices provides a promising path towards sustainable agriculture and resilient cropping systems.

REFERENCES

Abdela AA, Barka GD, Degefu T. Co-inoculation effect of Mesorhizobiumciceri and Pseudomonas fluorescens on physiological and biochemical responses of Kabuli chickpea (Cicer arietinum L.) during drought stress. Plant Physiology Reports. 2020;25:359–369. https://doi.org/10.1007/s40502-020-00511-x

Agarwal N, Srivastava R, Verma A, Rai KM, Singh B, Verma PC. Unravelling cotton nonexpressor of pathogenesis-related 1(NPR1)-like genes family: Evolutionary analysis and putative role in fiber development and defense pathway. Plants. 2020;9:999. https://doi.org/10.3390/plants9080999

Ahmad HM, Fiaz S, Hafeez S, Zahra S, Shah AN, Gul B, Aziz O, Fakhar A, Rafique M, Chen Y, Yang SH. Plant growth-promoting rhizobacteria eliminate the effect of drought stress in plants: A review. Frontiers in Plant Science. 2022 Aug 11;13:875774.

Ahmad M, Zahir ZA, Asghar HN, Asghar M. Inducing salt tolerance in mung bean through coinoculation with rhizobia and plant-growth-promoting rhizobacteria containing 1-aminocyclopropane-1-carboxylate deaminase. Canadian Journal of Microbiology. 2011;57(7):578–589. https://doi.org/10.1139/w11-044

Ahmad M, Zahir ZA, Khalid M, Nazli F, Arshad M. Efficacy of Rhizobium and Pseudomonas strains to improve physiology, ionic balance and quality of mung bean under salt-affected conditions on farmer's fields. Plant Physiology and Biochemistry. 2013;63:170–176. https://doi.org/10.1016/j.plaphy.2012.11.024.

Alharbi NH, Alghamdi SS, Migdadi HM, El-Harty EH, Adhikari KN. Evaluation of frost damage and pod set in faba bean (Vicia faba L.) under field conditions. Plants. 2021 Sep 15;10(9):1925.

Bakhoum GS, Sadak MS, Thabet MS. Induction of tolerance in groundnut plants against drought stress and Cercospora leaf spot disease with exogenous application of arginine and sodium nitroprusside under field conditions. Journal of Soil Science and Plant Nutrition. 2023 Dec;23(4):6612–6631.

Barnawal D, Bharti N, Maji D, Chanotiya CS, Kalra A. ACC deaminase-containing Arthrobacter protophormiae induces NaCl stress tolerance through reduced ACC oxidase activity and ethylene production resulting in improved nodulation and mycorrhization in Pisum sativum. Journal of Plant Physiology. 2014;171:884–894. https://doi.org/10.1016/j.jplph.2014.03.007

Bhat BA, Tariq L, Nissar S, Islam ST, Islam SU, Mangral Z, Ilyas N, Sayyed RZ, Muthusamy G, Kim W, Dar TUH. The role of plant-associated rhizobacteria in plant growth, biocontrol and abiotic stress management. Journal of Applied Microbiology. 2022;133:2717–2741. https://doi.org/10.1111/jam.15796

Chen X, Richter AR, Stout MJ, Davis JA. Effects of induced plant resistance on soybean looper (Lepidoptera: Noctuidae) in soybean. Arthropod-Plant Interactions. 2018 Aug;12(4):543–551.

Correia VR, Khan MR, de Oliveira Lima FS, Ahamad F. Nematode problems in vegetable legumes and their sustainable management. In Nematode Diseases of Crops and Their Sustainable Management 2023 Jan 1 (pp. 311–336). Academic Press. 11;14(1):241.

Dubey NK, Srivastava R, Singh M, Bhutani S, Ranjan A. Genetic engineering for crop improvement against stresses and future agriculture. In Omics and Genome Editing: Revolution in Crop Improvement for Sustainable Agriculture 2025 (pp. 55–67). Springer.

Ejigui J, Savoie L, Marin J, Desrosiers T. Improvement of the nutritional quality of a traditional complementary porridge made of fermented yellow maize (Zea mays): Effect of maize-legume combinations and traditional processing methods. Food and Nutrition Bulletin. 2007;28:23–34.

Eswaran SU, Sundaram L, Perveen K, Bukhari NA, Sayyed RZ. Osmolyte-producing microbial biostimulants regulate the growth of Arachis hypogaea L. under drought stress. BMC Microbiology. 2024 May 15;24(1):165.

Ghezal N, Rinez I, Sbai H, Saad I, Farooq M, Rinez A, Zribi I, Haouala R. Improvement of Pisum sativum salt stress tolerance by bio-priming their seeds using Typha angustifolia leaves aqueous extract. South African Journal of Botany. 2016 Jul 1;105:240–250.

Graham PH, Vance CP. Legumes: Importance and constraints to greater use. Plant Physiology. 2003 Mar 1;131(3):872–877.

Grau CR, Dorrance AE, Bond J, Russin JS. Fungal diseases. Soybeans: Improvement, Production, and Uses. 2004 Jan 1;16:679–763.

Grigoras I, Gronenborn B, Vetten HJ. First report of a nanovirus disease of pea in Germany. Plant Disease. 2010 May;94(5):642.

Hollaway GJ, Bretag TW, Price TV. The epidemiology and management of bacterial blight (Pseudomonas syringaepv. pisi) of field pea (Pisum sativum) in Australia: A review. Australian Journal of Agricultural Research. 2007 Nov 26;58(11):1086–1099.

Jida M, Assefa F. Phenotypic and plant growth promoting characteristics of Rhizobium leguminosarumbv. viciae from lentil growing areas of Ethiopia. African Journal of Microbiology Research. 2011;5:4133–4142.

Kang S-M, Radhakrishnan R, Khan AL, Kim M-J, Park J-M, Kim B-R, et al. Gibberellin secreting rhizobacterium, Pseudomonas putida H-2-3 modulates the hormonal and stress physiology of soybean to improve the plant growth under saline and drought conditions. Plant Physiology and Biochemistry. 2014;84:115–124. https://doi.org/10.1016/j.plaphy.2014.09.001

Kanobe C, McCarville MT, O'Neal ME, Tylka GL, MacIntosh GC. Soybean aphid infestation induces changes in fatty acid metabolism in soybean. PloS One. 2015 Dec 18;10(12):e0145660.

Kaur G, Kumar S, Thakur P, Malik JA, Bhandhari K, Sharma KD, Nayyar H. Involvement of proline in response of chickpea (Cicer arietinum L.) to chilling stress at reproductive stage. Scientia Horticulturae. 2011 Apr 11;128(3):174–181.

Kharayat BS, Yadav S, Srivastava R, Singh UM. Molecular approaches for multiple resistance genes stacking for management of major rice diseases. In Biotic Stress Management in Rice 2017 (pp. 331–362). Apple Academic Press.

Kishore GK, Pande S, Podile AR. Biological control of late leaf spot of peanut (Arachis hypogaea) with chitinolytic bacteria. Phytopathology. 2005 Oct;95(10):1157–1165.

Kumar P, Sharma PK. Soil salinity and food security in India. Frontiers in Sustainable Food Systems. 2020;4:533781. https://doi.org/10.3389/fsufs.2020.533781

Kumari VV, Roy A, Vijayan R, Banerjee P, Verma VC, Nalia A, Pramanik M, Mukherjee B, Ghosh A, Reja MH, Chandran MA. Drought and heat stress in cool-season food legumes in sub-tropical regions: Consequences, adaptation, and mitigation strategies. Plants. 2021 May 21;10(6):1038.

Metwali EM, Abdelmoneim TS, Bakheit MA, Kadasa NM. Alleviation of salinity stress in faba bean ('Vicia faba'L.) plants by inoculation with plant growth promoting rhizobacteria (PGPR). Plant Omics. 2015;8(5):449–460.

Mishra DK, Srivastava R, Pandey BK, Verma PC, Sawant SV. Identification and validation of the wound and insect bite early inducible promoter from Arabidopsis thaliana. 3 Biotech. 2022;12(3):74.

Mukherjee A, Gaurav AK, Chouhan GK, Singh S, Sarkar A, Abeysinghe S, Verma JP. Chickpea seed endophyte Enterobacter sp. mediated yield and nutritional enrichment of chickpea for improving human and livestock health. Frontiers in Nutrition. 2024 Apr 25;11:1387130.

Nadeem M, Li J, Yahya M, Wang M, Ali A, Cheng A, Wang X, Ma C. Grain legumes and fear of salt stress: Focus on mechanisms and management strategies. International Journal of Molecular Sciences. 2019 Feb 13;20(4):799.

Ohlson EW, Timko MP. Mapping and validation of Alectra vogelii resistance in the cowpea landrace B301. Agronomy. 2022 Oct 27;12(11):2654.

Pande S, Galloway J, Gaur PM, Siddique KH, Tripathi HS, Taylor P, MacLeod MW, Basandrai AK, Bakr A, Joshi S, Kishore GK. Botrytis grey mould of chickpea: A review of biology, epidemiology, and disease management. Australian Journal of Agricultural Research. 2006 Oct 27;57(11):1137–1150.

Pande S, Sharma SB, Ramakrishna A. Biotic stresses affecting legumes production in the Indo-Gangetic Plain. In Legumes in Rice and Wheat Cropping Systems of the Indo-Gangetic Plain – Constraints and Opportunities 2000 (pp. 129–155). Patancheru, Andhra Pradesh, India: International Crops Research Institute for the Semi-Arid Tropics.

Panwar M, Tewari R, Nayyar H. Native halo-tolerant plant growth promoting rhizobacteria Enterococcus and Pantoea sp. improve seed yield of Mungbean (Vigna radiata L.) under soil salinity by reducing sodium uptake and stress injury. Physiology and Molecular Biology of Plants. 2016 Oct;22(4):445–459. https://doi.org/10.1007/s12298-016-0376-9

Parmar RD, Varsani VG, Parmar V, Vyas S, Dudhagara D. Multidimensional evaluation of salt tolerance in groundnut genotypes through biochemical responses. Oil Crop Science. 2024 May 1;9(2):102–110.

Patel D, Jha CK, Tank N, Saraf M. Growth enhancement of chickpea in saline soils using plant growth-promoting rhizobacteria. Journal of Plant Growth Regulation. 2012;31:53–62.

Patel RR, Thakkar VR, Subramanian BR. A Pseudomonas guariconensis strain capable of promoting growth and controlling collar rot disease in Arachis hypogaea L. Plant and Soil. 2015;390(1):369–381.

Radl V, Sim˜oes-Ara´ujo JL, Leite J, Passos SR, Martins LM, Xavier GR, Rumjanek NG, Baldani JI, Zilli JE. Microvirgavignae sp. nov., a root nodule symbiotic bacterium isolated from cowpea grown in semi-arid Brazil. International Journal of Systematic Evolutionary Microbiology. 2014;64:725–730.

Rodríguez-Navarro DN, Dardanelli MS, Ruíz-Saínz JE. Attachment of bacteria to the roots of higher plants. FEMS Microbiology Letters. 2007 Jul 1;272(2):127–136.

Rubiales D, Mikic A. Introduction: Legumes in sustainable agriculture. Critical Reviews in Plant Sciences. 2015 Jun 1;34(1–3):2–3.

Saikia J, Sarma RK, Dhandia R, Yadav A, Bharali R, Gupta VK, et al. Alleviation of drought stress in pulse crops with ACC deaminase producing rhizobacteria isolated from acidic soil of Northeast India. Scientific Reports. 2018;8:3560. https://doi.org/10.1038/s41598-018-21921-w

Sandhu R, Bangarwa SK, Attri M, Tiwari S, Kohli S, Fayaz S, Chaudhary N. Effects of biotic stresses and their mitigation strategies in legumes: A review. Legume Research: An International Journal. 2025 Feb 1;48(2).

Sarma RK, Saikia R. Alleviation of drought stress in mung bean by strain Pseudomonas aeruginosa GGRJ21. Plant Soil 2014;377:111–126. https://doi.org/10.1007/s11104-013-1981-9

Sher A, Noor MA, Li HX, Nasir B, Manzoor MA, Hussain S, Zhang J, Riaz MW, Hussain S. Heat stress effects on legumes: Challenges, management strategies and future insights. Plant Stress. 2024 Sep 1;13:100537.

Sindhu SS, Dua S, Verma MK, Khandelwal A. Growth promotion of legumes by inoculation of rhizosphere bacteria. In Microbes for Legume Improvement 2010 Jul 12 (pp. 195–235). Vienna: Springer.

Singh V, Pandey PC, Jain DK. Botany Flowering Plants and Aesthetic Characteristics for B.Sc. Third Semester (B-96) (According to NEP For U.P. State Universities) 2022, ISBN: 978-93-92924-13-2, Page No. 95–97.

Srivastava R, Bajpai R, Khan Z, Singh SP, Mehrotra R, Dubey NK. Insight into strigolactone hormone functions in plant parasitic weeds: A regulatory perspective. Indian Journal of Experimental Biology. 2022;60:659–666.

Srivastava R, Srivastava R, Singh UM. Understanding the patterns of gene expression during climate change. In Climate Change Effect on Crop Productivity 2014 (pp. 279–328). Boca Raton, FL, USA: CRC Press, Taylor & Francis Group.

Verma V, Ravindran P, Kumar PP. Plant hormone-mediated regulation of stress responses. BMC Plant Biology. 2016 Apr 14;16:86. https://doi.org/10.1186/s12870-016-0771-y

Wobeng NB, Banfield CC, Megueni C, Mapongmetsem PM, Dippold MA. Impact of legumes on soil microbial activity and C cycle functions in two contrasting Cameroonian agro-ecological zones. Pedobiologia. 2020 Sep 1;81:150662.

8 Metagenomic Insights to Bacterial Communities and Functional Traits in Legume Crops

Alok Ranjan, Shruti Kaushik, and Anil Kumar Singh

8.1 INTRODUCTION

Leguminous plants are essential to global agriculture, playing key role in food production and environmental sustainability. They include important crops such as soybean (*Glycine max*), chickpea (*Cicer arietinum*), lentil (*Lens culinaris*), pea (*Pisum sativum*), cowpea (*Vigna unguiculata*), and common bean (*Phaseolus vulgaris*), which serve as vital sources of plant-based protein, essential amino acids, and micronutrients for both human and animal diets. Beyond their nutritional value, legumes significantly contribute to agroecosystem sustainability through their ability to biologically fix atmospheric nitrogen via symbiosis with rhizobia, reducing dependency on chemical fertilisers and improving soil fertility (Kebede 2021). Soil microbial communities are important component to legume growth and productivity. They influence nutrient availability, phytohormone regulation, disease resistance, and overall plant growth and development (Vasconcelos et al. 2020). However, the traditional focus on rhizobial nitrogen fixation has gradually expanded to include a broader understanding of the entire microbiome associated with legumes including rhizospheric, endophytic, and phyllospheric microorganisms (Anand et al. 2023). These microbes interact dynamically with their host plants, shaping nutrient fluxes, stress responses, and gene expression patterns under changing environmental conditions. Amid the growing challenges of climate change and soil degradation, maintaining legume productivity poses a growing challenge. Abiotic stresses such as drought, salinity, and heat, as well as biotic stresses including pests and pathogens, adversely impact yield and soil microbiota (Anand et al. 2023). The resilience of legumes under such stress conditions often depends on the adaptability and functional diversity of their microbial partners. Therefore, understanding these plant microbe associations at the molecular and community levels has become a research priority. Metagenomics defined as the direct genetic analysis of microbial genomes (whole DNA) from environmental samples has transformed microbial ecology by providing comprehensive insights into microbial diversity, function, and metabolic potential without the need for cultivation (Eloe-Fadrosh et al. 2016). This approach

DOI: 10.1201/9781003733324-9

employs next-generation sequencing (NGS) and advanced bioinformatics analysis to reconstruct taxonomic and functional profiles of complex microbial assemblages. In legume crops, metagenomic investigations have enabled the discovery of novel microbial taxa and key functional genes associated with nutrient cycling, stress adaptation, and symbiotic efficiency (Nwachukwu and Babalola 2022). The integration of metagenomics with complementary omics platforms such as metatranscriptomics, metaproteomics, and metabolomics offers a multidimensional view of the legume microbiome. Such integrative analyses link microbial composition with activity, unveiling the underlying mechanisms driving plant–microbe–soil interactions. Metagenomic studies support the development of microbial consortia or bioinoculants tailored to specific legume species and agroclimatic conditions, ultimately enhancing sustainable and climate-resilient agriculture (Anand et al. 2023). This chapter described current metagenomic research on legume-associated bacterial communities, highlighting their structural diversity, functional characteristics, and potential applications in enhancing crop performance and ecosystem stability.

8.2 LEGUME CROPS, MICROBES BIODIVERSITY, AND SOIL HEALTH

Leguminous crops play a key role in global agricultural systems because of their unique ability to perform biological nitrogen fixation. The family *Fabaceae* encompasses over 25,000 species distributed across varied ecological niches, from arid savannas to humid tropics (Sońta and Rekiel 2020). These plants are cultivated worldwide for food, feed, fibre, and green manure, contributing substantially to agricultural sustainability and global food security. Legumes not only enrich soils with bioavailable nitrogen but also improve soil structure, promote biodiversity, and support subsequent non-leguminous crops in rotation systems (Akshit et al. 2024). Their symbiotic nitrogen fixation mediated primarily through interaction with rhizobia in root nodules converts atmospheric nitrogen (N_2) into ammonia (NH_3), which plants can assimilate into amino acids and nucleotides. This process contributes between 70 and 80 million tonnes of nitrogen annually to agricultural soils, accounting for about 65% of biologically fixed nitrogen globally (Jensen and Hauggaard-Nielsen 2003). The residual nitrogen left after harvest benefits subsequent crops, reducing fertiliser input and associated greenhouse gas emissions. The rhizosphere of legumes is one of the most dynamic interfaces between plants and soil microorganisms. Through root exudates comprising sugars, amino acids, phenolics, and flavonoids legumes which attract and shape diverse microbial communities that influence soil structure and biochemical activity (Upadhayay and Ranjan 2025). These microbial populations, including both symbiotic and free-living bacteria, participate in critical ecological functions such as decomposition, nutrient cycling, and disease suppression (Upadhayay and Ranjan 2025). The best-known example is the symbiotic relationship between legumes and rhizobia, which leads to the formation of specialised root nodules. Inside these nodules, rhizobia differentiate into bacteroids that fix atmospheric nitrogen through the nitrogenase enzyme complex, while the plant supplies carbohydrates and energy (Anand et al. 2023). This mutualistic association benefits the host plant and also enhances soil organic matter and supports a cascade of beneficial microbial interactions. Recent metagenomic studies have revealed that the legume rhizosphere and nodules harbour a much wider microbial diversity than previously recognised. Alongside *Rhizobium*, *Bradyrhizobium*, and *Mesorhizobium*, genera such as *Pseudomonas*, *Bacillus*, *Burkholderia*, *Streptomyces*, and *Micromonospora* contribute significantly to nutrient solubilisation, phytohormone production, and pathogen inhibition (Abdellatif et al. 2024). These findings highlight the critical role of non-rhizobial endophytes and free-living bacteria in determining the functional dynamics of legume-associated microbiomes. Traditional microbiological approaches, limited to culturable organisms, captured only a fraction of this diversity. With the advent of next-generation sequencing (NGS) and metagenomics, researchers can now explore the complete microbial landscape associated with legume crops. It allows scientists to identify key functional traits such as nitrogenase activity, phosphate solubilisation, siderophore production, and stress response mechanisms directly from community DNA. These insights deepen our understanding of legume microbiome and also open avenues for developing microbial inoculants, synthetic consortia, and microbiome-based biofertilisers that enhance soil fertility while minimising environmental impact (Nwachukwu and Babalola 2022)

8.3 METAGENOMIC APPROACHES FOR LEGUME CROPS

Metagenomics is now a key tool for understanding microbial communities, enabling comprehensive exploration of microbial diversity and functional potential without the need for culturing. In the context of legume research, metagenomics approach has provided unprecedented insights into the composition, structure, and dynamics of bacterial communities inhabiting diverse plant-associated niches such as the rhizosphere, endosphere, and nodules. By analysing the total DNA extracted directly from environmental samples, metagenomics allows researchers to uncover both the taxonomic composition and metabolic capabilities of microbial consortia that play essential roles in nutrient cycling, plant growth promotion, and stress resilience (Nwachukwu and Babalola 2022). Traditional microbiological investigations relied mainly on culture-dependent techniques, which constrained microbial discovery to those species capable of growing under laboratory conditions. Research results suggest that less than 1% of environmental microorganisms are cultivable, leaving the vast majority of microbial diversity unexplored (Vitorino and Bessa 2018). The plate count anomaly revealed that most microbes cannot be cultured, leading to DNA-based metagenomic approaches. Early metagenomic studies employed clone libraries of environmental DNA fragments to identify novel enzymes and metabolic pathways (Lorenz and Schleper 2002). However, the advent of next-generation sequencing (NGS) technologies has enhanced sequencing depth, throughput, and cost-effectiveness, enabling large-scale analyses of complex microbial communities. These advancements have allowed researchers to characterise entire microbial ecosystems associated with legumes, uncovering novel symbiotic genes and revealing functional redundancy within bacterial populations that sustain soil fertility and plant productivity (Tsiknia et al. 2021). Metagenomic strategies are broadly classified into two main categories amplicon-based metagenomics and shotgun metagenomics both methods have unique perspectives and advantages in the study of microbial communities. Amplicon-based metagenomics, often referred to as marker gene sequencing, is a targeted approach that typically focuses on conserved genes such as the 16S rRNA gene in bacteria (Ramazzotti and Bacci 2017). This method is cost-effective and widely used for characterising microbial diversity across different environments. It involves PCR amplification of conserved genetic regions followed by high-throughput sequencing and subsequent taxonomic classification through bioinformatic pipelines like QIIME2, Mothur, and DADA2 (Callahan et al. 2016; Terrón-Camero et al. 2022). In studies of legume-associated microbiomes, this strategy has been particularly valuable for identifying dominant bacterial taxa in the rhizosphere, endosphere, and nodules, elucidating how community composition varies with plant genotype, soil type, and environmental factors. However, while amplicon sequencing provides detailed taxonomic profiles and relative abundance estimates, it has limitations in functional analysis since it targets only a small fraction of the genome. In contrast, shotgun metagenomics offers a comprehensive, untargeted view by sequencing all genomic DNA extracted from a sample (Tyson et al. 2004). This approach provides deeper insights into both the taxonomic composition and the functional potential of microbial communities. It enables the recovery of metagenome-assembled genomes (MAGs), offering near-complete reconstructions of individual microbial genomes, including those of uncultured species. Shotgun metagenomics can identify functional genes related to nutrient cycling, such as nitrogen fixation and phosphorus solubilisation, which are crucial for legume productivity. Despite being computationally demanding and costlier than amplicon sequencing, shotgun metagenomics provides unparalleled resolution in functional and ecological analyses. Together, these two metagenomic strategies complement each other amplicon sequencing efficiently profiles community structure, while shotgun sequencing elucidates the functional capacities and ecological roles of microbial populations associated with legumes.

8.4 DIVERSITY OF BACTERIAL COMMUNITIES IN LEGUME NICHES

The legume microbiome is one of the most diverse and functionally versatile microbial ecosystems in terrestrial environments. It comprises a complex consortium of bacteria, archaea, fungi, and other microorganisms that inhabit distinct plant-associated niches, including the rhizosphere, rhizoplane, endosphere, and phyllosphere. Among these, bacteria are the most abundant and functionally influential members, profoundly affecting nutrient dynamics, plant growth, and stress resilience (Hartmann and Six 2023). The rhizosphere is a hot spot for microbial activity and biochemical interactions. Root exudates released by legumes, including amino acids, organic acids, sugars, and flavonoids, act as chemical cues that recruit specific microbial taxa and shape community composition (Badri and Vivanco 2009). Metagenomic analyses have demonstrated that legume rhizospheres are dominated by members of *Proteobacteria, Actinobacteria, Firmicutes, Acidobacteria, and Bacteroidetes* (Sugiyama et al. 2014). In soybean rhizospheres, *Bradyrhizobium* species often predominate due to their symbiotic nitrogen-fixing capacity, but non-rhizobial groups such as *Burkholderia*, *Pseudomonas*, *Bacillus*, and *Streptomyces* are also consistently abundant (Liu et al. 2019). These bacteria perform essential ecological functions such as phosphate solubilisation, siderophore production, and hormone synthesis. Comparative metagenomic studies have further revealed that rhizosphere bacterial composition is influenced by multiple factors, including soil pH, organic carbon content, legume genotype, and seasonal variation (Sharma et al. 2005). In degraded or nutrient poor soils, legumes are often colonised by stress-tolerant bacterial taxa capable of producing osmoprotectants and antioxidant enzymes. For example, the relative abundance of *Bacillus* and *Microbacterium* increases under drought and salinity stress, highlighting the adaptive plasticity of rhizosphere microbial communities (Yaghoubi Khanghahi et al. 2022). Such dynamic microbial associations underline the rhizosphere's role as both a nutrient interface and a resilience hub for legumes in challenging environments.

8.5 ENDOPHYTIC AND PHYLLOSPHERE MICROBIOTA

Endophytic bacteria reside within plant tissues and contribute to plant growth and defence. These microorganisms colonise root cortical tissues, stems, leaves, and even reproductive organs. Metagenomic studies have uncovered a wide diversity of endophytes belonging to *Enterobacter*, *Pantoea*, *Bacillus*, *Paenibacillus*, *Streptomyces*, and *Micromonospora* genera(Santoyo et al. 2016). These bacteria enhance plant performance by producing phytohormones such as indole-3-acetic acid (IAA), gibberellins, and cytokinins; solubilising minerals; and inducing systemic resistance against pathogens (Weilharter et al. 2011; Wisniewski-Dyé et al. 2011). In soybean and chickpea, metagenomic studies have shown that endophytic bacterial communities are shaped by host genotype and tissue type. Leaf and stem tissues often harbour higher proportions of proteobacteria, particularly *Pseudomonas* and *Sphingomonas*, which are known for their antifungal and stress-mitigating properties (D.P.Singh 2011). The phyllosphere contains metabolically active bacterial assemblages that protect plants by competing with pathogens, modulating stomatal behaviour, and producing antimicrobial metabolites. These bacteria also contribute to nutrient recycling through the degradation of organic residues and atmospheric nitrogen assimilation.

8.6 ROOT NODULE MICROBIOMES

Legume root nodules are specialised organs harbour rhizobia that perform symbiotic nitrogen fixation. However, metagenomic and culture-independent analyses have revealed that nodules are not exclusively occupied by rhizobia; rather, they host complex bacterial communities that include diverse non-rhizobial endophytes. These non-rhizobial endophytes belonging to genera such as *Bacillus*, *Micromonospora*, *Pseudomonas*, *Burkholderia*, *Stenotrophomonas*, and *Paenibacillus* contribute to multiple functions, including nitrogen fixation, phosphate solubilisation, and pathogen inhibition (Martinez-Hidalgo and Hirsch 2017; Gopalakrishnan et al. 2018). Metagenomic studies in chickpea and lentil nodules have identified genes encoding nitrogenase reductase and stress-related proteins in non-rhizobial taxa, suggesting their potential involvement in associative nitrogen fixation and plant adaptation under suboptimal conditions (Dobbelaere et al. 2003). Moreover, non-rhizobial endophytes often exhibit synergistic interactions with classical rhizobia, improving nodule formation efficiency and nutrient assimilation. The discovery of these multispecies interactions within nodules highlights the need to redefine symbiosis as a multifaceted network rather than a binary host symbiont relationship.

8.7 DYNAMICS OF BACTERIAL COMMUNITIES

Microbial community composition is dynamic; it evolves throughout plant development and across spatial gradients. Seedling, vegetative, and reproductive stages of legumes each harbour distinct bacterial assemblages shaped by shifts in root exudation patterns and nutrient demand (Chaparro et al. 2014). Temporal metagenomic analyses of soybean and common bean indicate that rhizosphere diversity peaks during flowering, coinciding with maximum nitrogen fixation and carbon allocation to roots (Ciampitti et al. 2021). Spatial variation, from bulk soil to rhizoplane and endosphere, also influences microbial diversity. Generally, microbial abundance decreases from bulk soil to the endosphere, but functional specialisation increases, with endophytic bacteria exhibiting enriched genes related to nutrient transport, stress defence, and hormone signalling (Allison and Martiny 2008; Rodriguez et al. 2019). Such spatial partitioning ensures that different microbial groups complement each other's functions, collectively supporting the plant's physiological needs.

8.8 FUNCTIONAL TRAITS OF BACTERIAL COMMUNITIES

The legume-associated microbiome performs a wide array of ecological and biochemical functions essential for plant growth, soil fertility, and ecosystem sustainability. Metagenomic studies have illuminated the vast genetic potential of these bacterial communities, revealing genes and pathways involved in nutrient cycling, hormone biosynthesis, stress tolerance, and plant defence (Ciampitti et al. 2021). Beyond nitrogen fixation, the legume microbiome contributes significantly to phosphorus and sulphur mobilisation, carbon turnover, and the production of secondary metabolites that influence plant physiology and disease suppression (Martinez-Hidalgo and Hirsch 2017; Gopalakrishnan et al. 2018).

8.8.1 Nitrogen Fixation

Nitrogen fixation remains the hallmark functional trait of legume-associated bacteria. The symbiotic interaction between rhizobia and legumes converts atmospheric nitrogen (N_2) into ammonia (NH_3), a form readily assimilable by plants. Metagenomic sequencing of nodules, rhizospheres, and root-associated soils has identified a range of nitrogenase gene clusters (*nifH*, *nifD*, *nifK*, *fixABCX*, etc.) distributed among both canonical rhizobia (*Rhizobium*, *Bradyrhizobium*, *Ensifer*) and non-rhizobial diazotrophs such as *Azospirillum*, *Burkholderia*, *Klebsiella*, and *Paenibacillus* (Sánchez-Cañizares et al. 2018). Moreover, non-symbiotic nitrogen fixation by free-living diazotrophs in the rhizosphere contributes substantially to nitrogen inputs, particularly under conditions where nodulation efficiency is low due to soil acidity or salinity (Yang et al. 2020). Interestingly, metatranscriptomic analyses reveal that nitrogenase gene expression is dynamically regulated by oxygen concentration, carbon availability, and the host's developmental stage. Rhizobia within nodules express high-affinity cytochrome oxidases and leghemoglobin-binding proteins to maintain microaerobic conditions optimal for nitrogenase activity (Haskett et al. 2022).

8.8.2 Phosphate Solubilisation

Phosphorus (P) is a critical macronutrient often present in insoluble forms in soil. Legume-associated bacteria contribute to P mobilisation through enzymatic solubilisation and mineralisation processes. Metagenomic studies of legume rhizospheres consistently identify genes encoding phosphatases (*phoA, phoD*), phytases, and gluconic acid-producing enzymes (gcd) that facilitate the release of bioavailable phosphate from mineral and organic sources (Maldonado et al. 2020). Certain bacterial taxa such as *Pseudomonas*, *Bacillus*, *Enterobacter*, and *Burkholderia* harbour multiple P-solubilising genes, reflecting metabolic redundancy within the microbiome. Additionally, metagenomic data reveal the presence of sulphur oxidation (*sox*) and potassium transporter (*ktr*) genes, indicating a broader role of bacterial communities in mineral nutrient cycling (Malviya et al. 2022). These bacteria not only improve nutrient acquisition but also enhance soil fertility and reduce dependency on chemical fertilisers. Furthermore, phosphorus-solubilising bacteria often engage in synergistic interactions with rhizobia. Co-inoculation experiments demonstrate that phosphorus-solubilising bacteria can enhance nodulation and nitrogen fixation efficiency by providing adequate phosphorus supply, which is essential for ATP production during nitrogenase activity. Thus, P solubilisation and N fixation represent interconnected functional traits within the legume microbiome's metabolic network.

8.9 PRODUCTION OF PLANT GROWTH–PROMOTING SUBSTANCES

Legume-associated bacteria synthesise a different types of plant growth promoting substances (PGPS) that modulate root architecture, nutrient uptake, and stress responses (Zhang et al. 2022). Metagenomic analyses have uncovered gene clusters responsible for the biosynthesis of phytohormones, including indole-3-acetic acid (IAA), cytokinins, gibberellins, and ethylene-modulating enzymes (such as ACC deaminase). IAA biosynthetic genes such as *ipdC*, *amiE*, and *aldH* are widespread among *Pseudomonas*, *Bacillus*, *Azospirillum*, and *Streptomyces* species in legume rhizospheres (Spaepen and Vanderleyden 2011). This hormone promotes lateral root formation and enhances nutrient absorption. Similarly, genes encoding ACC deaminase (*acdS*) degrade the ethylene precursor ACC, reducing ethylene-induced inhibition of root elongation under stress conditions (Glick 2015). Additionally, metagenomic datasets reveal the enrichment of siderophore biosynthesis genes (*entABCDEF*) in the legume rhizosphere microbiome. Siderophores chelate iron and other micronutrients, making them available to plants while simultaneously limiting pathogen proliferation. These multifaceted growth-promoting traits position legume-associated bacteria as vital components of sustainable agricultural systems. Legume-associated bacterial communities frequently detect genes encoding antifungal enzymes such as chitinases, β-1,3-glucanases, and proteases, as well as nonribosomal peptide synthetase (NRPS) and polyketide synthase (PKS) gene clusters involved in antibiotic production (Raaijmakers and Mazzola 2012). Species of *Pseudomonas*, *Bacillus*, and *Streptomyces* are known to produce secondary metabolites such as phenazines, lipopeptides (surfactin, fengycin), and streptomycin-like compounds, which suppress soilborne pathogens including *Fusarium*, *Rhizoctonia*, and *Pythium*. Moreover, these bacteria trigger plant immune responses through microbe-associated molecular patterns (MAMPs) such as flagellin and lipopolysaccharides, leading to the activation of systemic resistance pathways (Yu et al. 2019).

8.10 CARBON AND SULPHUR CYCLING

Legume-associated bacteria also play crucial roles in the global biogeochemical cycling of carbon (C) and sulphur (S). Through decomposition of organic matter and enzymatic degradation of root exudates, they regulate soil organic carbon turnover. Metagenomic datasets frequently detect carbohydrate-active enzyme (CAZy) families such as glycoside hydrolases (GHs), polysaccharide lyases (PLs), and carbohydrate esterases (CEs), which participate in cellulose, hemicellulose, and lignin degradation (Cui et al. 2019). In sulphur metabolism, genes encoding *sox* enzymes, *cysK*, *sat*, and *aprAB* are prevalent among *Thiobacillus*, *Burkholderia*, and *Desulfovibrio* species in the legume rhizosphere (Berben et al. 2019; Chaudhary et al. 2022). These bacteria oxidise or reduce sulphur compounds, contributing to sulphur availability for plants. By regulating these elemental cycles, legume-associated bacteria maintain nutrient homeostasis and support long-term soil fertility.

8.11 MICROBIOME LEGUME INTERACTIONS

The advent of metagenomics has fundamentally transformed our understanding of microbiome and legume interactions, unveiling the molecular mechanisms that govern symbiosis, nutrient cycling, and environmental adaptation. Metagenomic research has revealed that the legume microbiome extends far beyond rhizobia. Diverse bacterial taxa such as *Pseudomonas*, *Burkholderia*, *Mesorhizobium*, *Bacillus*, and *Paenibacillus* participate in synergistic networks that collectively support plant nutrition and defence (Hartman et al. 2017). Metagenomic analyses have identified nodulation (nodABC) and nitrogenase (nifHDK) gene clusters within both canonical rhizobia and non-rhizobial endophytes. These findings suggest that horizontal gene transfer (HGT) has played a critical role in disseminating symbiotic traits across phylogenetically distinct bacterial lineages (Remigi et al. 2016). Furthermore, genes involved in secretion systems (T3SS, T4SS, T6SS), exopolysaccharide (EPS) biosynthesis, and signal molecule synthesis (AHLs, flavonoid receptors) are frequently detected in legume metagenomes (De Sousa et al. 2021). These components orchestrate microbial colonisation, root hair curling, and infection thread formation key steps in establishing functional nodules. The rhizosphere is a dynamic and chemically complex microenvironment that mediates nutrient and signal exchange between plants and microbes. Root exudates comprising sugars, amino acids, phenolics, and secondary metabolites act as selective filters that shape microbial community composition. Metagenomic studies have revealed that legume rhizospheres harbour higher functional diversity compared to non-legumes, largely due to enhanced carbon fluxes that sustain microbial metabolism (Zhalnina et al. 2018). Functional gene enrichment analyses of rhizosphere metagenomes indicate elevated representation of pathways related to carbon metabolism (glycolysis, TCA cycle), nutrient transporters, and biofilm formation. Genes involved in chemotaxis, and motility suggest that rhizosphere bacteria actively respond to plant-derived signals. Comparative metagenomics between nodulated and non-nodulated legumes further shows shifts in microbial functional genes associated with ammonium assimilation, denitrification, and organic acid degradation, implying that nitrogen-fixation status influences the entire rhizosphere metabolic network (Li et al. 2020). Metagenomic sequencing of legume tissues including roots, stems, leaves, and nodules has uncovered vast diversity of endophytic bacteria. These microbes reside within plant tissues without causing disease and often enhance growth and stress tolerance. Common endophytes include *Bacillus*, *Enterobacter*, *Arthrobacter*, and *Sphingomonas*, which harbour genes for auxin synthesis, siderophore production, and antioxidative enzymes (Lodewyckx et al. 2002). Moreover, metagenomic reconstructions of nodule microbiomes show the presence of non-rhizobial bacteria with complementary functions, such as phosphate solubilisation and stress protection. These findings redefine nodules as microbial consortia rather than binary symbioses, where multiple taxa contribute to nutrient exchange and host resilience.

8.12 STRESS ADAPTATION

Climate change and various environmental stressors such as drought, salinity, and heat pose significant challenges to legume productivity worldwide, reducing yield potential and threatening food security. Legumes, being vital sources of plant-based protein and soil nitrogen enrichment, are particularly sensitive to these abiotic stress factors. However, recent metagenomic analyses have revealed that the plant-associated microbiome plays an essential buffering role in mitigating the adverse effects of environmental stresses. The microbial communities inhabiting the rhizosphere and endosphere of legumes contribute to the host plant's resilience by conferring stress tolerance through diverse metabolic and physiological mechanisms. Functional metagenomic profiling of legume-associated microbiomes has identified several key stress-related genes that are instrumental in enhancing plant tolerance. Among these are genes involved in osmoprotectant synthesis, such *as otsA*, *proP*, and *betB*, which are crucial for the accumulation of compatible solutes that help maintain cellular osmotic balance under water-deficit or high-salinity conditions. Similarly, antioxidative enzyme genes like *katG*, *sodA*, and *ahpC* encode catalase, superoxide dismutase, and alkyl hydroperoxide reductase, respectively, which collectively function to scavenge reactive oxygen species generated during stress, thereby protecting plant cells from oxidative damage. In addition, heat-shock protein genes such as groEL and dnaK play vital roles in protein folding and stabilisation under elevated temperature conditions, preventing denaturation and maintaining cellular function. Ion transporters and efflux pump genes, including *nhaA*, *arsB*, and *czcD*, contribute to ionic homeostasis by regulating sodium, arsenic, and zinc transport, thus alleviating ion toxicity that typically accompanies saline or contaminated environments. Under conditions of salinity or drought stress, metagenomic datasets consistently show an enrichment of these stress-responsive genes, reflecting both microbial adaptation and the establishment of cooperative interactions with the host plant's defence machinery (Yadav et al. 2020). This enrichment suggests that the microbial community composition dynamically shifts in response to environmental pressures, favouring taxa that can support plant survival and productivity under adverse conditions. For example, species belonging to the genera Halomonas and Bacillus, which have been isolated from chickpea (*Cicer arietinum*) rhizospheres, demonstrate remarkable osmotic stress tolerance and exhibit multiple plant growth–promoting traits, including phosphate solubilisation, production of indole-3-acetic acid (IAA), and nitrogen fixation. These properties not only enhance plant physiological performance under saline stress but also improve soil fertility and root development. The dual functionality of these microbes as both stress mitigators and growth enhancers makes them promising candidates for the development of bioinoculants tailored for saline or arid agricultural systems. By harnessing the potential of such beneficial microbial consortia, sustainable strategies can be developed to reduce dependency on chemical fertilisers and improve crop resilience in the face of escalating climate challenges. Overall, metagenomic insights into legume-associated microbiomes provide a foundation for understanding the molecular mechanisms of plant microbe interactions under stress and open new avenues for microbiome-based agricultural innovations aimed at securing legume productivity in changing environments.

8.13 FUTURE PERSPECTIVES AND CHALLENGES

The rapid progress in metagenomics has profoundly reshaped our understanding of microbial ecology in legume systems, unlocking the functional and genetic diversity of bacterial communities that sustain plant growth and soil health. However, while this revolution has generated an unprecedented volume of data, translating these insights into practical agricultural innovations remains a formidable challenge. Most current legume microbiome studies remain descriptive, focusing primarily on taxonomic composition and relative abundance of microbial communities. While such studies provide foundational information, they often fail to establish causal links between microbial genes and their ecological functions in planta. The future lies in functional metagenomics where cloned environmental DNA fragments are expressed and screened for specific phenotypes such as phosphate solubilisation, nitrogen fixation, or antifungal activity. Functional metagenomics will allow direct validation of gene functions discovered in silico, bridging the gap between genetic potential and biological performance. For instance, screening metagenomic libraries for genes encoding ACC deaminase or siderophore biosynthesis can reveal novel mechanisms of plant growth promotion. Similarly, high-throughput expression systems can be used to characterise unannotated open reading frames, potentially leading to the discovery of new biofertiliser or biocontrol molecules. Future studies must combine functional metagenomics with stable isotope probing (SIP), metabolomics, and metaproteomics to connect microbial activity to specific biochemical transformations in the rhizosphere and nodule microenvironments. This shift from "who is there" to "what they are doing" will mark a critical evolution in the application of metagenomics to legume science. Legume microbiome interactions are inherently complex, influenced by multiple layers of molecular regulation involving both host and microbial genomes. A single omics approach cannot capture the full dynamics of these interactions. The future will thus depend on integrating multi-omics platforms metagenomics, metatranscriptomics, metaproteomics, metabolomics, and host genomics into a coherent systems biology framework. Such integrative approaches can uncover how microbial gene expression changes in response to host signalling, stress, or developmental stages. For example, coupling metagenomics with transcriptomics can identify active members of the microbiome during nitrogen fixation or drought stress, while metabolomics can reveal plant–microbe communication via root exudates and signalling molecules. Advanced bioinformatics pipelines and network modelling will help visualise gene-to-ecosystem linkages, identifying keystone microbial taxa and metabolic hubs critical for legume productivity. This holistic view can eventually lead to the rational design of microbiome-based management strategies tailored to specific soils, climates, and crop genotypes. A major gap in current research is the translation of metagenomic findings into measurable agronomic outcomes. While metagenomic data provide insights into microbial functions, direct evidence of their contribution to yield, nutrient use efficiency, or stress tolerance remains limited. Future studies must adopt integrated experimental designs that combine metagenomic profiling with physiological and agronomic measurements. For example, correlating the abundance of *nifH* and *phoD* genes with nitrogen and phosphorus uptake efficiency can validate microbial contributions to plant nutrition. Similarly, metagenomic biomarkers of drought resistance can be linked to physiological parameters such as relative water content, chlorophyll fluorescence, and root hydraulic conductivity. These interdisciplinary approaches will bridge the gap between microbiome science and practical crop improvement. Collaborative networks between genomic researchers, agronomists, and farmers will be crucial to translate laboratory insights into field-scale applications. Such partnerships can lead to microbiome-informed agronomy, where microbial data guide decisions on crop rotation, fertilisation, and soil amendment.

REFERENCES

Abdellatif AAM, Gebily DAS, Elmaghraby MMK, et al (2024) Transforming roles of actinobacteria in sustainable agriculture: From soil health and plant productivity perspective. Metabolomics, Proteomics Gene Ed Approaches Biofertilizer Ind 2:299–338

Akshit, KS, Sheoran N, et al (2024) Legumes in cropping systems: A way toward agricultural sustainability and diversification. Commun Soil Sci Plant Anal 55(4):596–608

Allison SD, Martiny JBH (2008) Resistance, resilience, and redundancy in microbial communities. Proc Natl Acad Sci U S A. https://doi.org/10.1073/pnas.0801925105

Anand U, Pal T, Yadav N, et al (2023) Current scenario and future prospects of endophytic microbes: Promising candidates for abiotic and biotic stress management for agricultural and environmental sustainability. Microb Ecol. https://doi.org/10.1007/s00248-023-02190-1

Badri DV, Vivanco JM (2009) Regulation and function of root exudates. Plant, Cell Environ. https://doi.org/10.1111/j.1365-3040.2009.01926.x

Berben T, Overmars L, Sorokin DY, Muyzer G (2019) Diversity and distribution of sulfur oxidation-related genes in thioalkalivibrio, a genus of chemolithoautotrophic and haloalkaliphilic sulfur-oxidizing bacteria. Front Microbiol. https://doi.org/10.3389/fmicb.2019.00160

Callahan BJ, McMurdie PJ, Rosen MJ, et al (2016) DADA2: High-resolution sample inference from Illumina amplicon data. Nat Methods. https://doi.org/10.1038/nmeth.3869

Chaparro JM, Badri DV, Vivanco JM (2014) Rhizosphere microbiome assemblage is affected by plant development. ISME J. https://doi.org/10.1038/ismej.2013.196

Chaudhary S, Dhanker R, Kumar R, Goyal S (2022) Importance of legumes and role of sulphur oxidizing bacteria for their production: A review. Legum Res. https://doi.org/10.18805/LR-4415

Ciampitti IA, de Borja Reis AF, Córdova SC, et al (2021) Revisiting biological nitrogen fixation dynamics in soybeans. Front Plant Sci. https://doi.org/10.3389/fpls.2021.727021

Cui J, Mai G, Wang Z, et al (2019) Metagenomic insights into a cellulose-rich niche reveal microbial cooperation in cellulose degradation. Front Microbiol. https://doi.org/10.3389/fmicb.2019.00618

De Sousa BFS, Castellane TCL, Tighilt L, et al (2021) Rhizobial exopolysaccharides and type VI secretion systems: A promising way to improve nitrogen acquisition by legumes. Front Agron 3:661468. https://doi.org/10.3389/fagro.2021.661468

Dobbelaere S, Vanderleyden J, Okon Y (2003) Plant growth-promoting effects of diazotrophs in the rhizosphere. CRC Crit Rev Plant Sci 22(2):107–149. https://doi.org/10.1080/713610853

D.P.Singh BBS (2011) Breeding for tolerance to abiotic stresses in mungbean. J Food Legum 24(2):83–90

Eloe-Fadrosh EA, Ivanova NN, Woyke T, Kyrpides NC (2016) Metagenomics uncovers gaps in amplicon-based detection of microbial diversity. Nat Microbiol. https://doi.org/10.1038/NMICROBIOL.2015.32

Glick B (2015) Stress control and ACC deaminase. In Lugtenberg B (eds) Principles of Plant-Microbe Interactions. Springer, Cham. https://doi.org/10.1007/978-3-319-08575-3_27

Gopalakrishnan S, Srinivas V, Vemula A, et al (2018) Influence of diazotrophic bacteria on nodulation, nitrogen fixation, growth promotion and yield traits in five cultivars of chickpea. Biocatal Agric Biotechnol. https://doi.org/10.1016/j.bcab.2018.05.006

Hartman K, van der Heijden MGA, Roussely-Provent V, et al (2017) Deciphering composition and function of the root microbiome of a legume plant. Microbiome. https://doi.org/10.1186/s40168-016-0220-z

Hartmann M, Six J (2023) Soil structure and microbiome functions in agroecosystems. Nat Rev Earth Environ 4:4–18

Haskett TL, Paramasivan P, Mendes MD, et al (2022) Engineered plant control of associative nitrogen fixation. Proc Natl Acad Sci U S A. https://doi.org/10.1073/pnas.2117465119

Jensen ES, Hauggaard-Nielsen H (2003) How can increased use of biological N2 fixation in agriculture benefit the environment? Plant Soil 252:177–186. https://doi.org/10.1023/A:1024189029226

Kebede E (2021) Contribution, utilization, and improvement of legumes-driven biological nitrogen fixation in agricultural systems. Front Sustain Food Syst 5:767998. https://doi.org/10.3389/fsufs.2021.767998

Li F, Zhang S, Wang Y, et al (2020) Rare fungus, Mortierella capitata, promotes crop growth by stimulating primary metabolisms related genes and reshaping rhizosphere bacterial community. Soil Biol Biochem. https://doi.org/10.1016/j.soilbio.2020.108017

Liu F, Hewezi T, Lebeis SL, et al (2019) Soil indigenous microbiome and plant genotypes cooperatively modify soybean rhizosphere microbiome assembly. BMC Microbiol. https://doi.org/10.1186/s12866-019-1572-x

Lodewyckx C, Vangronsveld J, Porteous F, et al (2002) Endophytic bacteria and their potential applications. CRC Crit Rev Plant Sci 21(6):583–606. https://doi.org/10.1080/0735-260291044377

Lorenz P, Schleper C (2002) Metagenome: A challenging source of enzyme discovery. In: Journal of Molecular Catalysis B: Enzymatic 19(20):13–19. https://doi.org/10.1016/S1381-1177(02)00147-9

Maldonado S, Rodríguez A, Ávila B, et al (2020) Enhanced crop productivity and sustainability by using native phosphate solubilizing rhizobacteria in the agriculture of arid zones. Front Sustain Food Syst. https://doi.org/10.3389/fsufs.2020.607355

Malviya D, Varma A, Singh UB, et al (2022) Sulfur-oxidizing bacteria from coal mine enhance sulfur nutrition in pigeonpea (Cajanus cajan L.). Front Environ Sci. https://doi.org/10.3389/fenvs.2022.932402

Martinez-Hidalgo P, Hirsch AM (2017) The nodule microbiome: N2-fixing rhizobia do not live alone. Phytobiomes J 1:70–82

Nwachukwu BC, Babalola OO (2022) Metagenomics: A tool for exploring key microbiome with the potentials for improving sustainable agriculture. Front Sustain Food Syst 6:886987. https://doi.org/10.3389/fsufs.2022.886987

Pascale A, Proietti S, Pantelides IS, Stringlis IA (2020) Modulation of the root microbiome by plant molecules: The basis for targeted disease suppression and plant growth promotion. Front Plant Sci 10:1741. https://doi.org/10.3389/fpls.2019.01741

Raaijmakers JM, Mazzola M (2012) Diversity and natural functions of antibiotics produced by beneficial and plant pathogenic bacteria. Annu Rev Phytopathol 50:403–424

Ramazzotti M, Bacci G (2017) 16S rRNA-based taxonomy profiling in the metagenomics era. In Metagenomics: Perspectives, Methods, and Applications. Academic Press, 103–119. https://doi.org/10.1016/B978-0-08-102268-9.00005-7

Remigi P, Zhu J, Young JPW, Masson-Boivin C (2016) Symbiosis within symbiosis: Evolving nitrogen-fixing legume symbionts. Trends Microbiol 24(1):63–75. https://doi.org/10.1016/j.tim.2015.10.007

Rodriguez PA, Rothballer M, Chowdhury SP, et al (2019) Systems biology of plant-microbiome interactions. Mol Plant 12(6):804–821

Sánchez-Cañizares C, Jorrín B, Durán D, et al (2018) Genomic diversity in the endosymbiotic bacterium rhizobium leguminosarum. Genes (Basel). https://doi.org/10.3390/genes9020060

Santoyo G, Moreno-Hagelsieb G, del Carmen Orozco-Mosqueda M, Glick BR (2016) Plant growth-promoting bacterial endophytes. Microbiol Res 183:92–99.

Sharma S, Aneja MK, Mayer J, et al (2005) Characterization of bacterial community structure in rhizosphere soil of grain legumes. Microb Ecol. https://doi.org/10.1007/s00248-004-0041-7

Sońta M, Rekiel A (2020) Legumes: Use for nutritional and feeding purposes. J Elem 25(3):835–849. https://doi.org/10.5601/jelem.2020.25.1.1953

Spaepen S, Vanderleyden J (2011) Auxin and plant-microbe interactions. Cold Spring Harb Perspect Biol. https://doi.org/10.1101/cshperspect.a001438

Sugiyama A, Ueda Y, Zushi T, et al (2014) Changes in the bacterial community of soybean rhizospheres during growth in the field. PLoS One. https://doi.org/10.1371/journal.pone.0100709

Terrón-Camero LC, Gordillo-González F, Salas-Espejo E, Andrés-León E (2022) Comparison of metagenomics and metatranscriptomics tools: A guide to making the right choice. Genes (Basel). https://doi.org/10.3390/genes13122280

Tsiknia M, Tsikou D, Papadopoulou KK, Ehaliotis C (2021) Multi-species relationships in legume roots: From pairwise legume-symbiont interactions to the plant – Microbiome – soil continuum. FEMS Microbiol Ecol

Tyson GW, Chapman J, Hugenholtz P, et al (2004) Community structure and metabolism through reconstruction of microbial genomes from the environment. Nature. https://doi.org/10.1038/nature02340

Upadhayay K, Ranjan A (2025) The biochemistry of plant-microbe interactions in the rhizosphere. Plant-Microbiome Interact Clim Agric 149

Vasconcelos MW, Grusak MA, Pinto E, et al (2020) The Biology of Legumes and their Agronomic, Economic, and Social Impact. Springer, Singapore. https://doi.org/10.1007/978-981-15-4752-2_1

Vitorino LC, Bessa LA (2018) Microbial diversity: The gap between the estimated and the known. Diversity

Weilharter A, Mitter B, Shin MV., et al (2011) Complete genome sequence of the plant growth-promoting endophyte burkholderia phytofirmans strain PsJN. J Bacteriol 193(13):3383–3384. https://doi.org/10.1128/JB.05055-11

Wisniewski-Dyé F, Borziak K, Khalsa-Moyers G, et al (2011) Azospirillum genomes reveal transition of bacteria from aquatic to terrestrial environments. PLoS Genet. https://doi.org/10.1371/journal.pgen.1002430

Yadav VK, Raghav M, Sharma SK, Bhagat N (2020) Rhizobacteriome: Promising candidate for conferring drought tolerance in crops. J Pure Appl Microbiol 14(1):73–92. https://doi.org/10.22207/JPAM.14.1.10

Yaghoubi Khanghahi M, Crecchio C, Verbruggen E (2022) Shifts in the rhizosphere and endosphere colonizing bacterial communities under drought and salinity stress as affected by a biofertilizer consortium. Microb Ecol. https://doi.org/10.1007/s00248-021-01856-y

Yang Y, Liu L, Singh RP, et al (2020) Nodule and root zone microbiota of salt-tolerant wild soybean in coastal sand and saline-alkali soil. Front Microbiol. https://doi.org/10.3389/fmicb.2020.523142

Yu K, Pieterse CMJ, Bakker PAHM, Berendsen RL (2019) Beneficial microbes going underground of root immunity. Plant Cell Environ 42(10):2860–2870. https://doi.org/10.1111/pce.13632

Zhalnina K, Louie KB, Hao Z, et al (2018) Dynamic root exudate chemistry and microbial substrate preferences drive patterns in rhizosphere microbial community assembly. Nat Microbiol. https://doi.org/10.1038/s41564-018-0129-3

Zhang X, Yan J, Khashi u Rahman M, Wu F (2022) The impact of root exudates, volatile organic compounds, and common mycorrhizal networks on root system architecture in root-root interactions. J Plant Interact 17(1):685–694. https://doi.org/10.1080/17429145.2022.2086307

9 Stress-Responsive Legume Microbiomes

Metagenomic Approaches for Functional and Applied Insights

Niloy Gain, Md. Zahidur Rahman, Shreyasi Bhowmik Tonton, Md. Sajedur Rahman, and Jamilur Rahman

9.1 INTRODUCTION

Climate-induced stressors threaten global food production and raise the food demand by 70% in the 2050 fiscal year (FAO, 2019). In addition, continuous crop production significantly reduces soil nutrients and organic matter status, leading to long-term reduction in soil fertility. In sustainable crop production, legume species simultaneously act as a vital source of health-beneficial proteins and enrich soil fertility through the synergistic interaction effects of soil microbes within legume root nodules (Goyal et al., 2021). However, legume cultivation is severely hampered by several stressors, particularly drought, salinity effects, irregular rainfall, extreme heat, heavy metals, nutrient deficiencies, pathogen invasion, and so on. These stressors impede the beneficial symbiotic relationship between legumes and microbiomes by altering plant physiological and biochemical functions, thereby diminishing grain yield (Yadav et al., 2025). Consequently, the potential of metagenomic decoding of the legume–microbiome interaction under stress stems from the limitations of culture-based approaches. It exposes the keystone taxa that contain potential candidate genes and their functions responsible for developing stress-responsive microbial traits (Ma et al., 2020). Furthermore, metagenomic profiling also paves the way for incorporating multi-omics approaches, including metagenomics, metatranscriptomics, and metabolomics. Together, these approaches reveal the gene-to-metabolite pathways of microbial consortia and predict the phenotype expression of microbial traits under stress (Zhang et al., 2023). Recently, the discovery of horizontal gene transfer hotspots, encoding genes responsible for stress tolerance, microbial plasticity, and genetic fluidity through metagenomics, has opened new windows for legume improvement (Cotta et al., 2025). However, considerable gaps remained, since most of the previous studies focus on single stressors or

 DOI: 10.1201/9781003733324-10

microbial taxa that regulate the beneficial symbiotic associations between legume and soil microbes. Moreover, the roles of non-rhizobial sensing in legume stress adaptation are still unknown, whereas microbial interactions are genotype-specific or host-dependent under stress (Schaedel et al., 2021). To harness the full potential of these interactions, microbial co-occurrence network analysis that is greatly influenced by the genotype × microbiome × environment interactions needs to be performed by employing advanced genomic tools (Niraula et al., 2022; Yang et al., 2024). Considering the significance of legume–microbiome interplays for sustainable ecosystems, this chapter focuses on the mechanistic view of legume–microbiome interaction through the metagenomic lens to understand the functional gene repertoires that produce unique metabolites involved in stress-responsive microbial traits. Additionally, this study also provides better insights into the joint effects of metagenomics, multi-omics data, AI modules, metadata accessibility, and bioinformatics tools that enable the identification of key genes showing adaptability under stress. We consider that this understanding would be rewarded to fabricate climate-smart legume systems that harness microbial diversity and legume cultivation for sustainable agriculture.

9.2 STRESS FACTORS AND ADAPTATION STRATEGIES IN LEGUME MICROBIOME THROUGH METAGENOMIC LENS

The legume–microbiome mutual relationship is highly susceptible to several environmental stressors. By identifying stress-responsive genes and metabolite-linked networks, metagenomic approaches have revolutionised the ability to move beyond cataloguing microbial taxa to understanding their functional potential in response to stressors (Figure 9.1). The abiotic and biotic stressors affecting the legume and microbiome symbiosis are discussed under the following subheadings.

9.2.1 Abiotic Stresses

The abiotic stress factors primarily disrupt the growth of legumes. Metagenomics has been pivotal in revealing how microbiomes adapt to support their host.

- **Drought:** Under drought conditions, beneficial microbes are significantly reduced in the soil rhizosphere, leading to lower N_2 fixation and nodule formation (Santos-Medellín et al., 2017). Metagenomic investigations identified some microbes, including *Actinobacteria* and *Proteobacteria* species, that are enriched in osmolyte biosynthesis (e.g., trehalose, proline), root colonisation traits, and exopolysaccharide synthesis-encoded genes, resulting in a higher stability of nodule formation within the roots, accelerating nitrogen fixation, soil fertility, and plant growth. This alteration in drought tolerance enhances the ecological adaptability of legume microbiomes under stress conditions.
- **Saline stress:** Higher salt concentrations cause osmotic imbalance, as well as water and nutrient deficits, which alter physiological and metabolic processes in microorganisms. However, halotolerant plant growth–promoting

rhizobacteria, such as *Pseudomonas* and *Bacillus spp.*, are enriched in Na^+/H^+ antiporters, osmoprotectants, and ion transporter genes (Ren et al., 2023). These taxa help both microorganisms and the host plant handle ionic stress through cellular membrane stability, alleviate salt-induced damage, and improve the legume's nutrient absorption.

- **Heat stress:** Raising temperatures in the root rhizosphere not only disrupts root colonisation of soil microbes but also impairs plant physiological functions. Metagenomic profiling of legumes identified several heat shock protein-encoding genes (e.g., chaperonins *GroEL/GroES*), fatty acid desaturases, and membrane fluidity regulators (Domingo-Serrano et al., 2025). These genes are functionally capable of alleviating thermal damage, resulting in a higher root cellular membrane stability, which increases nutrient absorption, biofilm development, and legume production under stress.
- **Heavy metal toxicity:** Symbiotic interactions within the legume microbiome remain functionally active and stable during heat stress by using metal ion retention and detoxifying pathways. Abundant genes for metal efflux, chelation (e.g., siderophores), redox enzymes, and siderophore biosynthesis mitigate microbial toxicity by exporting harmful ions and simultaneously reducing metal uptake in plants (Singh et al., 2023). Consequently, the adaptability and symbiotic relationships of legumes within the root rhizosphere under stress are improved by minimising phytotoxicity and metal ion bioavailability.
- **Nutrient deficiencies:** The scarcity of optimum nutrient uptake often retarded plants' growth. Metagenomic profiling of soil microbes under nutrient-deficient conditions reveals that several microbes contain *gcd* or *nifhs* genes, which facilitate gluconic acid synthesis that enhances phosphorus solubilisation, as well as siderophore biosynthetic gene clusters that are involved in iron acquisition (Wu et al., 2022). These biosynthetic gene clusters accelerate root modulation and unique metabolite production, triggering stress-adaptive signalling in the host plant.

9.2.2 Biotic Stresses

The symbiotic interactions of the legume–microbiome are also hampered by the intervention of several pathogens.

- **Pathogen invasion:** Pathogens in the root rhizosphere alter the microbial networks by disrupting the microbial colonisation and inducing nutrient depletion. Metagenomic profiling of legume–microbiome identified certain microbes like *Streptomyces* and *Bacillus* species that contain genes responsible for antibiotic production and immune modulation (Radhakrishnan et al., 2017). The antagonistic characteristics of these microbes indicate a natural defensive reaction that stimulates the resistance capacity and improves the host–pathogen relationship.

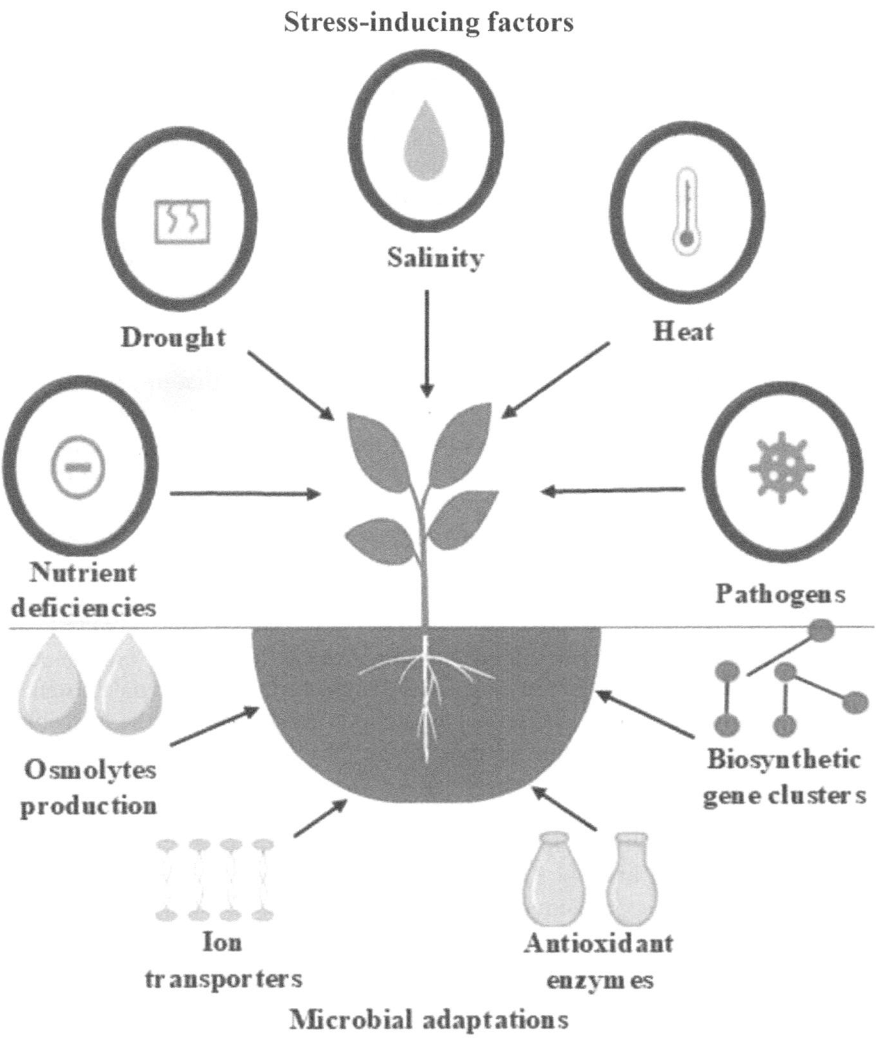

FIGURE 9.1 Factors affecting legume–microbiome interaction and microbial-resilient traits.

9.3 LEGUME MICROBIOME COMPOSITION AND ECOLOGICAL NICHES

The legume microbiome consists of multiple communities, like a cohesive organism, that occupy various ecological niches within the root system based on their distinct functions (Figure 9.2). Based on their ecological localisation, functional roles, and host-specific interactions, the legume microbiome may be categorised into several compartments, as detailed below.

- **Rhizosphere versus root endosphere versus nodule microbiota:** The soil microbial zone largely influences the legume–microbe symbiotic relationships. In the primary rhizosphere zone, root exudates and host microbial communities, such as PGPRs and saprophytic taxa, influence microbial diversity in root nodules. The endosphere zone is generally located in the root tissues and contains microorganisms encoding hormone modulation, immune system modulation, and cell wall disintegration genes (Kumar et al., 2023). Furthermore, in the legume nodule microbiome, N_2-fixing rhizobia such as *Bradyrhizobium* and *Mesorhizobium* coexist with non-rhizobial microbes. The synergistic functions of these microbes greatly enhance legume cultivation.
- **Core versus variable microbiome:** The microbiome's functions and stability are heavily reliant on the microorganisms that are frequently linked with host species. The core microbiome group is essential for host nutrition uptake and defence in harsh situations, while variable microbes possess certain genes that are responsible for adaptation under diverse ecological niches (Custer et al., 2023). The metagenomic approaches of co-occurrence network analysis distinguish the core and variable microbiome and provide a credible opportunity to uncover critical activities of these taxa and their predicted significance in stress adaptation.
- **Functional redundancy and keystone taxa:** In legume–microbiome interactions, functional redundancy and keystone taxa refer to the existence of two or more microbial taxa that maintain the stability of the microbial ecosystem. Since multiple microbes share similar genes or pathways, stress may reduce the core microbiome yet enrich beneficial taxa with antioxidant, osmolyte, and biofilm genes that sustain functional stability (Schaedel et al., 2021). Metagenomic tools offer precise opportunities in identifying the patterns of microbial balance between resilience *via* redundancy and reliance on keystone activities.

9.3.1 Cross-Tolerance and Multiple Stress Scenarios

In native conditions, legumes are frequently subjected to combined stressors. Metagenomics provides a systems-level view of microbiome resilience mechanisms under abiotic and biotic stresses by identifying the microbial taxa and gene networks. For instance, a microbiome enriched in osmoprotectant and antioxidant biosynthetic gene clusters simultaneously showed well-adaptability to drought and resistance against pathogen attack due to their antimicrobial capacity (Ali et al., 2022). Several microbes, including *Bacillus*, *Streptomyces*, and *Pseudomonas species,* show stress-adaptive responses under multiple stress conditions and enhance legume growth by improving nutrient cycling and root colonisation (Liu et al., 2025). This co-occurrence analysis of the metagenome harnesses sustainable legume cultivation by revealing the microbiome's multiple stress-resilient capabilities.

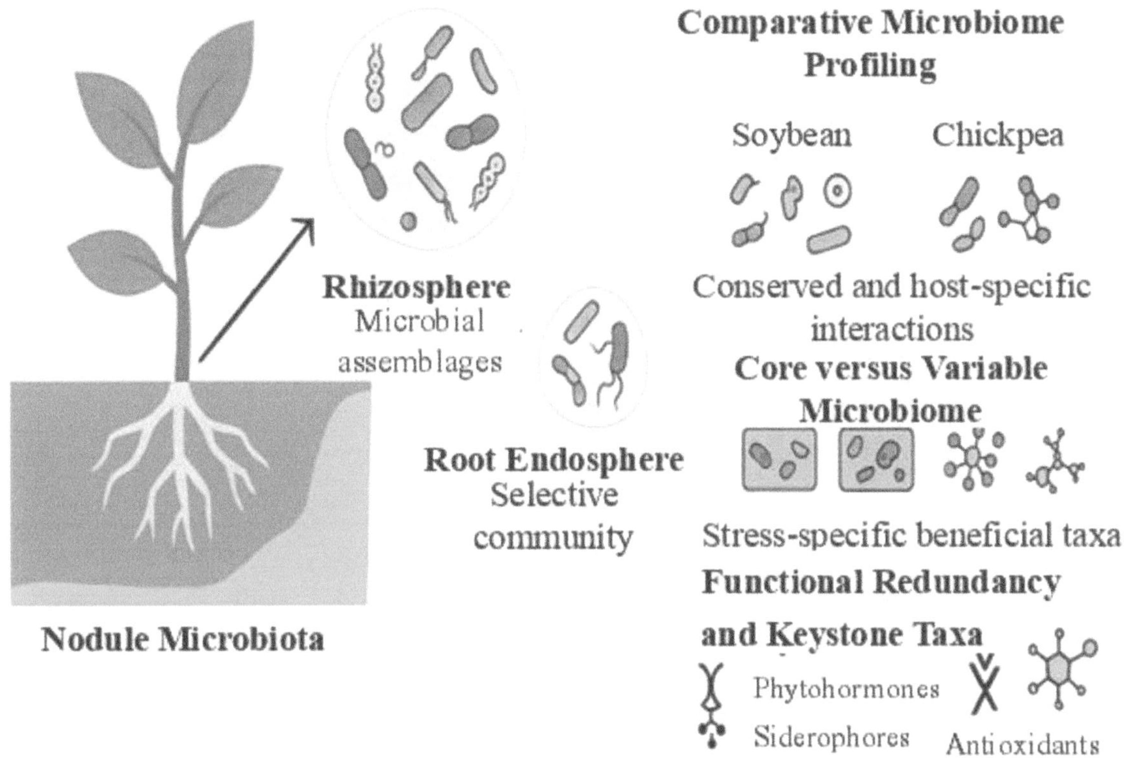

FIGURE 9.2 Functional stratification of legume root microbiomes under stress.

9.4 METAGENOMIC APPROACHES FOR LEGUME MICROBIOMES

Legume–microbiome possesses a complicated relationship, influenced by several factors under stress conditions. Metagenomics tools are frequently employed in identifying potential genes or gene clusters and metabolites responsible for generating stress-responsive microbial traits (Figure 9.3). A brief description of metagenomic approaches that are successfully applied in legume–microbe interaction analysis is presented in the following sections.

9.4.1 Amplicon Sequencing versus Shotgun Metagenomics

Employing amplicon sequencing of 16S rRNA in bacteria and ITS in fungi is frequently used for genome-level microbial community identification. Despite being cost-effective, this technique faces challenges in detecting microbial species at the genus level and revealing the functional role of various microbes within the rhizosphere and nodular environments. For instance, strain-level differentiation of *Bradyrhizobium* species that influence the level of nitrogen fixation (Yang et al., 2024). Compared to other techniques, shotgun metagenomic sequencing gives a more detailed functional representation of the microbiomes in the legumes, including genes responsible for phosphorus solubilisation and stress protection, as well as associating the microbes with plant drought tolerance (Duan et al., 2024). Nonetheless, this technique needs costlier analysis, comprehensive sequencing, and sophisticated computational tools to sift through host DNA and constraints on interpreting the data because of assembly mistakes and chimeric contigs.

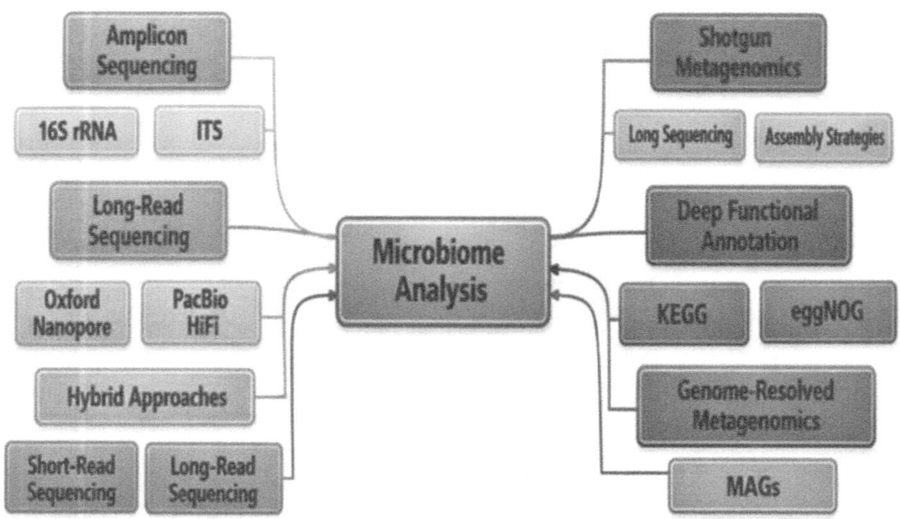

FIGURE 9.3 Methodological flow of different techniques used for microbiome analysis in legume crops.

9.4.2 Long-Read Sequencing Technologies

Long-read sequencing platforms (Oxford Nanopore, PacBio HiFi) generate repetitive sequences of genes that are shared by symbiotic plasmids and genomic islands, enhancing the entire assembly quality and sequencing continuity of the complex microbial genomes found in legume systems. Indicatively, enabling the comparison of nodulation (*nod, nif, fix*) gene clusters across an environmental gradient, the full resolution of symbiotic megaplasmids of *Rhizobium leguminosarum* or *Ensifer meliloti* populations has become the norm (Goyal et al., 2021). Moreover, long-read data plays a significant role in separating closely related strains within the same legume–microbiome zone, since minor genomic alterations can influence host adaptability, effectiveness of nitrogen-fixation, or competition for nodule occupancy. These long-read data also enable the study of mobile genetic elements and prophage regions, facilitating horizontal gene transfer that influences adaptive evolution in rhizobia and other advantageous groups (Rahimlou et al., 2025). In recent applications in soybean and pigeon pea metagenomes, these technologies yielded adaptive stress alleles associated with salinity and oxidative defence.

9.4.3 Hybrid Sequencing and Bioinformatics Workflows

The hybrid pipelines between short- and long-read metagenomic data, where short reads provide higher accuracy rates of bases, and long reads eliminate structural ambiguities, allow the reconstruction of complete genomes to virtually detect major areas such as plant growth-stimulating loci, biosynthetic gene clusters, and nutrient cycling pathways. For example, the detection of nitrogenase operons and *ACC deaminase* genes in peanut rhizosphere employing hybrid assemblies is an important aspect of plant–microbe stress interactions (Lai et al., 2022). Bioinformatics workflows can be additionally used to reconstruct genomes by relying on stress-specific reference databases and metagenome binning algorithms (MetaBAT, MaxBin, CONCOCT), to assess the genome more easily with integrative platforms such as metaWRAP and ATLAS (Peng et al., 2025). These computational pipelines play crucial roles in microbiome restructuring during abiotic stress (drought) by connecting taxonomic and functional changes in legume root microbiomes (osmoprotectants, aquaporins, and heat-shock proteins) and provide an understanding of how plant–microbe co-adaptation may occur.

9.4.4 Deep Functional Annotation Pipelines

The metagenomic data requires the presence of effective functional annotation systems (KEGG, eggNOG, and MetaCyc) to match genes to metabolic pathways, and antiSMASH to discover secondary metabolite biosynthetic gene clusters (BGCs). Such resources in legume microbiomes disclose microbial functions that affect the functioning of phytohormone regulation (e.g., indole-3-acetic acid) and antagonistic metabolic products against pathogens (Wu et al., 2022; Sahil et al., 2025). Furthermore, integrating multiple databases improves coverage and minimises bias, for example, co-annotation of KEGG and MetaCyc, whereas AntiSMASH demonstrated the presence of varied non-ribosomal peptide synthetases (NRPS) and polyketide synthases (PKS) clusters in rhizosphere actinobacteria, indicating an unexplored domain of chemical ecology with implications for root health. These multi-database annotations determine stress resilience genes among microbial genotypes, such as oxidative stress regulators for drought-stimulated nodulation, or genes for siderophore against pathogens (Wang et al., 2025). Such sequencing information transforms biological assumptions into predictive frameworks for sustainable legume crop enhancement.

9.4.5 Genome-Resolved Metagenomics

Genome-resolved metagenomics now enables recovery of near-complete genomes of uncultured legume symbionts, advancing insights into symbiosis and environmental adaptation (Peng et al., 2025). Particularly, uncultured *Bradyrhizobium spp.* in African cowpea have acquired nitrogenase genes, leading to improved nitrogen fixation of low-nitrogen soils. Moreover, novel *Streptomyces* and *Pseudomonas* isolates in the rhizosphere of lentils and chickpeas have acquired biocontrol and plant-derived compounds, respectively (Pramanik et al., 2020). The reconstructions of metagenome assembled genomes (MAGs) enable cross-environment comparisons, revealing microbial genome shifts under different agronomic regimes. Additionally, the meta-transcriptomics and meta-proteomics are now employed in tapping microbial gene expressions at the rhizosphere zone and their impact on plant performance level, facilitating the species and environment-specific bioinoculants production (Hewitt et al., 2023; Kazarina et al., 2025). These bioinoculants not only reduce the use of chemical inoculants and culture-based methods, but also enhance the advantageous microbial range. Collectively, these genomic approaches now permit accurate reconstructions of genetic and metabolic networks that regulate interactions between legumes and microbes, thereby enhancing legume productivity and resilience to climate change.

9.5 MULTI-OMICS INTEGRATION AND MICROBIAL CO-OCCURRENCE NETWORK ANALYSIS

The high-throughput metagenomic approach revolutionised researchers' ability to identify potential microbial networks; however, it does not fully reveal the dynamic gene expression or functions of metabolites responsible for host physiological responses of legumes under stress. The microbial co-occurrence analysis through the metagenomic lens often identifies the presence of microbial taxa and their dimensions of co-occurrence, but fails to express the biological mechanisms. In recent studies of legume–microbiome interactions, multi-omics approaches, such as metagenomics, metatranscriptomics, metabolomics, and proteomics, are now employed to overcome these limitations (Ali et al., 2022). These genomic tools establish the linkage between microbial activity and stress-responsive traits (Figure 9.4). Furthermore, metabolomics profiling reveals the biochemical interaction among the microbial communities by identifying potential metabolites (Mohamedikbal et al., 2025). Such metabolomics-based legume–microbe symbiotic associations can

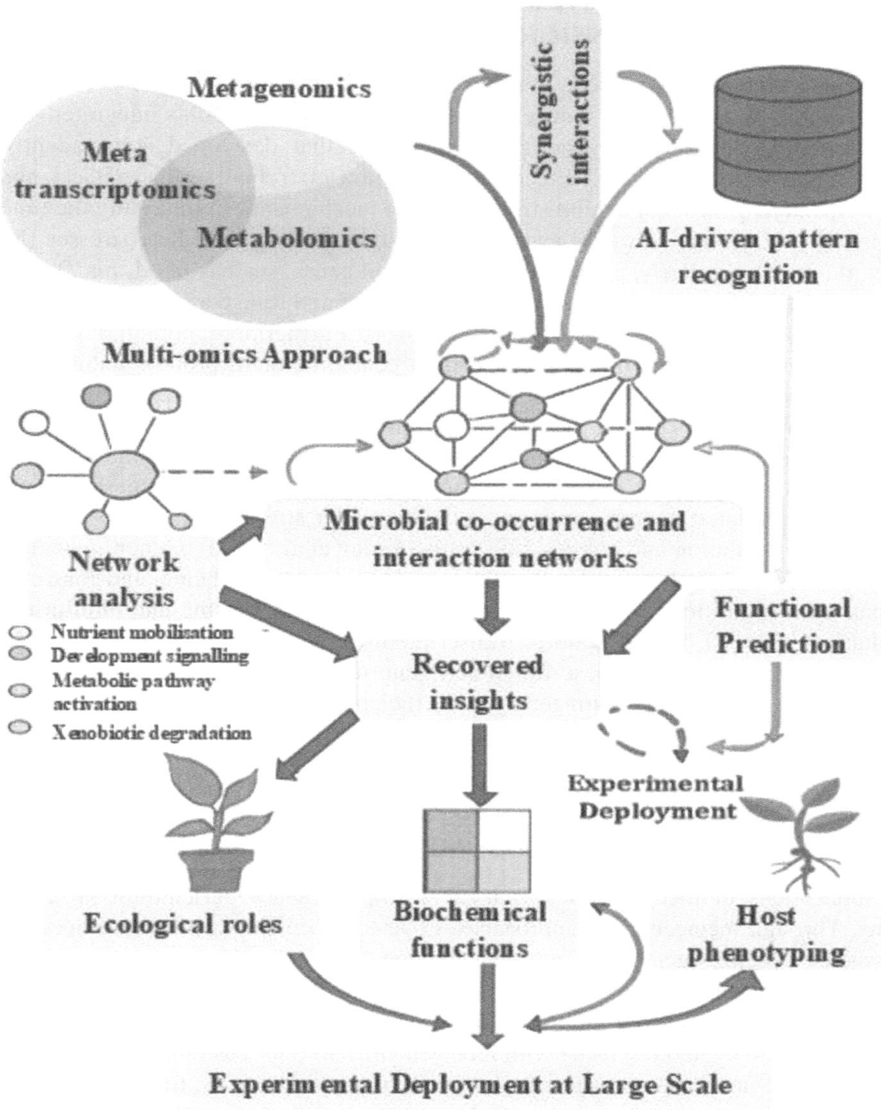

FIGURE 9.4 Predictive modelling of microbial ecosystems via multi-omics and AI-driven network analysis.

be used to predict the synergistic or antagonistic effects of metabolites regarding nutrient cycling, stress-responsive microbial traits, and host phenotype expressions. Coupling these multilayered datasets with AI modelling yields functionally annotated networks that identify hidden hubs of keystone taxa and predict stress-tolerant consortia for sustainable legume cultivation.

9.6 FUNCTIONAL GENE REPERTOIRES AND MECHANISTIC INSIGHTS

Legume microbe interactions develop a complex network of microbes and their associated hosts. In stress-induced conditions, metagenomic tools integrated with multi-omics identified several gene repertoires that developed stress-adaptive microbial traits (Table 9.1). For example, certain microbial strains possess *nod*, *nifH*, and *fix* genes that exhibit stress-response mechanisms, maintaining the functionality of nitrogen fixation even under drought, salinity, and heat stresses (Ma et al., 2020). Similarly, phosphate solubilisation genes, such as *phoA*, *phoD*, *pstS*, and *siderophore*, allow microorganisms to chelate and transport phosphorus, iron, and other micronutrients under abiotic stressors. Furthermore, potential microbes with osmoprotectant (trehalose or proline) genes maintain protein stability and cellular structure development under osmotic stress (Gaied et al., 2023), thereby regulating cellular osmotic adjustment and transferring stress-responsive metabolites to host tissues and improving legume cultivation under arid and semi-arid agroecological niches. Furthermore, phytohormone-modulated signalling genes or biofilm-related operon genes are activated under stress conditions by stabilising root nodulation and colonisation traits (Yadav et al., 2025). Genomic analysis revealed soil microbes as a biochemical shield that enables efficient and controlled pathogen inhibition in legumes by coordinated quorum sensing and biofilm production. Overall, metagenomics, transcriptomics, and metabolomics harness the mechanistic insights of these functional gene repertoires and pathways, offering a possible microbiome engineering and efficient stress-tolerant legume breeding strategies.

9.7 TRANSLATIONAL APPLICATIONS OF METAGENOMICS IN LEGUME CROPS

Applications of metagenomics in legume crops enhance agricultural productivity. Through metagenomics approaches, targeted inoculants can be developed that employ specific bacteria for nitrogen fixation, while SynComs combine multiple microorganisms for enhanced resilience. Additionally, AI-based prediction tools are frequently employed for forecasting microbial shifts, while precision agriculture integrates remote sensing with metagenomics to offer customised solutions for handling leguminous crops to enhance growth and productivity, thereby promoting sustainable agricultural practices (Table 9.2). The following tables describe these applications and their benefits and limitations while explaining the key focus points and providing examples.

9.8 CHALLENGES AND FUTURE ROADMAP

Despite recent advances that underscore the possibility of leveraging legume microbiome stress-adaptation capacity, several important challenges and knowledge gaps need to be addressed through the metagenomic lens to sustain legume cultivation.

TABLE 9.1
Case Studies of Microbial Gene Repertoires and Mechanistic Insights across Stress Conditions in Legume Hosts

Gene Clusters	Functions	Stressors	Mechanistic Insight	Legumes	References
nifh, *acdS*, *treA*, *otsA*	Biosynthesis of trehalose	Drought, Heat	N_2 fixation, ethylene suppression, Osmoprotection, and desiccation tolerance	Soybean, Lentil, Pigeon pea, Lupin	Valente et al. (2025) and Dempsey and Thavarajah (2024)
ipdC, *phoA*, *phoD*, *siderophore*	Solubilisation of phosphate ions	Salinity, Nutrient Deficiency	Improve Fe and P uptake and root elongation	Faba bean, Chickpea, Cowpea, Groundnut	Panigrahi et al. (2024) and Sharma et al. (2024)
sod, *katG*, *czcABC*, *copA*	Heavy metal efflux pumps and resistance	Heavy Metals	Detoxification of Zn, Cu, and Cd in the rhizosphere	Chickpea, Common bean, Mung bean	Singh et al. (2023) and Liu et al. (2024)
NRPS, *PKS* clusters	Antimicrobial biosynthesis	Pathogens	Suppression of soil-borne pathogens	Groundnut, Lupin, Lotus japonicus	Khater et al. (2017) and Sun et al. (2024)
katA, *SOD*, *GR*	Antioxidant defence	Heat, Drought, Heavy Metals	ROS detoxification and redox homeostasis	Soybean, Cowpea, Lentil, Lupin	Burchardt et al. (2024) and Aleem et al. (2022)
ipdC, *iaaM*	Ethylene modulation and auxin biosynthesis	Drought, Salinity	Root elongation and stress signalling modulation	Chickpea, Cowpea, Faba bean	Orozco-Mosqueda et al. (2023) and Ma et al. (2020)
bph	Biphenyl degradation	Herbicides	Microbial detoxification of xenobiotics in the rhizosphere	Groundnut, Soybean, Cowpea	Sharma et al. (2024) and Aleem et al. (2022)
cry genes	Cold shock proteins	Cold	Stabilisation of microbial membranes, proteins, and chlorophyll contents	Pea, Lentil, Lupin, Faba bean	Panigrahi et al. (2024), Sharma et al. (2024), and Gondalia et al. (2022)
photolyase, *uvrABC*	DNA repair and UV protection	UV radiation	Microbial survival and rhizosphere stability under high UV exposure	Lupin, Lotus, Pea	Burchardt et al. (2024) and Sun et al. (2024)
nirK, *norB*	Denitrification enzymes	Flooding	Maintenance of nitrogen cycling under anaerobic conditions	Mung bean, Common bean,	Tariq et al. (2022) and Caldas et al. (2016)

TABLE 9.2
Comparative Translating Metagenomic Insights into Legume |Crop Management

Section	Focus Area	Key Points	Examples/Applications	Benefits	Limitations	References
Designing targeted microbial inoculants	Metagenomic-based inoculants	-Legumes fix nitrogen via rhizobia symbiosis -Metagenomics identifies beneficial microbes -Inoculants improve N_2-fixation, soil health, and yield	-*Bradyrhizobium japonicum* (soybean) -*Rhizobium leguminosarum* (peas, lentils) -PGPB isolation	-Highly crop-specific -Reduces chemical fertiliser use -Enhances soil fertility	-Narrow host range -Limited adaptability -Requires careful strain selection	Ren et al. (2023), Gondalia et al. (2022), and Pramanik et al. (2020)
Synthetic microbial communities	Multi-strain microbial consortia	-SynComs integrate multiple microbes synergistically -Enhance nutrient cycling, pathogen resistance, and stress tolerance	-*Bacillus spp.* and *Trichoderma spp.* (pathogen resistance) -Mycorrhizal fungi (phosphorus uptake). -SynComs for N_2-fixation	-Broader functional coverage -Synergistic effects improve resilience -Species customisable	-Complex to design -Microbial competition risk -Regulatory, scalability challenges	You et al. (2024) and Sahil et al. (2025)
AI-based microbiome prediction	AI-assisted stress management	-Predicts microbial shifts under stress -Enables site-specific inoculant recommendations -Supports timely disease prevention and irrigation decisions	-Forecast drought-resistant soybean inoculants -Prevent root rot in lentils/chickpeas -Predictive models for microbial shifts	-Site-specific precision -Early warning for stress/disease -Data-driven decision-making	-Metagenomic datasets dependent -Requires computational infrastructure -Costly for smallholders	D'Elia et al. (2023) and Clagnan et al. (2024)
Precision agriculture integration	Metagenomics and remote sensing	-Combines GPS, UAVs, and satellites with microbiome data. -Identifies nutrient cycling and microbial activity zones. -Enables site-specific management.	-Pea production: phosphorus recycling -Remote sensing + inoculation for nutrient/disease hotspots -Customised crop management strategies	-Optimises resources -Reduces waste -Boosts the yield's sustainability	-High initial investment -Requires technical expertise -Data integration challenges	Nwachukwu and Babalola (2022) and Peng et al. (2025)

9.8.1 Microbial Insights for Understudied Legume Microbiomes

Many wild and underutilised *Fabaceae* members possess excellent stress-tolerant genes and metabolites, but the scarcity of phenotypical or molecular characterisation slows down the advancement of these neglected taxa. Recently, meta-pangenome studies have opened a new window to reveal microbial functions across diverse ecological niches under stress (Ma et al., 2020). Nonetheless, employing more advanced communication technology and analytical techniques, namely mass spectrometry and nuclear magnetic resonance, and deploying CRISPR-Cas tools will expose the full potential of legume–microbiome symbiosis (Sharma et al., 2025). Revealing the functions of stress-responsive genes, physiological responses of metabolites, and precise genome modification of the host or microbes will aid in sustaining legume production under stress.

9.8.2 Functional Verification of Metagenomic Predictions

Metagenomic profiling demonstrates the functions of stress-responsive microbial genes linked to the host phenotypes. Addressing the reconstruction of metagenomic profiling challenged by gene misannotation with limited genomic data of the microbiome members (Nwachukwu & Babalola, 2022), multi-omics datasets provide a dynamic picture of active gene expressions, metabolic reactions, and regulatory responses in stress adaptations. Furthermore, combining omics data with AI/ML and the open access of Synthetic Microbial Consortia (SynComs), Genome Expression Atlases (GEAs), Plant Regulomics Portal (PRP), and Plant Metabolome Hub (PMhub) will open the way for predictive models, genomic breeding, and QTL mining (Sharma et al., 2025), and aid in designing efficient stress-resilient legume breeding programs.

9.8.3 Linking Lab-Scale Metagenomic Findings to Field-Scale Production

The benefits of legume–microbe symbiotic interactions, particularly for stress resistance, are largely influenced by the complex interactions among microbe × microbe, host × microbe, and environment × microbe factors (Ullah et al., 2025). For translating controlled laboratory findings into on-the-ground practice, the development of robust delivery systems needs to be optimised for long-term trials across agroecological systems. Moreover, applying precision tools (IoT sensors, drones, and remote sensing) will ensure stability of the microbially mediated ecosystem, soil health, and yield response (Yousuf et al., 2025). Transferring the laboratory to field findings will necessitate the validation of both in vitro and in vivo experiments for improving the legume–microbiome mechanistic insights.

9.8.4 Concerns About Ethics, Biosafety, and Policy in Engineering Microbiomes

Microbiome engineering through gene editing tools poses significant moral, biosafety, and policy challenges. The strict legislation restricts scientific progress by virtue of the delayed approvals of genetically modified microorganisms. Furthermore, technology patenting could slow down open research and data sharing. For example, the modification of novel microbial communities using biocontrol agents requires strict biosafety assessment and containment strategies (Patra et al., 2025). Collectively, public acceptance and regulatory status of microbiome-based products remain the key obstacle to their uptake. To promote and gain trust among stakeholders, analysing economic feasibility and transparency information about biosafety is required.

9.9 CONCLUSION

The advent of metagenomics has now revealed functional insights into legume–microbiome synergistic effects that enhance root architecture, nutrient acquisition, and systemic resistance capabilities to mitigate the adverse impacts of abiotic and biotic stressors. Moreover, the coupling of omics approaches and AI-driven metadata opens a new avenue in exploring microbial activity and ecological function that enhances the precision of legume-based breeding strategies. However, the limitation of genomic data of many legume species hinders the identification of stress-adaptive host genes and the resolution of microbial trait mapping, which reduces the accuracy in predictive modelling that links microbial genotypes to field-relevant phenotypes. Unlocking the functional diversity of legume-associated microbiomes through metagenomics, multi-omics, AI data, and bioinformatics may aid in designing targeted microbial consortia for sustainable, stress-resilient agriculture.

REFERENCES

Aleem, M., Aleem, S., Sharif, I., et al. 2022. "Characterization of SOD and GPX Gene Families in Soybeans in Response to Drought and Salinity Stresses." *Antioxidants* 11(3): 460. https://doi.org/10.3390/antiox11030460

Ali, A., Altaf, M.T., Nadeem, M.A., et al. 2022. "Recent Advancements in OMICS Approaches to Enhance Abiotic Stress Tolerance in Legumes." *Frontiers in Plant Science* 13: 952759. https://doi.org/10.3389/fpls.2022.952759

Burchardt, S., Czernicka, M., Kućko, A., et al. 2024. "Exploring the Response of Yellow Lupine (*Lupinus luteus*) Root to Drought via Phytohormone and Redox Pathways." *BMC Plant Biology* 24: 1049. https://doi.org/10.1186/s12870-024-05748-4

Caldas, D.G.G., Konzen, E.R., Recchia, G.H., et al. 2016. "Functional Genomics of Biotic and Abiotic Stresses in *Phaseolus vulgaris*." In *Abiotic and Biotic Stress in Plants*, 1–20. IntechOpen. https://doi.org/10.5772/62051

Clagnan, E., Costanzo, M., Visca, A., et al. 2024. "Culturomics- and Metagenomics-Based Insights into the Soil Microbiome Preservation and Application for Sustainable Agriculture." *Frontiers in Microbiology* 15: 1473666. https://doi.org/10.3389/fmicb.2024.1473666

Cotta, S.R., Oliveira, A.P., Silva, M.F., et al. 2025. "Role of Horizontal Gene Transfer and Cooperation in Rhizosphere Microbiome Assembly." *Brazilian Journal of Microbiology* 56: 225–236. https://doi.org/10.1007/s42770-024-01583-9

Custer, G.F., Gans, M., van Diepen, L.T.A., et al. 2023. "Comparative Analysis of Core Microbiome Assignments: Implications for Ecological Synthesis." *mSystems*, 8(2): e01066-22. https://doi.org/10.1128/msystems.01066-22

D'Elia, D., Truu, J., Lahti, L., et al. 2023. "Advancing Microbiome Research with Machine Learning: Key Findings from the ML4Microbiome COST Action." *Frontiers in Microbiology* 14: 1257002. https://doi.org/10.3389/fmicb.2023.1257002

Dempsey, M., and Thavarajah, D. 2024. "Low Molecular Weight Carbohydrates and Abiotic Stress Tolerance in Lentil (*Lens culinaris*)." *Frontiers in Plant Science* 15: 1408252. https://doi.org/10.3389/fpls.2024.1408252

Domingo-Serrano, L., Sanchis-López, C., Alejandre, C., et al. 2025. "A Microaerobically Induced Small Heat Shock Protein Contributes to *Rhizobium leguminosarum/Pisum sativum* Symbiosis and Interacts with a Wide Range of Bacteroid Proteins." *Applied and Environmental Microbiology* 91: e01385-24. https://doi.org/10.1128/aem.01385-24

Duan, H.-X., Luo, C.-L., Wang, X., et al. 2024. "Responses of Legumes to Rhizobia and Arbuscular Mycorrhizal Fungi under Abiotic Stresses: A Global Meta-Analysis." *Agronomy* 14(11): 2597. https://doi.org/10.3390/agronomy14112597

FAO. 2019. *How to Feed the World in 2050.* Rome: Food and Agriculture Organization of the United Nations.

Gaied, B.R., Brígido, C., Sbissi, I., and Tarhouni, M. 2023. "Sustainable Strategy to Boost Legumes Growth under Salinity and Drought Stress in Semi-Arid and Arid Regions." *Soil Systems* 8(3): 84. https://doi.org/10.3390/soilsystems8030084

Gondalia, N., Vashi, R., Barot, V., et al. 2022. "Genomic Designing for Abiotic Stress Tolerance in Pea (*Pisum sativum*)." In *Genomic Designing for Abiotic Stress Resistant Pulse Crops*, 45–113. Springer. https://doi.org/10.1007/978-3-030-91039-6_3

Goyal, R.K., Mattoo, A.K., and Schmidt, M.A. 2021. "Rhizobial–Host Interactions and Symbiotic Nitrogen Fixation in Legume Crops toward Agriculture Sustainability." *Frontiers in Microbiology* 12: 669404. https://doi.org/10.3389/fmicb.2021.669404

Hewitt, S., Ghogare, R., Troxel, W., et al. 2023. "Metatranscriptomic Analysis of Tomato Rhizospheres Reveals Insight into Plant Microbiome Molecular Response to Biochar-amended Organic Soil." *Frontiers in Analytical Science* 3: 1205583. https://doi.org/10.3389/frans.2023.1205583

Kazarina, A., Wiechman, H., Sarkar, S., et al. 2025. "Recovery of 679 Metagenome-assembled Genomes from Different Soil Depths along a Precipitation Gradient." *Scientific Data* 12: 521. https://doi.org/10.1038/s41597-025-04884-2

Khater, S., Gupta, M., Agrawal, P., et al. 2017. "SBSPKSv2: Structure-Based Sequence Analysis of PKS and NRPS Gene Clusters." *Nucleic Acids Research* 45(W1): W72–W79. https://doi.org/10.1093/nar/gkx344

Kumar, V., and Nautiyal, C.S. 2023. "Endophytes Modulate Plant Genes: Present Status and Future Perspectives." *Current Microbiology* 80: 353. https://doi.org/10.1007/s00284-023-03466-y

Lai, H., Gao, F., Su, H., et al. 2022. "Nitrogen Distribution and Soil Microbial Community Characteristics in a Legume–Cereal Intercropping System: A Review." *Agronomy* 12(8): 1900. https://doi.org/10.3390/agronomy12081900

Liu, K., Deng, F., Zeng, F., et al. 2025. "Plant Growth-Promoting Rhizobacteria Improve Drought Tolerance of Crops: A Review." *Plant Growth Regulation* 105: 567–581. https://doi.org/10.1007/s10725-025-01300-y

Liu, L., Gao, Y., Yang, W., et al. 2024. "Community Metagenomics Reveals Nutrient Cycling Regulated by Microbial Functions in Soils with P Fertilizer Input." *Plant and Soil* 499: 139–154. https://doi.org/10.1007/s11104-023-05875-1

Ma, Y., Dias, M.C., and Freitas, H. 2020. "Drought and Salinity Stress Responses and Microbe-Induced Tolerance in Plants." *Frontiers in Plant Science* 11: 591911. https://doi.org/10.3389/fpls.2020.591911

Mohamedikbal, S., Al-Mamun, H.A., Bestry, M.S., et al. 2025. "Integrating Multi-Omics and Machine Learning for Disease Resistance Prediction in Legumes." *Theoretical and Applied Genetics* 138: 163. https://doi.org/10.1007/s00122-025-04948-2

Niraula, S., Rose, M., and Chang, W.S. 2022. "Microbial Co-occurrence Network in the Rhizosphere Microbiome: Its Association with Physicochemical Properties and Soybean Yield at a Regional Scale." *Journal of Microbiology* 60: 986–997. https://doi.org/10.1007/s12275-022-2363-x

Nwachukwu, B.C., and Babalola, O.O. 2022. "Metagenomics: A Tool for Exploring Key Microbiomes with the Potential for Improving Sustainable Agriculture." *Frontiers in Sustainable Food Systems* 6: 886987. https://doi.org/10.3389/fsufs.2022.886987

Orozco-Mosqueda, M.C., Santoyo, G., and Glick, B.R. 2023. "Recent Advances in the Bacterial Phytohormone Modulation of Plant Growth." *Plants* 12(3): 606. https://doi.org/10.3390/plants12030606

Panigrahi, S., Kumar, U., Swami, S., et al. 2024. "Meta QTL Analysis for Dissecting Abiotic Stress Tolerance in Chickpea." *BMC Genomics* 25: 439. https://doi.org/10.1186/s12864-024-10336-9

Patra, D., Mishra, S., Mohanty, J., et al. 2025. "Harnessing the Plant Microbiome: A Genomic and Metagenomic Perspective for Sustainable Agriculture." *Academia Molecular Biology and Genomics* 2(3). https://doi.org/10.20935/AcadMolBioGen7892

Peng, H., Ruiz-Moreno, A.J., and Fu, J. 2025. "Multi-Dimensional Metagenomics." *Nature Reviews Bioengineering*. https://doi.org/10.1038/s44222-025-00346-x

Pramanik, K., Das, A., Banerjee, J., et al. 2020. "Metagenomic Insights into Rhizospheric Microbiome Profiling in Lentil Cultivars Unveils Differential Microbial Nitrogen and Phosphorus Metabolism under Rice-Fallow Ecology." *International Journal of Molecular Sciences* 21(23): 8895. https://doi.org/10.3390/ijms21238895

Radhakrishnan, R., Hashem, A., and Abd Allah, E.F. 2017. "*Bacillus*: A Biological Tool for Crop Improvement through Bio-Molecular Changes in Adverse Environments." *Frontiers in Physiology* 8: 667. https://doi.org/10.3389/fphys.2017.00667

Rahimlou, S., Hosseyni Moghadam, M.S., Gazis, R., et al. 2025. "Unveiling Root Nodulation in *Tribulus terrestris* and *Roystonea regia* via Metagenomics Analysis." *Molecular Genetics and Genomics* 300(1): 9. https://doi.org/10.1007/s00438-024-02218-2

Ren, H., Zhang, F., Zhu, X., et al. 2023. "Manipulating Rhizosphere Microorganisms to Improve Crop Yield in Saline-Alkali Soil: A Study on Soybean Growth and Development." *Frontiers in Microbiology* 14. https://doi.org/10.3389/fmicb.2023.1233351

Sahil, R., Pal, V., Kharat, A.S., and Jain, M. 2025. "A Multi-Omics Meta-Analysis of Rhizosphere Microbiome Reveals Growth-Promoting Marker Bacteria at Different Stages of Legume Development." *Plant, Cell & Environment*. https://doi.org/10.1111/pce.15429

Santos-Medellín, C., Edwards, J., Liechty, Z., et al. 2017. "Drought Stress Results in a Compartment-Specific Restructuring of the Rice Root-Associated Microbiomes." *mBio* 8(4): e00764-17. https://doi.org/10.1128/mBio.00764-17

Schaedel, M., Hidrobo, G., and Grossman, J. 2021. "From Microns to Meters: Exploring Advances in Legume Microbiome Diversity for Agroecosystem Benefits." *Frontiers in Sustainable Food Systems* 5: 668195. https://doi.org/10.3389/fsufs.2021.668195

Sharma, A., Das, N., Pandey, P., and Shukla, P. 2025. "Plant-Microbiome Responses under Drought Stress and Their Metabolite-Mediated Interactions towards Enhanced Crop Resilience." *Current Plant Biology*: 100513.

Sharma, V., Mahadevaiah, S.S., Latha, P., et al. 2024. "Dissecting Genomic Regions and Underlying Candidate Genes in Groundnut MAGIC Population for Drought Tolerance." *BMC Plant Biology* 24: 1044. https://doi.org/10.1186/s12870-024-05749-3

Singh, H.W., Creamer, K.E., Chase, A.B., et al. 2023. "Metagenomic Data Reveals Type I Polyketide Synthase Distributions across Biomes." *mSystems* 8(3): e00012-23. https://doi.org/10.1128/msystems.00012-23

Sun, Z., Liu, Z., Zhi, M., et al. 2024. "Comparative Genomics of *Lotus japonicus* Reveals Insights into Proanthocyanidin Accumulation and Abiotic Stress Response." *Plants* 13(8): 1151. https://doi.org/10.3390/plants13081151

Tariq, R., Hussain, A., Tariq, A., et al. 2022. "Genome-Wide Analyses of the Mung Bean NAC Gene Family Reveals Orthologs and Expression Profiling under Abiotic Stresses." *BMC Plant Biology* 22: 343. https://doi.org/10.1186/s12870-022-03716-4

Ullah, F., Ali, S., Siraj, M., et al. 2025. "Plant Microbiomes Alleviate Abiotic Stress-Associated Damage in Crops and Enhance Climate-Resilient Agriculture." *Plants* 14(12): 1890. https://doi.org/10.3390/plants14121890

Valente, I.M., Soares, C., and Trindade, H. 2025. "Editorial: Integrating Omics to Unveil Legume Stress Response Mechanisms." *Frontiers in Plant Science* 16: 1583412. https://doi.org/10.3389/fpls.2025.1583412

Wang, H., Chen, J., Wang, Y., et al. 2025. "Metagenomic Study on Microbial Function in Increasing Total Nitrogen Content in Soil under Crop Rotation Systems." *Applied Soil Ecology* 213: 106331. https://doi.org/10.1016/j.apsoil.2025.106331

Wu, X., Cui, Z., Peng, J., Zhang, F., and Liesack, W. 2022. "Genome-Resolved Metagenomics Identifies the Particular Genetic Traits of Phosphate-Solubilizing Bacteria in Agricultural Soil." *ISME Communications* 2(1): 17. https://doi.org/10.1038/s43705-022-00100-z

Yadav, S., Gaikwad, A.B., Kalwan, G., et al. 2025. "The Emerging Trends and Prospects of DNA Methylomics for Abiotic Stress Tolerance in Legumes." *Journal of Plant Biochemistry and Biotechnology* 34(1): 62–78. https://doi.org/10.1007/s13562-024-00942-9

Yang, Y., Xu, N., Zhang, Z., et al. 2024. "Deciphering Microbial Community and Nitrogen Fixation in the Legume Rhizosphere." *Journal of Agricultural and Food Chemistry* 72(11): 5659–5670. https://doi.org/10.1021/acs.jafc.3c09160

You, X., Wang, S., Du, L., et al. 2024. "Metagenomics Reveals the Variations in Functional Metabolism Associated with Greenhouse Gas Emissions during Legume-Vegetable Rotation Process." *Ecotoxicology and Environmental Safety* 275: 116268. https://doi.org/10.1016/j.ecoenv.2024.116268

Yousuf, S., Wang, Y., Yu, P., et al. 2025. "Metagenomic Analysis for Unveiling Agricultural Microbiome-2nd Edition." *Agronomy* 15: 1419. https://doi.org/10.3390/agronomy15061419

Zhang, Y., Wang, X., Liu, Y., et al. 2023. "A Multi-Omics Approach for Microbiome Data Analysis in Legumes." In *Legume Genomics: Methods in Molecular Biology*, 2977. Springer. https://doi.org/10.1007/978-1-0716-4820-9_13

10 Effect of Chemical Fertilisers on Microbiome in Legumes

Shruti Kaushik, Alok Ranjan, Ashwani Kumar, and Geetika Sirhindi

10.1 INTRODUCTION

The rapid expansion of global agriculture, driven by rising demands for food, feed, and biofuel production, has resulted in a marked increase in fertiliser inputs to maintain high crop productivity and support food security. At present, approximately half of the world's habitable land is utilised for agricultural purposes, illustrating the growing dependence on intensive cultivation systems. Chemical fertilisers also referred to as mineral or inorganic fertilisers serve as concentrated sources of essential macronutrients, including nitrogen (N), phosphorus (P), and potassium (K), as well as important secondary nutrients such as calcium (Ca), magnesium (Mg), and sulphur (S). These inputs play a critical role in sustaining plant growth, enhancing soil nutrient availability, and supporting high-yield production systems (Dincă et al., 2022). These fertilisers are predominantly produced through industrial processes: nitrogen fertilisers are synthesised from atmospheric N_2 and hydrogen via the energy-intensive Haber–Bosch process; phosphorus fertilisers are manufactured from mined phosphate rock through chemical beneficiation; and potassium fertilisers are extracted and purified from potash deposits. Although the widespread use of these inputs has substantially increased global crop productivity, the growing reliance on chemical fertilisers has raised critical concerns regarding long-term soil health, nutrient imbalances, and the overall sustainability of soil biological functioning (Thapa et al., 2021). In legume-based systems, fertilisers interact intricately with soil microbial communities that play a vital role in nutrient cycling and plant productivity. Soil biota comprising bacteria, fungi, archaea, and other microorganisms represent the largest and most functionally diverse component of the soil ecosystem (Gouda et al., 2018). These microbial communities orchestrate key ecological processes, including biological nitrogen fixation, denitrification, phosphorus solubilisation, organic matter mineralisation, and the biodegradation of xenobiotic compounds. Their functional activity and taxonomic diversity serve as critical indicators of soil quality, nutrient status, and overall ecosystem productivity. Consequently, a well-structured and diverse soil microbiome is essential for supporting optimal legume growth, sustaining soil functional integrity, and maintaining long-term ecological stability (Gouda et al., 2018). However, excessive

 DOI: 10.1201/9781003733324-11

chemical fertilisation can disrupt the natural equilibrium of soil-plant–microbe interactions. Changes in soil pH, nutrient availability, and organic carbon content induced by fertiliser inputs can alter microbial community composition, reduce enzymatic activities (e.g. dehydrogenase activity), and diminish beneficial symbiotic associations such as Rhizobium-legume and arbuscular mycorrhizal fungi (AMF) networks (Bai et al., 2020). Over time, such perturbations can result in reduced microbial biomass and functional diversity, ultimately diminishing soil fertility and ecological resilience. A comprehensive understanding of the multifaceted interactions between chemical fertilisers and the legume-associated microbiome is therefore crucial for designing sustainable nutrient management strategies. The integration of microbial inoculants and biofertilisers with judicious chemical fertilisation offers a promising framework for improving nutrient use efficiency, preserving microbial diversity, and promoting long-term soil health within legume-based agroecosystems.

10.2 OVERVIEW OF CHEMICAL FERTILISERS

Chemical fertilisers, also known as mineral or inorganic fertilisers, are widely used in modern agriculture to replenish essential nutrients in the soil and enhance crop growth and productivity. These fertilisers supply nutrients in readily available forms that plants can absorb quickly, thereby supporting rapid vegetative growth, efficient photosynthesis, and higher yields (Geisseler and Scow, 2014). The primary nutrients supplied by most chemical fertilisers include nitrogen (N), phosphorus (P), and potassium (K) collectively known as macronutrients. In addition, several fertilisers supply secondary nutrients such as calcium (Ca), magnesium (Mg), and sulphur (S). Nitrogen fertilisers like urea, ammonium sulphate, and calcium ammonium nitrate are critical for promoting vigorous growth, chlorophyll formation, and protein synthesis (Verma et al., 2023). Phosphatic fertilisers such as diammonium phosphate (DAP), monoammonium phosphate (MAP), single superphosphate, and triple superphosphate are integral for promoting root development, facilitating cellular energy transfer through ATP synthesis, and supporting early seedling establishment. Potassic fertilisers particularly muriate of potash (potassium chloride) enhances key physiological functions, including enzyme activation, stomatal regulation, and abiotic stress tolerance. Balanced NPK compound fertilisers deliver multiple macronutrients in a single formulation, making them suitable across diverse cropping systems and soil types (Chianu et al., 2012). Fertiliser application rates are typically calibrated according to crop-specific nutrient demands, inherent soil fertility, and prevailing environmental conditions. Although chemical fertilisers substantially enhance crop productivity, their excessive or unregulated use can induce soil degradation, disrupt nutrient balance, and exert negative impacts on the soil microbial community structure and function (Kumar et al., 2024).

10.3 ROLE OF LEGUMES AND THEIR RHIZOSPHERE MICROBIOME

Legumes occupy a central position in global agriculture owing to their capacity to form specialised associations with soil microorganisms that enhance nutrient cycling,

TABLE 10.1
Common Chemical Fertilisers

Chemical Fertiliser	Primary Nutrients
Urea	Nitrogen (N)
Diammonium Phosphate (DAP)	Nitrogen (N), Phosphorus (P)
Monoammonium Phosphate (MAP)	Nitrogen (N), Phosphorus (P)
Potassium Chloride (Muriate of Potash)	Potassium (K)
Calcium Ammonium Nitrate	Nitrogen (N), Calcium (Ca)
Ammonium Sulphate	Nitrogen (N), Sulphur (S)
Superphosphate	Phosphorus (P)
Magnesium Sulphate	Magnesium (Mg), Sulphur (S)
Triple Superphosphate	Phosphorus (P)
NPK (Compound Fertilisers)	Nitrogen (N), Phosphorus (P), Potassium (K)

soil fertility, and overall crop productivity. Beyond their importance as a major dietary protein source, legumes contribute to soil enrichment through biological nitrogen fixation, stimulation of organic matter accumulation, and support of productive crop rotations, thereby playing an essential role in sustainable agroecosystem functioning (Ren et al., 2020). A hallmark feature of legumes is their symbiotic interaction with nitrogen-fixing rhizobia, which colonise root tissues and initiate nodule formation wherein atmospheric nitrogen is reduced to ammonia. This symbiosis provides a substantial proportion of the plant's nitrogen requirement, decreases reliance on synthetic nitrogen fertilisers, and promotes vigorous plant growth under nutrient-limited conditions (Goyal and Habtewold, 2023). In addition to nitrogen fixation, rhizobial associations enhance plant stress tolerance, nutrient mobilisation, and defence responses, further contributing to crop resilience (Adedayo and Babalola, 2023). The legume rhizosphere harbours a dynamic and functionally diverse microbiome shaped by root exudation patterns, soil physicochemical properties, and environmental variables. Legume roots secrete a broad array of organic compounds including amino acids, sugars, flavonoids, organic acids, and other signalling molecules that selectively recruit beneficial microorganisms and modulate microbial activity within the root-associated zone (Wankhade et al., 2025). These compounds serve as chemical cues for rhizobia during nodule initiation and also shape the recruitment of other plant growth–promoting microorganisms. Flavonoids in particular act as key regulatory molecules, triggering the expression of nod genes in rhizobia and initiating the symbiotic interaction required for nodule formation and nitrogen fixation (Goyal and Habtewold, 2023). Alongside rhizobia, plant growth–promoting rhizobacteria (PGPR) represent another important group of microorganisms associated with legumes. These beneficial bacteria including species of Bacillus, Pseudomonas, Azospirillum, Enterobacter, and Burkholderia contribute to nutrient solubilisation, hormone production, siderophore release, and pathogen suppression. Many PGPR strains enhance root elongation, improve nutrient uptake, and help plants manage oxidative stress through the production of antioxidant enzymes and metabolites (Ren et al., 2020). Genomic studies show that PGPR associated with legumes often carry functional genes related to nitrogen metabolism, phosphorus mobilisation, and biocontrol activities, which help them thrive in the competitive rhizosphere environment (Ren et al., 2020). The synergy between PGPR and rhizobia frequently results in enhanced nodulation efficiency, greater nitrogen fixation capacity, and improved yields in many legume species. Arbuscular mycorrhizal fungi (AMF) form another essential component of the legume microbiome. These fungi form mutualistic associations with roots, increasing the functional surface area of the root system and enhancing the uptake of phosphorus, micronutrients, and water. Phosphorus is one of the most limiting nutrients in many soils, and AMF play a critical role in mobilising and delivering it to the plant, thereby supporting nitrogen fixation and overall plant growth (Shah et al., 2025). AMF also improve soil structure by producing glomalin, a glycoprotein that helps stabilise soil aggregates thereby enhancing soil aeration, water retention, and root penetration. Several studies show that co-inoculation of legumes with AMF and rhizobia results in synergistic effects, improving nutrient-use efficiency, biomass accumulation, stress tolerance, and soil microbial balance (Zhu et al., 2024). This tripartite interaction is particularly important under conditions of drought, salinity, and nutrient-poor soils, where AMF help alleviate stress by improving water uptake, regulating ionic balance, and supporting antioxidative pathways. The legume rhizosphere microbiome plays a pivotal role in maintaining soil fertility by mediating organic matter decomposition, nutrient cycling, and soil enzymatic activities. Microbial assemblages associated with legume roots participate in key biogeochemical processes, including nitrogen transformation pathways, phosphorus solubilisation, carbon stabilisation, and the secretion of critical soil enzymes such as phosphatases, ureases, and dehydrogenases. These functional contributions not only enhance nutrient acquisition and growth of the legume crop but also improve soil fertility for subsequent crops within rotational systems. Many legumes are widely incorporated as green manure crops because the microbial biomass and organic residues they contribute substantially enhance soil structure, nutrient availability, and microbial community diversity (Shah et al., 2025).

10.4 SOIL PROPERTIES IMPACTED BY CHEMICAL FERTILISERS

The application of inorganic fertilisers is widely regarded as an effective approach for enhancing soil fertility and increasing crop productivity (Chen et al., 2017). Owing to their high solubility, these fertilisers supply nutrients in readily available forms, resulting in rapid and visibly pronounced effects on plant growth. Their concentrated nutrient composition allows relatively small application rates to achieve substantial improvements in crop performance. When applied at appropriate levels, inorganic fertilisers can also indirectly contribute to the buildup of soil organic matter by promoting greater root biomass production and increasing the return of crop residues to the soil (Han et al., 2016). Consequently, chemical fertilisers have become an important part of modern agriculture, with many farmers depending on them to enhance yield (Basel and Sami, 2014). Chemical fertilisers influence several soil chemical properties by increasing the availability of both macro- and micronutrients required for plant growth (Prasetyo et al., 2013). Nitrogen, phosphorus, and potassium fertilisers in particular play a key role in improving nutrient supply and supporting crop productivity. However, their effects on soil health can be both beneficial and detrimental, depending on rate, type, and frequency of application. Excessive application of inorganic fertilisers accelerates the decomposition of soil organic matter, leading to deterioration of soil structure, reduced aggregation, and increased nutrient losses through leaching, volatilisation, and fixation (Alimi et al., 2007). Overuse also harms decomposers and beneficial soil organisms, reduces mycorrhizal colonisation, and compromises soil health (Hernandez et al., 2021). Nutrient imbalance caused by high fertiliser inputs can lead to soil acidification, which in turn reduces crop yield. Soil acidity lowers phosphate uptake, increases toxic ion concentrations, and adversely affects plant growth (Hernandez et al., 2021). Soil physical properties also exhibit measurable responses to chemical fertiliser application. Nitrogen inputs have been shown to improve soil physical and chemical attributes by enhancing soil organic carbon content, which in turn contributes to higher infiltration rates (Azizi et al., 2016). Soil organic carbon is strongly and positively associated with improved infiltration capacity and overall structural stability (Brar et al., 2015). Likewise, phosphorus fertilisers particularly when applied in combination with organic amendments enhance soil hydraulic conductivity, porosity, and pore continuity (Brar et al., 2015). The improvement in soil aggregation observed with phosphorus supplementation is attributed to increased root density and root exudation, as well as the formation of phosphate-associated hydroxide and carbonate precipitates that promote particle binding and structural refinement. Some studies report no significant change because chemical fertilisers do not directly enhance soil aggregation (Bhatt et al., 2019a), whereas others observed slight reductions in bulk density with NPK application due to increased biomass production and associated organic matter inputs (Bhatt et al., 2019b). Soil aggregation, infiltration, flocculation, porosity, and water-holding capacity are therefore influenced indirectly through changes in soil organic carbon and root activity. Continuous or excessive fertiliser use, however, can lead to long-term chemical degradation of soil. Overuse contributes to soil acidification, loss of organic matter, decline in beneficial microflora, increased soil crusting, shifts in pH, pest outbreaks, and enhanced greenhouse gas emissions. Loss of humus reduces nutrient-holding capacity, while excessive nitrogen disrupts the balance of N, P, and K, often resulting in micronutrient deficiencies and reduced yield potential, particularly in sandy soils that lack buffering capacity. Chemical fertilisers may also introduce toxic heavy metals such as cadmium, arsenic, and uranium especially from sources like triple superphosphate, and these contaminants can accumulate in plants and enter the food chain (Hernandez et al., 2021). Inorganic fertilisers also add salts to the soil, increasing salinity and electrical conductivity (Azizi et al., 2016). Nitrogen-induced salinity can inhibit nitrification and cause NO_2 buildup. Phosphorus fertilisers such as DAP increase soluble salt concentration, and at high salinity levels, nitrogen remains largely in ammonium form due to reduced conversion to nitrate (Irshad et al., 2018). Potassium fertilisers based on chloride or sulphate salts further contribute to salinity because of their high salt index. The biological properties of soil are strongly modulated by fertiliser application. Soil microorganisms play essential roles in nutrient cycling, organic matter decomposition, and the formation of soil structure. Chemical fertilisers influence microbial diversity and community composition in ways that are contingent on soil pH, fertiliser type, and the chemical form of applied nutrients (Alzamel et al., 2022). Fertilisers that increase soil pH tend to suppress acidophilic microbial taxa, whereas acidic fertilisers inhibit microorganisms adapted to neutral or alkaline conditions (Lin et al., 2019). Elevated phosphorus availability has been shown to reduce mycorrhizal fungal colonisation, with the magnitude of suppression depending on fungal species and prevailing soil P concentrations. Fertiliser inputs also generate nutrient-rich microsites or "hotspots" that strongly influence microbial spatial distribution and functional diversity (Wu et al., 2012). Nitrogen gradients created by fertilisers such as ammonium sulphate or urea induce shifts in microbial biomass and restructure communities involved in nitrogen transformation processes (Li et al., 2013).

10.5 EFFECTS OF NITROGEN FERTILISERS ON LEGUME MICROBIOME

Nitrogen is a key macronutrient that governs plant productivity, vegetative growth, and yield formation across diverse cropping systems. Its continuous supplementation through chemical fertilisers generally enhances biomass accumulation, improves yield-determining components, and strengthens overall crop performance. Increased nitrogen availability has been associated with greater plant height, enhanced tillering, increased spikelet or pod numbers, and improved kernel or seed development in cereals, legumes, and oilseed crops. In sugarcane, elevated nitrogen supply promotes vigorous vegetative growth and enhances sucrose accumulation. For instance, nitrogen fertilisation has been reported to significantly increase the grain yield of wheat, maize, sesame, and soybean by improving essential yield attributes such as spikelet number, boll weight, kernel number, and straw biomass (Imran et al., 2015). Studies in common bean demonstrate that increasing external nitrogen significantly reduces the percentage of nitrogen derived from atmospheric fixation (%Ndfa), thereby diminishing the plant's reliance on symbiotic nitrogen fixation (Singh et al., 2023). Application of nitrogen has been reported to reduce nodulation activity by interfering with both the initiation and development of root nodules, thereby lowering nitrogenase activity and the contribution of BNF to plant nutrition. Genotype-dependent responses observed in common bean further reveal that certain varieties exhibit a stronger suppression of nodulation and nitrogen fixation under high-N fertilisation, indicating variation in N-sensitivity within legume species (Singh et al., 2023). In addition to reducing nodulation, nitrogen fertilisers exert direct effects on rhizobial physiology, signalling, and competitive dynamics within the rhizosphere. Elevated concentrations of nitrate or ammonium have been shown to inhibit the synthesis of Nod factors, key lipochitooligosaccharide signalling molecules required for initiating the legume–rhizobium symbiosis. Suppression of these molecular cues weakens rhizobial infection processes and significantly reduces nodule initiation (Abd-Alla et al., 2023). Such interference in early signalling steps alters the recruitment of beneficial microorganisms and induces shifts in the overall microbial community structure. Nitrogen-rich environments also promote the proliferation of fast-growing, fertiliser-responsive microbial taxa that can outcompete symbiotic rhizobia, leading to a decline in the abundance of efficient nitrogen-fixing strains (Abd-Alla et al., 2023). This displacement has long-term implications for soil fertility, as the natural biological pathways responsible for replenishing nitrogen become progressively impaired. The inhibitory influence of nitrogen fertilisers on biological nitrogen fixation (BNF) extends beyond alterations in rhizobial population dynamics or disruptions in molecular signalling. Classical studies employing nitrate-reductase-deficient Rhizobium mutants demonstrate that exogenous nitrate directly suppresses nodulation and nitrogenase activity, even in strains incapable of metabolising nitrate internally. These findings indicate that nitrate itself functions as a potent negative regulator of symbiotic establishment and N_2-fixation efficiency (Marc et al., 2024). This suggests that the suppression of legume–rhizobium symbiosis occurs through multiple independent pathways, including metabolic feedback, hormonal regulation, and downregulation of symbiosis-related genes. Nitrogen fertilisation also modifies the microbial diversity surrounding legume roots. High N levels typically reduce microbial richness by favouring groups that thrive in nutrient-rich environments while diminishing beneficial microbial consortia necessary for sustainable soil function (Marc et al., 2024). This shift in community composition alters carbon and nutrient cycling dynamics, further affecting plant–microbe interactions. At the same time, inadequate nitrogen supply can also reduce rhizobial populations due to limited substrate availability, implying that optimising, rather than maximising, nitrogen fertilisation is essential for maintaining symbiotic efficiency. Despite these suppressive effects, moderate N application at early vegetative stages is sometimes beneficial for legumes by stimulating initial growth before the onset of active nitrogen fixation. However, surpassing threshold levels can abruptly inhibit nodule formation and reduce BNF contribution. Therefore, balanced nutrient management is vital for sustaining legume productivity without jeopardising the biological processes that enhance soil fertility (Kebede, 2021). Research across several cropping systems clearly indicates that excessive nitrogen fertilisation disrupts legume microbiomes, restricts nodulation, reduces nitrogenase activity, and diminishes the ecological and agronomic benefits of biological nitrogen fixation (Huang, 2024). By understanding these interactions, modern agricultural practices aim to optimise nitrogen input to balance crop yield improvement with environmental sustainability and microbiome health. Strategies such as selecting rhizobial strains with higher competitiveness under nitrogen-rich conditions, breeding for genotypes less sensitive to nitrate inhibition, and adopting integrated nutrient management approaches can mitigate the negative impacts of nitrogen fertilisation on the legume microbiome (Abd-Alla et al., 2023).

10.6 EFFECTS OF PHOSPHORUS FERTILISERS ON LEGUME MICROBIOME

Phosphorus (P) is widely regarded as one of the most limiting macronutrients in global agriculture, second only to nitrogen in the frequency of deficiency across cropping systems. Although P constitutes only 0.1–0.4% of plant dry matter, it performs vital functions in root morphogenesis, energy transfer through ATP, photosynthetic efficiency, nodulation, reproductive development, and overall plant vigour. Adequate P availability facilitates rapid seedling establishment and enhances key yield components, including grain number, grain mass, and uniformity of maturation. Numerous studies have demonstrated substantial yield improvements in crops such as wheat, barley, cotton, sesame, and sugarcane with increased P fertilisation, primarily attributed to enhanced root proliferation, improved tillering capacity, and more efficient cellular energy metabolism (Dikr and Abayechaw, 2022). In legume cropping systems, the impact of P fertilisation extends beyond plant physiology and directly influences the microbial community inhabiting the rhizosphere (Dikr and Abayechaw, 2022). Legume roots release carbon-rich exudates that shape complex microbial networks, including beneficial groups such as phosphate-solubilising bacteria (PSB) and AMF. These microorganisms collectively determine how efficiently P fertilisers are transformed, mobilised, and supplied to plants. Studies have shown that phosphorus fertilisation alters the abundance and metabolic activity of PSB, often stimulating populations capable of organic acid secretion, mineral dissolution, and enzymatic mobilisation of insoluble P forms (Pan and Cai, 2023). This study emphasises that PSB respond dynamically to external P inputs, with certain taxa becoming dominant under high-P conditions because they are more efficient at converting fertiliser-derived P into plant-available forms (Silva et al., 2023). These shifts are particularly relevant in legumes, where PSB activity can influence nodulation efficiency, nitrogen fixation, and early seedling vigour.

The interactions between phosphorus (P) fertilisation and arbuscular mycorrhizal fungi (AMF) are similarly critical for soil–plant–microbe dynamics. Evidence indicates that AMF symbiosis is highly responsive to soil P availability, with moderate P inputs promoting fungal colonisation, whereas excessive P fertilisation can markedly suppress AMF abundance and diversity (García et al., 2017). García et al. (2017) further emphasise the synergistic relationship between AMF and phosphate-solubilising bacteria (PSB), wherein PSB increase P solubility while AMF facilitate its uptake through extensive extraradical hyphal networks (Etesami, 2021). This synergism is particularly important in legumes, which require adequate P for nodule initiation and efficient biological nitrogen fixation. Under optimal P supply, AMF colonisation enhances root surface area, nutrient transfer efficiency, and rhizosphere stability, collectively fostering a microbiome that supports improved plant productivity. Phosphorus fertilisation also triggers broader shifts in microbial community structure. Experimental studies demonstrate that P amendments can alter both taxonomic composition and functional diversity within the rhizosphere, enriching microbial groups involved in P mineralisation while reducing populations adapted to low-P environments (Nacoon, 2020). In legumes, these shifts often favour bacteria associated with organic acid production, phosphatase activity, and root growth promotion. Isolations from legume rhizospheres revealed that many PSB strains exhibit enhanced solubilisation potential under P fertilisation, indicating that phosphorus inputs can modify microbial behaviour and functional capabilities (Janati et al., 2022). Co-inoculation studies provide additional evidence of P-driven microbial responses. Dual inoculation with AMF and PSB has been reported to significantly improve P mobilisation and uptake in crop plants, particularly under moderate P regimes (Jatana et al., 2021). These findings support that balanced P fertilisation not only supplies nutrients directly but also enhances beneficial plant–microbe interactions. In legumes, where P demand is high due to nodulation processes, these synergistic effects can translate into improved biomass accumulation, nitrogen-fixing efficiency, and yield performance (Jatana et al., 2021).

10.7 EFFECTS OF POTASSIUM FERTILISERS ON LEGUME MICROBIOME

Potassium (K) is a fundamental macronutrient required for optimal plant growth, serving as a key regulator of osmotic homeostasis, enzyme activation, and abiotic stress tolerance. Its agronomic significance has been well recognised for over a century (Rehanul, 2002). Numerous field studies in crops such as sugar beet, wheat, and rice demonstrate that potassium fertilisation enhances biomass production, improves nutrient-use efficiency, and increases major yield components (Singh and Wanjari, 2012). These improvements arise from K's central role in energy transfer processes, stomatal regulation, carbohydrate translocation, and root system development all of which collectively shape the biochemical and physiological properties of the rhizosphere. In legume-based production systems, where root exudation patterns, nodule development, and biological nitrogen fixation depend on a balanced nutrient regime, fluctuations in K availability can markedly influence rhizosphere microbial community structure and functional dynamics (Bhatt et al., 2019b). A growing body of evidence highlights that potassium fertilisation alters microbial activity in legume rhizospheres by modifying nutrient availability and root-mediated carbon inputs. Potassium-solubilising microorganisms (KSMs) play a central role in microbial K cycling by releasing plant-available K from silicate minerals through organic acid production, chelation and acidolysis (Olaniyan et al., 2022). These microbial communities exhibit high sensitivity to potassium fertilisation regimes, with their abundance often increasing when moderate K inputs enhance nutrient fluxes and carbon availability in the rhizosphere. Soumare (2023) further demonstrated that not only the application rate but also the chemical form of K fertiliser can drive significant shifts in microbial community composition, suggesting that specific potassium sources interact with microbial metabolic pathways and niche preferences. Several studies have shown that K fertilisation modulates soil enzymatic systems, which act as sensitive indicators of microbial activity and nutrient turnover. In legume-based systems, the combined application of phosphorus and potassium markedly enhances key soil enzyme activities including β-glucosidase, urease, phosphatase, and dehydrogenase reflecting increases in microbial biomass, metabolic intensity, and nutrient mobilisation (Solangi et al., 2024). Findings by Zielewicz et al. (2022) also confirm that microbial oxidative and hydrolytic capacities expand under improved K availability. Given that legumes release substantial quantities of high-quality carbon substrates into the rhizosphere through root exudates and nodule turnover, these K-induced enzymatic responses tend to be even more pronounced in legume systems. Furthermore, potassium availability interacts with the ecological stoichiometry of legume–microbe associations, influencing nutrient balance, microbial energetics, and the overall functional performance of the rhizosphere microbiome. Kanté et al. (2021) reported that legumes substantially alter soil enzyme stoichiometry and microbial resource demand, suggesting that changes in K availability may influence how microbes allocate resources among C, N and P acquisition pathways. This correlate with findings from Yu et al. (2021), who showed that rotations including legumes significantly increased microbial functional diversity and enzyme activity, implying that legumes establish a microbial environment that is particularly responsive to nutrient inputs such as potassium.

Beyond effects on microbial activity and enzyme dynamics, potassium fertilisers shape microbial nutrient cycling pathways. Olaniyan et al. (2022) emphasised that KSMs not only release mineral-bound K but also interact with other nutrient cycles by producing growth-promoting compounds that affect nitrogen fixation and phosphorus solubilisation, two processes central to legume performance. Moreover, Messaoudi et al. (2023) showed that potassium-induced improvements in plant water relations and root growth are tightly linked with secondary changes in microbial functioning, indicating that physiological changes in the host plant mediate microbiome responses. Oliveira et al. (2025) demonstrated that organomineral potassium fertilisers reshape soil microbial community structure differently than conventional KCl, highlighting that fertiliser composition influences microbial assemblage, activity and enzyme regulation.

10.8 CONCLUSION

This chapter highlights that chemical fertilisers exert substantial influence on the legume rhizosphere microbiome by modifying soil chemical, physical, and biological attributes. Although nitrogen, phosphorus, and potassium inputs enhance plant growth, nodulation, and yield by improving nutrient availability, their long-term or unbalanced application can disrupt natural soil–plant–microbe interactions. Excessive nitrogen fertilisation suppresses nodule formation, diminishes biological nitrogen fixation, and drives microbial community shifts away from beneficial rhizobial populations. Phosphorus inputs regulate key microbial functional groups such as phosphate-solubilising bacteria and arbuscular mycorrhizal fungi, with moderate P availability supporting beneficial symbioses, whereas excessive P reduces AMF colonisation and rhizosphere cooperation. Potassium fertilisation influences microbial activity, enzymatic dynamics, and nutrient cycling, particularly through shifts in potassium-solubilising microorganisms. Across nutrient types, judicious fertiliser application can stimulate microbial biomass, enhance soil enzymatic activities, and improve nutrient turnover. However, overuse leads to soil acidification, organic matter decline, reduced microbial diversity, and weakened legume–microbe symbioses. Sustaining microbial functionality in legume-based agroecosystems therefore requires balanced fertilisation integrated with microbial inoculants and ecologically informed nutrient management. Combining chemical fertilisers with biofertilisers, organic amendments, and improved agronomic practices provides a viable strategy to maintain soil microbial diversity, enhance nutrient-use efficiency, and safeguard long-term soil fertility while sustaining high crop productivity.

REFERENCES

Abd-Alla, M.H., Al-Amri, S.M., and El-Enany, A.W.E. 2023. “Enhancing rhizobium-legume symbiosis and reducing nitrogen fertilizer use are potential options for mitigating climate change.” Agriculture, 13(11): 2092.

Adedayo, A.A., and Babalola, O.O. 2023. “Genomic mechanisms of plant growth-promoting bacteria in the production of leguminous crops.” Frontiers in Genetics, 14: 1276003.

Alimi, T., Ajewole, O.C., Awosola, O., and Idowu, E.O. 2007. “Organic and inorganic fertilizer for vegetable production under tropical conditions.” Journal of Agriculture and Rural Development, 1: 120–136.

Alzamel, N.M., Taha, E.M., Bakr, A.A., and Loutfy, N. 2022. "Effect of organic and inorganic fertilizers on soil properties, growth yield, and physiochemical properties of sunflower seeds and oils." Sustainability, 14(19): 12928.

Azizi, S., Thomas, T., and Rao, S. 2016. "Effect of different levels of chemical fertilizers on soil physico-chemical properties of inceptisols." International Journal of Multidisciplinary Research and Development, 3(8): 29–32.

Bai, Y.C., Chang, Y.Y., Hussain, M., Lu, B., Zhang, J.P., Song, X.B., Lei, X.S., and Pei, D. 2020. "Soil chemical and microbiological properties are changed by long-term chemical fertilizers that limit ecosystem functioning." Microorganisms, 8(5): 694.

Basel, N., and Sami, M. 2014. "Effect of organic and inorganic fertilizers application on soil and cucumber (Cucumis sativa L.) plant productivity." International Journal of Agriculture and Forestry, 4: 166–170.

Bhatt, M.K., Labanya, R., and Joshi, H.C. 2019a. "Influence of long-term chemical fertilizers and organic manures on soil fertility: A review." Universal Journal of Agricultural Research, 7(5): 177–188.

Bhatt, M.K., Singh, A.P., Singh, V., Kala, D.C., and Kumar, V. 2019b. "Long-term effect of organic and inorganic fertilizers on soil physico-chemical properties of a silty clay loam soil under rice–wheat cropping system in Tarai region of Uttarakhand." Journal of Pharmacognosy and Phytochemistry, 8(1): 2113–2118.

Brar, B.S., Singh, J., Singh, G., and Kaur, G. 2015. "Effects of long-term application of inorganic and organic fertilizers on soil organic carbon and physical properties in maize–wheat rotation." Agronomy, 5: 220–238.

Chen, D., Yuan, L., Liu, Y., Ji, J., and Hou, H. 2017. "Long-term application of manures plus chemical fertilizers sustained high rice yield and improved soil chemical and bacterial properties." European Journal of Agronomy, 90: 34–42.

Chianu, J.N., Chianu, J.N., and Mairura, F. 2012. "Mineral fertilizers in the farming systems of sub-Saharan Africa: A review." Agronomy for Sustainable Development, 32(2): 545–566.

Dikr, W., and Abayechaw, D. 2022. "Effects of phosphorus fertilizer on agronomic, grain yield and other physiological traits of some selected legume crops." Journal of Biology, Agriculture and Healthcare, 12: 1–13.

Dincă, L.C., Grenni, P., Onet, C., and Onet, A. 2022. "Fertilization and soil microbial community: A review." Applied Sciences, 12(3): 1198.

Etesami, H., Jeong, B.R., and Glick, B.R. 2021. "Contribution of arbuscular mycorrhizal fungi, phosphate-solubilizing bacteria, and silicon to P uptake by plants." Frontiers in Plant Science, 12: 699618.

García, S., Pezzani, F., and Rodríguez-Blanco, A. 2017. "Long-term phosphorus fertilization effects on arbuscular mycorrhizal fungal diversity in Uruguayan grasses." Journal of Soil Science and Plant Nutrition, 17(4): 1013–1027.

Geisseler, D., and Scow, K.M. 2014. "Long-term effects of mineral fertilizers on soil microorganisms: A review." Soil Biology and Biochemistry, 75: 54–63.

Gouda, S., Kerry, R.G., Das, G., Paramithiotis, S., Shin, H.S., and Patra, J.K. 2018. "Revitalization of plant growth–promoting rhizobacteria for sustainable development in agriculture." Microbiological Research, 206: 131–140.

Goyal, R.K., and Habtewold, J.Z. 2023. "Evaluation of legume–rhizobial symbiotic interactions beyond nitrogen fixation that help the host survival and diversification in hostile environments." Microorganisms, 11(6): 1454.

Han, S.H., Young, J., Hwang, J., Kim, S.B., and Park, B. 2016. "The effects of organic manure and chemical fertilizer on the growth and nutrient concentrations of yellow poplar (*Liriodendron tulipifera Lin.*) in a nursery system." Forest Science and Technology, 12: 137–143.

Hernandez, T., Berlanga, J.G., Tormos, I., and Garcia, C. 2021. "Organic versus inorganic fertilizers: Response of soil properties and crop yield." AIMS Geosciences, 7(3): 415–439.

Huang, W. 2024. "Nitrogen fixation in legumes: Genetic mechanisms and agricultural applications." Field Crop, 7.

Imran, S., Arif, M., Khan, A., Shah, W., Abdul, L., and Ali, M. 2015. "Effect of nitrogen level and plant population on yield and yield components of maize." Advances in Crop Technology, 3: 1–7.

Irshad, M., Hafeez, F., Naseem, M., Rizwan, M., and Al-Wabel, M.I. 2018. "Effect of coal and wood ash on phosphorus immobilization in different textured soils." Arabian Journal of Geosciences, 11(18): 536.

Janati, W., Mikou, K., El Ghadraoui, L., and Errachidi, F. 2022. "Isolation and characterization of phosphate-solubilizing bacteria naturally colonizing legumes rhizosphere in Morocco." Frontiers in Microbiology, 13: 958300.

Jatana, B.S., Kitchens, C., Ray, C., Gerard, P., and Tharayil, N. 2021. "Dual inoculation with arbuscular mycorrhizal fungi and phosphorus-solubilizing fungi synergistically enhances the mobilization and plant uptake of phosphorus from meat and bone meal." Frontiers in Soil Science, 1: 757839.

Kanté, M., Riah-Anglet, W., Cliquet, J.B., and Trinsoutrot-Gattin, I. 2021. "Soil enzyme activity and stoichiometry: Linking soil microorganism resource requirement and legume carbon rhizodeposition." Agronomy, 11(11): 2131.

Kebede, E. 2021. "Contribution, utilization, and improvement of legumes-driven biological nitrogen fixation in agricultural systems." Frontiers in Sustainable Food Systems, 5: 767998.

Kumar, A., Sheoran, P., Kumar, N., Devi, S., Kumar, A., Malik, K., Rani, M., Bhardwaj, A.K., and Mann, A. 2024. "Elucidating morphogenic and physiological traits of rice with nitrogen substitution through nano-nitrogen under salt stress conditions." BMC Plant Biology, 24(1): 908.

Li, F., Liu, M., Li, Z., Jiang, C., Han, F., and Che, Y. 2013. "Changes in soil microbial biomass and functional diversity with a nitrogen gradient in soil columns." Applied Soil Ecology, 64: 1–6.

Lin, W., Lin, M., Zhou, H., Wu, H., and Li, Z. 2019. "The effects of chemical and organic fertilizer usage on rhizosphere soil in tea orchards." PLoS One.

Marc, B., Alexandre, D., Natasha, H.A., Dominique, C., Marie, P., Yassine, E.F., Marc, L., Gilles, C., Renaud, B., and Alexandre, B. 2024. "Nitrate reductase activity is required in Medicago truncatula–Sinorhizobium meliloti nitrogen-fixing symbiosis." bioRxiv, 2024-05.

Messaoudi, A., Labdelli, F., Rebouh, N.Y., Djerbaoui, M., Kucher, D.E., Hadjout, S., Ouaret, W., Zakharova, O.A., and Latati, M. 2023. "Investigating the potassium fertilization effect on morphological and agrophysiological indicators of durum wheat under Mediterranean rain-fed conditions." Agriculture, 13(6): 1142.

Nacoon, S., Jogloy, S., Riddech, N., Mongkolthanaruk, W., Kuyper, T.W., and Boonlue, S. 2020. "Interaction between phosphate-solubilizing bacteria and arbuscular mycorrhizal fungi on growth promotion and tuber inulin content of *Helianthus tuberosus* L." Scientific Reports, 10(1): 4916.

Olaniyan, F.T., Alori, E.T., Adekiya, A.O., Ayorinde, B.B., Daramola, F.Y., Osemwegie, O.O., and Babalola, O.O. 2022. "The use of soil microbial potassium solubilizers in potassium nutrient availability in soil and its dynamics." Annals of Microbiology, 72(1): 45.

Oliveira, C.F., Mendes, L.W., and Alleoni, L.R.F. 2025. "Potassium organomineral fertilizer alters the microbiome of a sandy loam tropical soil." Applied Soil Ecology, 207: 105960.

Pan, L., and Cai, B. 2023. "Phosphate-solubilizing bacteria: Advances in their physiology, molecular mechanisms and microbial community effects." Microorganisms, 11(12): 2904.

Prasetyo, W., Santoso, M., and Wardiyanti, T. 2013. "Pengaruh beberapa macam kombinasi pupuk organik dan anorganik terhadap pertumbuhan dan hasil tanaman jagung manis (Zea mays saccharata Sturt)." Jurnal Produksi Tanaman, 1(3): 79–86.

Rehanul, H. 2002. "Potassium status of soils in India." Better Crops International, 16(2), November.

Ren, N., Wang, Y., Ye, Y., Zhao, Y., Huang, Y., Fu, W., and Chu, X. 2020. "Effects of continuous nitrogen fertilizer application on the diversity and composition of rhizosphere soil bacteria." Frontiers in Microbiology, 11: 1948.

Shah, S., Khan, Y., Wang, M., and Zhang, T. 2025. "Arbuscular mycorrhizal fungi as mediators of nitrogen, phosphorus, and carbon: Implications for plant growth and development." Journal of Plant Nutrition, 1–21.

Silva, L.I.D., Pereira, M.C., Carvalho, A.M.X.D., Buttrós, V.H., Pasqual, M., and Dória, J. 2023. "Phosphorus-solubilizing microorganisms: A key to sustainable agriculture." Agriculture, 13(2): 462.

Singh, A., Schöb, C., and Iannetta, P.P. 2023. "Nitrogen fixation by common beans in crop mixtures is influenced by growth rate of associated species." BMC Plant Biology, 23(1): 253.

Singh, M., and Wanjari, R.H. 2012. "Potassium response in Vertisols in long-term fertilizer experiment in India." e-ifc (International Potash Institute), 37: 10–31.

Solangi, F., Zhu, X., Solangi, K.A., Iqbal, R., Elshikh, M.S., Alarjani, K.M., and Elsalahy, H.H. 2024. "Responses of soil enzymatic activities and microbial biomass phosphorus to improve nutrient accumulation abilities in leguminous species." Scientific Reports, 14(1): 11139.

Soumare, A., Djibril, S.A.R.R., and Diedhiou, A.G. 2023. "Potassium sources, microorganisms and plant nutrition: Challenges and future research directions." Pedosphere, 33(1): 105–115.

Thapa, S., Bhandari, A., Ghimire, R., Xue, Q., Kidwaro, F., Ghatrehsamani, S., Maharjan, B., and Goodwin, M. 2021. "Managing micronutrients for improving soil fertility, health, and soybean yield." Sustainability, 13(21): 11766.

Verma, R.C., Singh, N.K., Gangavati, A.R., Ashoka, P., Kesarwani, A., Ali, I., and Pandey, S.K. 2023. "A review of long-term effects of mineral fertilizers on soil microorganisms." International Journal of Plant & Soil Science, 35(20): 1145–1155.

Wankhade, A., Wilkinson, E., Britt, D.W., and Kaundal, A. 2025. "A review of plant–microbe interactions in the rhizosphere and the role of root exudates in microbiome engineering." Applied Sciences, 15(13): 7127.

Wu, F., Gai, Y., Jiao, Z., Liu, Y., Ma, X., An, L., Wang, W., and Feng, H. 2012. "The community structure of microbes in arable soil under different long-term fertilization regimes in the Loess Plateau of China." African Journal of Microbiology Research, 6: 6152–6164.

Yu, H., Wang, F., Shao, M., Huang, L., Xie, Y., Xu, Y., and Kong, L. 2021. "Effects of rotations with legumes on soil functional microbial communities involved in phosphorus transformation." Frontiers in Microbiology, 12: 661100.

Zhu, Y., Xing, Y., Li, Y., Jia, J., Ying, Y., and Shi, W. 2024. "The role of phosphate-solubilizing microbial interactions in phosphorus activation and utilization in plant–soil systems: A review." Plants, 13(19): 2686.

Zielewicz, W., Swędrzyńska, D., Swędrzyński, A., Grzebisz, W., and Goliński, P. 2022. "The influence of calcium sulfate and different doses of potassium on soil enzyme activity and the yield of the sward with a mixture of alfalfa and grasses." Agriculture, 12(4): 475.

11 Metagenomics of Legumes Under Acidic and Alkaline Soil Conditions

Yashika Sarangle, Ritu, Dhruvi Vaish, and Ram Singh Purty

11.1 LEGUMES AS KEY COMPONENTS IN AGROECOSYSTEMS

Legumes are important for agriculture because of their capability to biologically fix atmospheric nitrogen, thereby enhancing the soil fertility, food production, and ecological sustainability (Jena et al. 2022). With over 20,000 species found in various habitats, legumes such as soybean (*Glycine max*), groundnut (*Arachis hypogaea*), common bean (*Phaseolus vulgaris*), chickpea (*Cicer arietinum*), lentil (*Lens culinaris*), and cowpea (*Vigna unguiculata*) are vital for both human and animal nutrition while also promoting soil health (Sandhu and Chaturvedi 2025). Their high-protein seeds help mitigate nutritional deficiencies, especially in developing countries where diets are predominantly plant-based (Kumar et al. 2019). Biological nitrogen fixation (BNF) is an essential ecosystem service facilitated by the symbiotic relationship between legumes and rhizobial bacteria, which transform atmospheric N_2 into ammonia through the action of nitrogenase enzymes (Huang 2024). This process significantly lessens reliance on synthetic nitrogen fertilisers, thereby reducing both economic expenses and environmental issues like groundwater contamination and greenhouse gas emissions (Tyagi et al. 2022). Additionally, legumes engage with beneficial microbial communities, such as arbuscular mycorrhizal fungi (AMF), phosphate-solubilising bacteria (PSB), and plant growth–promoting rhizobacteria (PGPR), which together enhance nutrient absorption, disease resistance, and resilience to stress. These characteristics establish legumes as foundational species in agricultural systems that are resilient to climate change and have low input requirements (Ben Gaied et al. 2024).

DOI: 10.1201/9781003733324-12

11.2 IMPORTANCE OF SOIL PH IN LEGUME GROWTH AND MICROBIAL SYMBIOSIS

Soil pH acts as a key factor that governs various biochemical and physiological mechanisms, including how roots absorb nutrients, the solubility of metals, enzyme functions, and the formation of microbial communities (Gondal et al. 2021). Worldwide, the acidification of soils is increasing due to the overuse of nitrogen fertilisers, acid rain, and nutrient leaching in regions with high rainfall, leading to the creation of acidic soils that heighten the toxicity of aluminium (Al^{3+}) and manganese (Mn^{2+}), both of which are significant impediments to the growth and nodulation of legume roots. Conversely, arid and semi-arid areas often experience alkaline soil conditions stemming from sodium carbonates, high salinity, and inadequate drainage, which results in a decreased availability of micronutrients like iron, zinc, and phosphates – crucial elements for nitrogenase activity within nodules (Sharma et al. 2025). Optimal nitrogen fixation occurs within a specific pH range (6.0–7.5), where legume signalling molecules, such as flavonoids and Nod factors, function efficiently to facilitate the creation of infection threads and the subsequent development of nodules. Extreme pH levels disrupt these signalling pathways, diminish the survival of rhizobia, and lower the efficiency of nodules – ultimately leading to reduced plant biomass and yield (Shumilina et al. 2023). Microbial communities in the rhizosphere react swiftly to changes in pH. Acidic soils generally favour the growth of acidophilic bacteria, such as *Acidobacteria,* while beneficial nitrogen-fixing groups like *Bradyrhizobium* see a significant decline. In contrast, alkaline soils often lead to a prevalence of *Actinobacteria* and alkaliphilic *Proteobacteria,* which frequently results in a decrease in functional diversity related to nitrogen fixation and the mobilisation of phosphorus (Bai et al. 2025). These shifts in microbial composition have cascading effects on the resilience of legumes, as maintaining microbiome integrity is essential for plant adaptation to environmental stress. Therefore, soil pH has a dual influence – alternately affecting legumes directly and regulating the microbial ecosystem indirectly – highlighting the necessity for further exploration of microbe-mediated adaptation strategies (Gupta 2024).

11.3 RATIONALE FOR METAGENOMIC APPROACHES IN STUDYING PH-STRESSED SOILS

Traditionally, studies on rhizosphere microorganisms predominantly depended on culture-dependent techniques, which only capture a limited subset of soil microbes, as many cannot be cultured due to specific environmental or metabolic conditions (Tate III 2020). This reliance led to significant gaps in understanding the microbial taxa and their functional roles that help legumes adapt to pH stress. Metagenomics has transformed soil microbiology by allowing the direct sequencing of environmental DNA, offering high-resolution insights into unculturable microbial communities and their various functional gene pools (Clagnan et al. 2024).

With metagenomic methods, scientists can track changes in microbial diversity across soil pH gradients, characterise metabolic pathways linked to nitrogen fixation and siderophore production, identify novel functional genes that enhance survival in acidic or alkaline environments, investigate the interactions within microbial groups that aid nutrient mobilisation, and correlate microbiome structure with plant health indicators like biomass and nodule nitrogen content (Yavitt et al. 2021). Recent progress in shotgun sequencing, long-read technologies, and multi-omics approaches has further improved the understanding of community-level functional alterations driven by soil degradation due to pH changes (Ullah et al. 2025).

11.4 METAGENOMIC APPROACHES IN RHIZOSPHERE STUDIES

Understanding the complexity of the legume rhizosphere requires molecular approaches that can investigate various microbial communities and their metabolic functions across different soil pH levels. Two primary metagenomic methods are commonly utilised: amplicon sequencing and shotgun metagenomics. Amplicon sequencing focuses on conserved marker genes like 16S rRNA for bacteria and archaea, ITS (Internal Transcribed Spacer) for fungi, as well as nod and nif genes for nitrogen-fixing symbionts (Gaby and Buckley 2014). It offers a cost-efficient way to assess microbial community composition and observe changes under conditions of acidity or alkalinity. Nevertheless, studies using amplicon sequencing do not provide direct insights into functional capabilities, especially regarding stress adaptation mechanisms such as pH homeostasis, siderophore production, or the regulation of nitrogenase (Meng et al. 2024). On the other hand, shotgun metagenomics randomly sequences all genetic material present in the rhizosphere. This allows for accurate taxonomic classification, identification of rare microbial taxa that endure pH extremes, and quantification of genes related to stress responses, along with the identification of plasmids and symbiotic islands that carry nod, nif, and effector genes. Shotgun sequencing is particularly effective in acidic soils where microbial diversity may be limited, yet functional redundancy is vital for ecosystem stability (Xun et al. 2021). Thus, many studies on legume microbiomes utilise multi-omics strategies that merge amplicon and shotgun sequencing data to uncover both the identities of the microorganisms present and their functional activities.

11.5 ACIDIC SOIL MICROBIOME ASSOCIATED WITH LEGUMES

The microbiome of acidic soils associated with legumes is influenced by acid-tolerant microorganisms that possess specialised genes to cope with acid stress, detoxify metals, and adapt to harsh soil environments. Prominent bacterial groups such as *Bradyrhizobium*, *Burkholderia,* and certain members of the *Acidobacteria* are vital for symbiotic nitrogen fixation, nutrient solubilisation, and the functionality of the rhizosphere in low pH environments (Mendes et al. 2013; Lu et al. 2017; Pérez-Jaramillo et al. 2019). Soil pH is commonly regarded as a key factor governing microbial communities, and acidic soils ($pH < 5.5$) create a distinct ecological environment that favours microorganisms that can tolerate acidity and resist aluminium. As a result, the composition and functional capabilities of the microbial community diverge significantly from those found in neutral or alkaline soils, typically exhibiting a reduction in bacterial diversity in highly acidic conditions (Zhou et al. 2024; Rousseau et al. 2024; Wu et al. 2017). Soils with acidic conditions support members of the phylum *Acidobacteria,* which flourish in nutrient-poor, low-pH environments and often prevail in the soil metagenome. In the rhizospheres of legumes, species of *Bradyrhizobium* are recognised as some of the most acid-resistant rhizobia and are commonly found in acidic tropical (Mendes et al. 2013; Lu et al. 2017; Pérez-Jaramillo et al. 2019). Other prominent groups include *Burkholderia,* acid-tolerant strains of *Mesorhizobium loti,* and various *Actinobacteria* (Vincze et al. 2024; Dludlu et al. 2017; Brígido et al. 2007). These microorganisms play key roles in decomposition and nutrient cycling, helping to maintain balance in the rhizosphere under challenging soil conditions. Although overall microbial diversity is decreased, acidophilic bacterial communities demonstrate significant metabolic flexibility and high functional efficiency. Fungal communities found in acidic soils predominantly consist of *Ascomycota* and specific lineages of *Basidiomycota,* exhibiting strong saprotrophic abilities and resilience to low pH levels (Vincze et al. 2024; Dludlu et al. 2017; Brígido et al. 2007). Acid-resistant AMF enhance phosphorus absorption and promote microbial partnerships that support the growth of stress-tolerant plant-promoting rhizobacteria (Wen et al. 2023; Alotaibi et al. 2021; Deng et al. 2025; Bonfante and Genre 2010). Although exhibiting reduced diversity, these fungal and bacterial communities create ecologically effective systems for decomposition, nutrient mobilisation, and detoxification in soils experiencing acidity stress.

Metagenomic analyses indicate an increase in genes associated with acid-stress tolerance, including those for proton pumps, Na+/H+ antiporters, and membrane transporters that help regulate cytoplasmic pH and mitigate metal toxicity. Common detoxification mechanisms for aluminium and manganese, such as citrate and malate transporters, have been identified. Under conditions of low pH and high Al^{3+} concentrations, the abundance and expression of nitrogen fixation genes (nifH, nifD, nifK) are often diminished due to their vulnerability to proton and metal stress (Jaiswal et al. 2018; Dai et al. 2014; Pereira e Silva et al. 2013; Hungria and Vargas 2000).

11.6 ALKALINE SOIL MICROBIOME ASSOCIATED WITH LEGUMES

Alkaline soils (pH > 8) are common in arid and semi-arid areas and are marked by low levels of organic matter, poor soil structure, and nutrient deficiencies, particularly in phosphorus (P), iron (Fe), and zinc (Zn). Such conditions create osmotic and nutritional stress for legumes, which negatively affects nodulation and the process of nitrogen fixation. The microbial communities found in alkaline and sodic soils are primarily composed of stress-resistant taxa that are efficient at nutrient mobilisation and osmotic regulation. Under these conditions, specific bacterial phyla including *Actinobacteria, Firmicutes, and Proteobacteria* are prevalent, with genera such as *Bacillus, Pseudomonas, Nocardiopsis,* and *Streptomyces* regularly identified for their resilience and abilities to promote plant growth (Bai et al. 2023; Hou et al. 2021; Yang et al. 2024; Ben Gaied et al. 2024; Abdalla Abdelshafy Mohamad et al. 2025). Halotolerant and alkaliphilic bacteria, such as those in the *Halomonadaceae, Micrococcaceae*, and *Nocardiaceae* families, are prevalent in alkaline soils that are arid. These bacterial groups promote legume growth by solubilising phosphate, producing siderophores, and synthesising phytohormones like indole-3-acetic acid. Species from the *Ensifer* and *Rhizobium* genera are primarily the symbiotic partners found in calcareous soils, though they frequently display reduced nitrogenase efficiency. Microbial species like *Pseudomonas fluorescens* and *Burkholderia cepacia* enhance the availability of iron and phosphorus through the secretion of siderophores and acidifying the rhizosphere, which helps alleviate nutrient stress (Pang et al. 2024; Keshri et al. 2013; Sharma et al. 2013). Alkaline soil fungal communities are generally dominated by genera such as *Aspergillus, Penicillium, Trichoderma,* and *Fusarium,* which are noted for their enzymatic diversity and ability to withstand osmotic stress (Cuadros-Orellana et al. 2013; Corbu et al. 2023; Gadd et al. 2024). These fungi release alkali-resistant extracellular enzymes and organic acids (like oxalic and citric acids) that enhance the solubilisation of phosphorus and increase the availability of micronutrients. The production of osmoprotectants, including trehalose and mannitol, boosts tolerance to ionic stress, while the synthesis of siderophores aids in iron acquisition. AMF, especially *Glomus* and *Rhizophagus,* are integral to improving phosphorus uptake and promoting soil aggregation through the production of glomalin. By altering root exudation and encouraging beneficial bacterial communities, AMF facilitate interactions that stabilise the rhizospheres of legumes in alkaline and sodic environments. Functional metagenomics reveals that alkaline soils favour a distinct set of genes associated with pH regulation, osmoprotection, and nutrient mobilisation. Genes responsible for alkaline phosphatases, such as phoD and phoA, are significantly enriched, aiding in the mineralisation of organic phosphorus in low phosphorus conditions. Additionally, genes involved in siderophore biosynthesis are upregulated to address iron scarcity, while genes that transform arsenic and metals are co-selected in high pH environments (Zhao et al. 2024). Other key adaptive systems include Na^+/H^+ antiporters, F_1F_0-ATPases, and genes for the biosynthesis of compatible solutes, which are essential for maintaining ionic balance and osmotic stability.

11.7 METAGENOMIC INSIGHTS INTO SYMBIOTIC EFFICIENCY UNDER PH STRESS

Metagenomic studies have demonstrated that soil pH acts as a vital ecological factor influencing both the taxonomical makeup and functional capabilities of diazotrophic communities, which in turn affects symbiotic efficacy under pH-related stress (Long et al. 2023). Extensive shotgun metagenomic assessments indicate that the presence of nitrogenase genes (nifH, nifD, nifK) and variety of key genes and corresponding cofactor biosynthesis pathways (such as Fe–S cluster assembly and molybdenum transport) significantly fluctuate with soil pH variations. In a worldwide meta-analysis encompassing 1,451 soil metagenomes, neutral pH soils were found to support a greater diversity of symbiotic lineages from *Bradyrhizobium, Rhizobium,* and *Mesorhizobium,* while acidic soils were primarily occupied by stress-tolerant groups like *Burkholderia* and *Geobacteraceae* (Garcia et al. 2020). Likewise, rhizosphere metagenomic profiles from legumes cultivated in acidic soils display diminished nif gene presence and reduced transcriptional activity, alongside an increase in genes associated with proton efflux, metal transport, and oxidative stress response. These results imply that pH stress may indirectly inhibit nitrogenase gene expression by affecting both the bioavailability of metal cofactors (Mo, Fe) and microbial redox balance. Additionally, network-based metagenomic investigations suggest that extremes in pH disturb the stability of microbial co-occurrence patterns in the rhizosphere, leading to a decrease in cooperative relationships that are essential for effective nodulation and nitrogen fixation. From a practical perspective, integrative metagenomic profiling facilitates the identification of pH-tolerant diazotrophic strains and the development of management practices – such as liming, micronutrient supplementation, and targeted inoculation – to restore symbiotic effectiveness and boost biological nitrogen fixation in agroecosystems experiencing pH stress (Niraula et al. 2022).

11.8 METAGENOME-GUIDED BIOTECHNOLOGICAL APPLICATIONS

The emergence of metagenome-guided discovery pipelines presents a robust method for identifying novel pH-tolerant microbial inoculants aimed at enhancing legumes in marginal soils, achieved by merging community profiling, genome reconstruction, and functional trait analysis. Shotgun metagenomes and metagenome-assembled genomes (MAGs) from acidic and alkaline soils reveal consistent genomic traits linked to pH adaptation, including a broader range of proton-export systems, resilient Fe–S cluster assembly pathways, and an abundance of metal-uptake and siderophore biosynthesis genes, which can serve as indicators for identifying potential strains in culture collections or enrichment cultures (Liu et al. 2023). The integration of metatranscriptomics further helps in prioritising taxa that actively express nitrogen-fixation (nif) genes and functions that mitigate stress in situ, thereby enhancing the probability that the isolated inoculants will thrive under field conditions with pH stress (Thapa et al. 2025). Recent research shows that genome-resolved trait screening speeds up the isolation of acid-tolerant Bradyrhizobium and non-canonical diazotrophs that possess complementary P-solubilisation and osmo-protection genes, which, whether tested as individual strains or in consortia, enhance nodulation and nitrogen fixation during greenhouse and field experiments. Complementary culturomics strategies, guided by metagenomic abundance data and metabolic models, improve the recovery of essential rhizosphere members that might otherwise be overlooked by conventional methods (Clagnan et al. 2024). However, it is essential to assess metagenome-guided candidates for ecological compatibility with local microbiomes and agronomic conditions through co-occurrence network analyses and targeted microcosm experiments to prevent niche displacement or functional loss after release. Ultimately, combining MAG-level trait catalogues with high-throughput phenotyping and formulation science facilitates the rational development of pH-resilient inoculants, a strategy that, when paired with soil amendments and micronutrient management, provides a feasible approach to maintaining legume productivity in marginal lands affected by pH stress.

11.9 ENGINEERING RHIZOBIA OR MICROBIAL CONSORTIA USING METAGENOMIC DATA

The incorporation of metagenomic data into microbial engineering workflows has transformed the creation of advanced rhizobial inoculants and synthetic microbial groups designed to enhance legume yield in stressful conditions. Through comparative metagenomics, it is possible to pinpoint candidate genes that provide stress resilience, nutrient mobilisation, quorum sensing, and specific nodulation suited to hosts, which can be transferred horizontally or upregulated in superior strains (Meaney 2025). For example, gene clusters that encode proton efflux pumps, membrane chaperones, and metal-ion transporters extracted from pH-resistant *Bradyrhizobium* and *Burkholderia* metagenome-assembled genomes (MAGs) have been integrated into symbiotic *Rhizobium* strains, leading to enhanced nodule occupancy and nitrogen fixation efficiency under acidic stress (Chandola et al. 2025). The reconstruction of synthetic consortia, guided by multi-omics, also allows for the intentional pairing of nitrogen-fixing microorganisms with phosphorus-solubilising *Bacillus, Azospirillum*, or arbuscular mycorrhizal fungi, forming cooperative networks that improve nutrient cycling and root colonisation in suboptimal soils (Timofeeva et al. 2023). Additionally, analyses of metagenomic co-occurrence and metabolic networks assist in screening compatibility among members of the community, reducing antagonism and ensuring functional stability after being released into the field. Recent systems-biology approaches further integrate genome-scale metabolic models alongside CRISPR-based genome editing techniques to optimise carbon-nitrogen flows and quorum-sensing mechanisms in rhizobia, laying a data-informed groundwork for the creation of resilient biofertilisers (Pantoja-Alonso et al. 2025).

11.10 CHALLENGES AND FUTURE DIRECTIONS

Although advancements in rhizosphere metagenomics have been rapid, there are still considerable gaps in understanding the genomic and ecological characteristics of microbiomes associated with legumes in extreme pH conditions. The majority of current datasets focus on neutral-pH agricultural soils and temperate agricultural systems, leaving acidic and alkaline environments – key regions for legume production in Asia, Africa, and South America – underrepresented (Ma et al. 2025). Comparative analyses of metagenomics indicate that soil pH is the main factor influencing the structure of microbial communities; however, the functional gene pools that support symbiotic efficiency, nodulation, and nitrogenase activity across different pH levels are not well understood. For example, acidic soils tend to favour strains of *Bradyrhizobium* and *Burkholderia* that are rich in genes for stress tolerance, while alkaline soils often contain *Mesorhizobium* that possess unique mechanisms for osmoprotection and metal-ion transport (Wen et al. 2023). Nevertheless, the absence of geographically diverse and high-depth shotgun metagenomic data, along with standardised metadata, impedes comparisons across systems and the establishment of global functional baselines ((Vargas 2024). Moreover, many studies on rhizospheres depend on 16S rRNA amplicon sequencing, which offers taxonomic information but provides minimal functional understanding. Gene-centric and genome-resolved metagenomics (MAGs) are still limited for legumes grown in the field, particularly for less common pulses like pigeon pea or lentil, which are crucial for smallholder farmers (Sampara et al. 2022). There is also a notable lack of metagenomic time-series studies that examine nodulation dynamics, rhizosphere acidification, and community progression throughout plant development. This deficiency restricts our capacity to model microbial adaptation influenced by pH and to develop pH-tolerant inoculants informed by the genomic characteristics of communities (Afkhami et al. 2025). Closing this gap will necessitate collaborative global efforts to integrate standardised sampling, metadata collection, and open-access metagenomic databases for legume systems across varied soil and climate conditions (Bauchet et al. 2019). Future studies should focus on conducting *in silico* simulations and validating microcosms to predict community-level dynamics prior to implementation. With the growth of metagenomic databases and advancements in computational tools, predictive microbiome modelling is expected to transition from merely correlational analyses to a deeper mechanistic understanding, facilitating precision microbiome engineering and promoting sustainable legume farming across various soil pH conditions (Imran and Gul 2025).

11.11 ACKNOWLEDGEMENTS

We like to thank Department of Biotechnology and University Grant Commission, India for granting Junior Research Fellowship to YS (Grant No.: DBTHRDPMU/JRF/BET-20/1/2020/AL/119) and Ritu (Grant No.:NTA Ref. No. 2116100801), respectively. We also thank GGS Indraprastha University, New Delhi for support and encouragement.

11.12 AUTHOR'S CONTRIBUTION

Ritu, DV, YS and RSP compiled the data and prepared the manuscript. All the authors have read and agreed to publish the version of the manuscript.

REFERENCES

Abdalla Abdelshafy Mohamad O, Liu Y-H, Elsamahy T, Li S, Govindan R, Kuchkarova N, Hatab S, Zhang Y, Li W-J (2025) Dual-functionality of Nocardiopsis alba B57 in biocontrol and plant growth: A metabolomic approach to agricultural sustainability. NPJ Biofilms and Microbiomes 11 (1):164

Afkhami ME, Classen AT, Dice CG, Hernandez DJ, Li VW, Rawstern AH, Rudgers JA, Stinchcombe JR, Crawford KM (2025) Unraveling complexity in climate change effects on beneficial plant–microbe interactions: Mechanisms, resilience, and future directions. New Phytologist 249 (1):93

Alotaibi MO, Saleh AM, Sobrinho RL, Sheteiwy MS, El-Sawah AM, Mohammed AE, Elgawad HA (2021) Arbuscular mycorrhizae mitigate aluminum toxicity and regulate proline metabolism in plants grown in acidic soil. Journal of Fungi 7 (7):531

Bai X, Wu J, Zhang B, Zhao H, Tian F, Wang B (2025) Metagenomics reveals functional profiles of soil nitrogen and phosphorus cycling under different amendments in saline-alkali soil. Environmental Research 267:120686

Bai Z, Jia A, Li H, Wang M, Qu S (2023) Explore the soil factors driving soil microbial community and structure in Songnen alkaline salt degraded grassland. Frontiers in Plant Science 14:1110685

Bauchet GJ, Bett KE, Cameron CT, Campbell JD, Cannon EK, Cannon SB, Carlson JW, Chan A, Cleary A, Close TJ (2019) The future of legume genetic data resources: Challenges, opportunities, and priorities. Legume Science 1 (1):e16

Ben Gaied R, Brígido C, Sbissi I, Tarhouni M (2024) Sustainable strategy to boost legumes growth under salinity and drought stress in semi-arid and arid regions. Soil Systems 8 (3):84

Bonfante P, Genre A (2010) Mechanisms underlying beneficial plant–fungus interactions in mycorrhizal symbiosis. Nature Communications 1 (1):48

Brígido C, Alexandre A, Laranjo M, Oliveira S (2007) Moderately acidophilic mesorhizobia isolated from chickpea. Letters in Applied Microbiology 44 (2):168–174

Chandola U, Manirakiza E, Maillard M, Lavier Aydat LJ, Camuel A, Trottier C, Tanaka A, Chaumier T, Giraud E, Tirichine L (2025) A Bradyrhizobium isolate from a marine diatom induces nitrogen-fixing nodules in a terrestrial legume. Nature Microbiology:1–12

Clagnan E, Costanzo M, Visca A, Di Gregorio L, Tabacchioni S, Colantoni E, Sevi F, Sbarra F, Bindo A, Nolfi L (2024) Culturomics-and metagenomics-based insights into the soil microbiome preservation and application for sustainable agriculture. Frontiers in Microbiology 15:1473666

Corbu VM, Gheorghe-Barbu I, Dumbravă AȘ, Vrâncianu CO, Șesan TE (2023) Current insights in fungal importance: A comprehensive review. Microorganisms 11 (6):1384

Cuadros-Orellana S, Leite LR, Smith A, Medeiros JD, Badotti F, Fonseca PL, Vaz AB, Oliveira G, Góes-Neto A (2013) Assessment of fungal diversity in the environment using metagenomics: A decade in review. Fungal Genomics & Biology 3 (2):1

Dai Z, Guo X, Yin H, Liang Y, Cong J, Liu X (2014) Identification of nitrogen-fixing genes and gene clusters from metagenomic library of acid mine drainage. PloS One 9 (2):e87976

Deng X, Zhao D, Li Y, Li K, Xia J, Yu S, Wang J, Lin J (2025) Arbuscular mycorrhizal fungi confer aluminum toxicity tolerance in Ricinus communis via modulating root metabolic mechanisms and the composition and quantity of root exudates. Plant Physiology and Biochemistry:110149

Dludlu MN, Chimphango SB, Stirton CH, Muasya AM (2017) Differential preference of Burkholderia and Mesorhizobium to pH and soil types in the Core Cape Subregion, South Africa. Genes 9 (1):2

Gaby JC, Buckley DH (2014) A comprehensive aligned nifH gene database: A multipurpose tool for studies of nitrogen-fixing bacteria. Database 2014:bau001

Gadd GM, Fomina M, Pinzari F (2024) Fungal biodeterioration and preservation of cultural heritage, artwork, and historical artifacts: Extremophily and adaptation. Microbiology and Molecular Biology Reviews 88 (1):e00200–00222

Garcia AK, McShea H, Kolaczkowski B, Kaçar B (2020) Reconstructing the evolutionary history of nitrogenases: Evidence for ancestral molybdenum-cofactor utilization. Geobiology 18 (3):394–411

Gondal AH, Hussain I, Ijaz AB, Zafar A, Ch BI, Zafar H, Sohail MD, Niazi H, Touseef M, Khan AA (2021) Influence of soil pH and microbes on mineral solubility and plant nutrition: A review. International Journal of Agriculture and Biological Sciences 5 (1):71–81

Gupta S (2024) Future directions in rhizosphere biology. In: Rhizosphere Revolution. CRC Press, pp 387–408

Hou Y, Zeng W, Hou M, Wang Z, Luo Y, Lei G, Zhou B, Huang J (2021) Responses of the soil microbial community to salinity stress in maize fields. Biology 10 (11):1114

Huang W (2024) Nitrogen fixation in legumes: Genetic mechanisms and agricultural applications. Field Crop 7

Hungria M, Vargas MA (2000) Environmental factors affecting N2 fixation in grain legumes in the tropics, with an emphasis on Brazil. Field Crops Research 65 (2–3):151–164

Imran H, Gul A (2025) Computational approaches to understand rhizomicrobiome community and its future implications. In: Rhizomicrobiome in Sustainable Agriculture and Environment. Elsevier, pp 391–402

Jaiswal SK, Naamala J, Dakora FD (2018) Nature and mechanisms of aluminium toxicity, tolerance and amelioration in symbiotic legumes and rhizobia. Biology and Fertility of Soils 54 (3):309–318

Jena J, Maitra S, Hossain A, Pramanick B, Gitari HI, Praharaj S, Shankar T, Palai JB, Rathore A, Mandal TK (2022) Role of legumes in cropping system for soil ecosystem improvement: Ecosystem services: Types, management and benefits. Nova Science Publishers, Inc 415:1–21

Keshri J, Mishra A, Jha B (2013) Microbial population index and community structure in saline–alkaline soil using gene targeted metagenomics. Microbiological Research 168 (3):165–173

Kumar S, Meena RS, Datta R, Verma SK, Yadav GS, Pradhan G, Molaei A, Rahman GM, Mashuk H (2019) Legumes for carbon and nitrogen cycling: An organic approach. In: Carbon and Nitrogen Cycling in Soil. Springer, pp 337–375

Liu Z, Zhao Y, Zhang B, Wang J, Zhu L, Hu B (2023) Deterministic effect of pH on shaping soil resistome revealed by metagenomic analysis. Environmental Science & Technology 57 (2):985–996

Long G, Chao W, Wenxing L, Zejiang C, Boren W, Juan C, Ren Fang S (2023) Fertilization practices affect biological nitrogen fixation by modulating diazotrophic communities in an acidic soil in southern China. Pedosphere 33 (2):301–311

Lu J, Yang F, Wang S, Ma H, Liang J, Chen Y (2017) Co-existence of rhizobia and diverse non-rhizobial bacteria in the rhizosphere and nodules of Dalbergia odorifera seedlings inoculated with Bradyrhizobium elkanii, Rhizobium multihospitium–like and Burkholderia pyrrocinia–like strains. Frontiers in Microbiology 8:2255

Ma J, Yin B, Gao T, He K, Huang X, Jiang T, Zhen W (2025) Legume–non-legume cover crop mixtures enhance soil nutrient availability and physical properties: A meta-analysis across Chinese agroecosystems. Agronomy 15 (8):1756

Meaney JS (2025) Engineering Rhizobia Endosymbionts Towards the Development of Synthetic Nitrogen-Fixing Organelles. The University of Western Ontario

Mendes R, Garbeva P, Raaijmakers JM (2013) The rhizosphere microbiome: Significance of plant beneficial, plant pathogenic, and human pathogenic microorganisms. FEMS Microbiology Reviews 37 (5):634–663

Meng W, Zhou Z, Tan M, Liu A, Liu S, Wang J, Sun Z, Tan Y, Liu Y, Wang B (2024) Integrated analysis of metatranscriptome and amplicon sequencing to reveal distinctive rhizospheric microorganisms of salt-tolerant rice. Plants 14 (1):36

Niraula S, Rose M, Chang W-S (2022) Microbial co-occurrence network in the rhizosphere microbiome: Its association with physicochemical properties and soybean yield at a regional scale. Journal of Microbiology 60 (10):986–997

Pang F, Li Q, Solanki MK, Wang Z, Xing Y-X, Dong D-F (2024) Soil phosphorus transformation and plant uptake driven by phosphate-solubilizing microorganisms. Frontiers in Microbiology 15:1383813

Pantoja-Alonso MA, Camas-Reyes JA, Cano-Segura R, Cárdenas-Aquino MDR, Martínez-Antonio A (2025) A comprehensive review of genomic-scale genetic engineering as a strategy to improve bacterial productivity. Microbiology 171 (11):001628

Pereira e Silva MC, Schloter-Hai B, Schloter M, van Elsas JD, Salles JF (2013) Temporal dynamics of abundance and composition of nitrogen-fixing communities across agricultural soils. PLoS One 8 (9):e74500

Pérez-Jaramillo JE, De Hollander M, Ramírez CA, Mendes R, Raaijmakers JM, Carrión VJ (2019) Deciphering rhizosphere microbiome assembly of wild and modern common bean (Phaseolus vulgaris) in native and agricultural soils from Colombia. Microbiome 7 (1):114

Rousseau M, Siegenthaler A, Skidmore AK, de Groot GA, Laros I (2024) Further reduction in soil bacterial diversity under severe acidification in European temperate forests. European Journal of Soil Science 75 (6):e70005

Sampara P, Luo Y, Lin X, Ziels RM (2022) Integrating genome-resolved metagenomics with trait-based process modeling to determine biokinetics of distinct nitrifying communities within activated sludge. Environmental Science & Technology 56 (16):11670–11682

Sandhu JS, Chaturvedi SK (2025) Legumes crops cultivation for food, feed and soil health. Legume Crops for Food Security-Cultivation and Benefits: Cultivation and Benefits:75

Sharma SB, Sayyed RZ, Trivedi MH, Gobi TA (2013) Phosphate solubilizing microbes: Sustainable approach for managing phosphorus deficiency in agricultural soils. SpringerPlus 2 (1):587

Sharma U, Datta M, Sharma V (2025) Global status and extent of acid soils. In: Soil Acidity: Management Options for Higher Crop Productivity. Springer, pp 49–119

Shumilina J, Soboleva A, Abakumov E, Shtark OY, Zhukov VA, Frolov A (2023) Signaling in legume–Rhizobia symbiosis. International Journal of Molecular Sciences 24 (24):17397

Tate III RL (2020) Soil Microbiology. John Wiley & Sons.

Thapa A, Hasan MR, Kabir AH (2025) Transcriptional reprogramming and microbiome dynamics in garden pea exposed to high pH stress during vegetative stage. Planta 261 (4):83

Timofeeva AM, Galyamova MR, Sedykh SE (2023) Plant growth-promoting soil bacteria: Nitrogen fixation, phosphate solubilization, siderophore production, and other biological activities. Plants 12 (24):4074

Tyagi J, Ahmad S, Malik M (2022) Nitrogenous fertilizers: Impact on environment sustainability, mitigation strategies, and challenges. International Journal of Environmental Science and Technology 19 (11):11649–11672

Ullah F, Ali S, Siraj M, Akhtar MS, Zaman W (2025) Plant microbiomes alleviate abiotic stress-associated damage in crops and enhance climate-resilient agriculture. Plants 14 (12):1890

Vargas R (2024) Microbial Community Composition and Assembly in Diverse Environmental Systems. Washington University in St. Louis

Vincze É-B, Becze A, Laslo É, Mara G (2024) Beneficial soil microbiomes and their potential role in plant growth and soil fertility. Agriculture 14 (1):152

Wen Z, Yang M, Han H, Fazal A, Liao Y, Ren R, Yin T, Qi J, Sun S, Lu G (2023) Mycorrhizae enhance soybean plant growth and aluminum stress tolerance by shaping the microbiome assembly in an acidic soil. Microbiology Spectrum 11 (2):e03310–03322

Wu Y, Zeng J, Zhu Q, Zhang Z, Lin X (2017) pH is the primary determinant of the bacterial community structure in agricultural soils impacted by polycyclic aromatic hydrocarbon pollution. Scientific Reports 7 (1):40093

Xun W, Liu Y, Li W, Ren Y, Xiong W, Xu Z, Zhang N, Miao Y, Shen Q, Zhang R (2021) Specialized metabolic functions of keystone taxa sustain soil microbiome stability. Microbiome 9 (1):35

Yang S, Xu Y, Tang Z, Jin S, Yang S (2024) The impact of alkaline stress on plant growth and its alkaline resistance mechanisms. International Journal of Molecular Sciences 25 (24):13719

Yavitt JB, Roco CA, Debenport SJ, Barnett SE, Shapleigh JP (2021) Community organization and metagenomics of bacterial assemblages across local scale pH gradients in northern forest soils. Microbial Ecology 81 (3):758–769

Zhao X-D, Gao Z-Y, Peng J, Konstantinidis KT, Zhang S-Y (2024) Various microbial taxa couple arsenic transformation to nitrogen and carbon cycling in paddy soils. Microbiome 12 (1):238

Zhou X, Tahvanainen T, Malard L, Chen L, Pérez-Pérez J, Berninger F (2024) Global analysis of soil bacterial genera and diversity in response to pH. Soil Biology and Biochemistry 198:109552

12 Metagenomics Insights of Major and Orphan Legumes under Mineral Deficiencies

Nishant, Riya Mary Cherian, and Ragini Sinha

12.1 INTRODUCTION

Legumes belong to the third-largest angiosperm family, Fabaceae (Roy *et al.*, 2020). Globally, 27% of the crop production and half of the total plant-mediated proteins humans consume are provided by legumes (Goyal *et al.*, 2021). Legumes are enriched with secondary metabolites such as flavonoids, which have antioxidant properties. *Phaseolus vulgaris,* a commonly consumed legume (known as common bean), is known for its low glycemic index, making it an appropriate choice for diabetic patients (Shohag *et al.*, 2024). They have immense nutritional value and also play an indispensable role in sustainable agriculture due to their ability to form symbiotic associations with *Rhizobia* and other microorganisms, enriching the soil with nitrogen (N) and other essential mineral nutrients (Sulieman and Tran, 2015). They can harness symbiotic microbial populations in their specialised root nodules, where atmospheric N_2 (the non-absorbable form to plants) is converted into ammonia (NH_3), making the mineral available for the plant's metabolic activities (Liu *et al.*, 2018). Crop rotation strategies in cropping systems explore the ability of legumes to enhance soil N and phosphorus (P) via symbiotic nitrogen fixation (SNF) (Pang *et al.*, 2018). The decomposition of the legume crop residues also enriches the soil with N, making the macronutrient available for the subsequent crops and thus reducing the dependency of farming systems on N fertilisers (Plaza-Bonilla *et al.*, 2017).

Legume-rhizobia symbiosis marks an evolutionary relationship of 60 million years or more (Wang *et al.*, 2017). Legume-rhizobium interaction and nodule formation are entirely dependent on chemical signals. Once the legume senses N-deprived conditions, it releases certain metabolites such as N-acetylglucosamine and stringolactone to attract the rhizobia in the rhizosphere (Chiurazzi *et al.*, 2025). After attaching to the root hair cells of the legume, the rhizobia release Nod factors, recognised by the LysM receptor kinase of the plant, which stimulates a Ca^{2+} spike in the root hair cell nuclei (Leng *et al.*, 2025). The whole process of SNF has been reviewed previously (Zhu *et al.*, 2023) and is also beyond the scope of this chapter. However, briefly it can be summarised as the bacteria colonising the rhizosphere, root hair infection, and subsequent initiation of the infection thread, leading to the formation of a symbiosome and finally differentiating into bacterioids capable of N_2 fixation (Speck *et al.*, 2019). The molecular mechanisms involved in the SNF and its regulation have been studied extensively for food security and agricultural stability (Ferguson *et al.*, 2019; Yang *et al.*, 2022). The process of SNF not only allows legumes to survive in N-deficient soil but also reduces the use of fertilisers (Zhang *et al.*, 2024). Two types of nodules exist in legumes – the indeterminate type reported in temperate legumes, where amides (glutamine, asparagine) are formed from N_2, and determinate nodules in tropical legumes, where ureides (allantoin, allantoic acid) are synthesised from the atmospheric N_2. Inside the nodules, NH_3 is assimilated from atmospheric N_2 by the enzyme nitrogenase, and the ammonium ions (NH_4^+) thus produced are further converted by the glutamine synthetase-glutamate synthase pathway into glutamine and glutamate (Ma *et al.*, 2024). Plant growth, development, and productivity are highly altered by nutrient supplementation and deficiencies. Among the nutrients essential for plant productivity, N represents an abundantly required mineral, whose efficiently absorbable forms (nitrate and ammonia) are found in lesser quantities in the soil, making it a major productivity-limiting nutrient in plants (Shumilina *et al.*, 2023). Legumes can sense the level of soil N and can monitor rhizobial infection, nodulation, and SNF accordingly. SNF is an energetically costly process for the plant, where it will have to provide the photosynthates to the bacterial symbiotic (Dagan *et al.*, 2023). So, in soils having sufficient N, nodule formation is inhibited by nodule inception-like protein (NLP). Nitrogen can be acquired via both non-symbiotic and symbiotic means in legumes. The direct uptake of reduced N forms (nitrate and ammonia) via nitrate and ammonia transporters present in the roots contributes to the major share of N in well-fertilised soil (Rahmat *et al.*, 2023). However, in N-deprived conditions, the plant triggers the rhizosphere for potent diazotrophs for the symbiotic association leading to SNF (Gu *et al.*, 2022). Nitrate uptake, availability, and distribution in plants also regulate SNF (Sol *et al.*, 2019). Nitrogen deprivation studies in *Pisum sativum* revealed a decrease in pigment production and growth and structural changes in the etioplast membrane (Kosa *et al.*, 2015). Furthermore, N-deficiency was found to induce alterations in the root microbiome in soybean (Yazaki *et al.*, 2021). In *Vigna radiata* a significant reduction in the activity of the enzymes involved in N-assimilation was observed during N-limited conditions (Zhou *et al.*, 2023a). The status of N nutrition and availability in the plant is a major factor in the regulation of nodulation and SNF, along with negative feedback regulation generated by the byproducts of N assimilation (Murray *et al.*, 2017). Phosphorus (P) is a mineral that is in higher requirement for SNF in legumes (Dokwal *et al.*, 2021). Nodular transport and homeostasis of P can also regulate SNF. Active nodules are found to contain up to three times more P than other plant parts, with an elevated nodulation rate under high P levels (Zhong *et al.*, 2023). Soil inorganic phosphorus (Pi) levels are considerably low. Soil pH and the concentration of metal ions that compete with and co-precipitate with Pi are also crucial in determining the bioavailability of soil Pi (Hinsinger, 2001). Cellular processes such as respiration, photosynthesis, nucleic acid synthesis, carbohydrate metabolism, membrane synthesis, SNF, cell signalling, and enzyme activity regulation require Pi (Vance *et al.*, 2003; Bechtaoui *et al.*, 2021). Carbon-N assimilation pathways can be severely affected by

Pi deficiency (Valentine *et al.*, 2017; Pang *et al.*, 2018). It is a very crucial macronutrient involved in several metabolic and regulatory functions in plants and is also a structural component in major biomolecules, including nucleotides, ATP, NADPH, phospholipids, and several secondary metabolites (Bechtaoui *et al.*, 2021). Pi deficiency in plants can be marked by enhanced root growth, improved expression of P transporters, higher carbon allocation to the roots, and improved release of acid phosphatases and organic acids, thereby increasing the availability of phosphates (Vance *et al.*, 2003). Pi deficiency in legumes can significantly reduce plant growth and productivity (Sulieman and Tran, 2015). Metabolomic studies depicted a drastic drop in organic acid content in *Medicago truncatula* under P-deprived conditions (Dokwal *et al.*, 2022). The genes involved in P-deficiency responses in plants are reported to be expressed in a coordinated manner and have a shared system of regulation (Vance *et al.*, 2003). Purple acid phosphatase in *Glycine max* (*GmPAP4*) is found to be overexpressed under P-deficient conditions in the infection zones of the nodules and can significantly enhance SNF, Pi levels in the nodule, plant growth, and yield (Sun *et al.*, 2024). Increased root hair initiation, leading to root surface area enhancement, is commonly observed as a response to low Pi content in plants (Ajmera *et al.*, 2019). Metabolomic studies during P-limited conditions revealed alterations in carbohydrate, lipid, amino acid. and other P-integrated metabolite levels. The accumulation of sugars (storage carbohydrates) and free amino acids has been observed in plants during Pi-limited conditions. Intense Pi deprivation in plants calls for the release of Pi from phospholipids in the plasma membrane for crucial cellular activities, thus replacing membrane phospholipids with other lipid components (Shohag *et al.*, 2024). Apart from being part of the respiratory electron transport mechanism, iron (Fe) is a cofactor for the N-fixing enzyme, nitrogenase, and is also required for the synthesis of leghaemoglobin, which makes it an absolutely relevant mineral for legume growth and SNF (Day *et al.*, 2021; Chikoti *et al.*, 2020). Deficiencies of minerals such as zinc (Zn), Fe, and molybdenum (Mo) are also known to reduce plant productivity, growth, and nodulation rates in legumes (Dhaliwal *et al.*, 2022). Genomic studies in *M. truncatula* revealed that F-bZIP transcription factors play a crucial role in Zn deficiency responses (Liao *et al.*, 2022). Calcium ions (Ca^{+2}) have significance in SNF signalling, enabling the legumes to distinguish between pathogenesis and symbiosis (Chiurazzi *et al.*, 2025). Due to the higher energy and photosynthate cost, legumes opt for SNF only under severe N starvation (Xu *et al.*, 2025). The availability of sulphur (S) is also critical in SNF. S-containing metabolites such as glutathione are vital in nodule formation. Apart from metabolites, the nitrogenase enzyme itself contains S as FeS clusters (Speck *et al.*, 2019). The role of the molybdate transporter, *MOT 1.2*, in nitrogenase activity and for Fe-Mo cofactor synthesis is an established fact, showing the relevance of Mo in SNF (Chiurazzi *et al.*, 2025).

12.2 ORPHAN LEGUMES

Legumes such as *G. max*, *Arachis hypogaea*, *M. truncatula*, *P. sativum* L, *P. vulgaris*, and *L. ustalisinum* are commonly cultivated and domesticated. However, there are several other legumes such as *Psophocarpus tetragonolobus*, *Vicia faba*, *V. mungo*, *V. umbelleta*, *V. unguiculata*, *V. subterranean*, *V. aconitifolia*, *Lathyrus sativus*,

Canavalia sp., *Cajanus cajan*, *P. lunatus*, *Lublub perpureous*, *Macrotyloma uniflorum*, *Tylosema esculentum*, and *Lens culinaris* that are regarded as underutilised, as they have been neglected and less cultivated even though they are grown widely in the tropics (Popoola *et al.*, 2022). Although these legumes stand underutilised, they have high nutrient as well as phytochemical content that can prevent malnutrition and even chronic diseases. They are also among staple food in many developing countries and are grown in traditional farming systems to meet nutritional requirements at the community scale. Despite their nutritional capacity for serving as "super food" they are poorly researched and documented and are still considered as orphan legumes (Cullis and Kunert, 2017). Compared to the majorly reported legumes, these so called orphan legumes can grow in harsh soil and adverse climatic conditions like drought, salinity, etc. These have immense tolerance potential towards majority of abiotic stresses and hence can play significant roles in breeding programs for bio-fortification of pulses. Hence, adding these legume species to the main food chain might be useful for improving human health and nutrition (Popoola *et al.*, 2022). Certain genomics and gene expression-based studies have been conducted on these orphan legumes identifying responsive genes towards abiotic stress conditions (Sinha *et al.* 2018, 2019, 2021). This chapter provides an overview of achievements in the current decade on the impact of mineral (macro and micronutrient) deficiency on a broad range of legume species, also providing an insight of orphan legumes. The scope of the chapter covers primarily the genes and pathways responding or differentially expressing during mineral-restricted conditions (Figure 12.1). The metagenomics insights covered details in legumes regarding efficient SNF during limited mineral availability and also depict the mechanisms, through altered gene profile. The insights regarding orphan legumes are also mentioned to emphasise the significant contribution that these legumes may have in raising potentially tolerant and nutritionally rich legume and also advocate the inclusion of these so called orphan legumes in mainstream agriculture and research.

12.3 APPLICATION OF METAGENOMIC-BASED STUDIES IN REVEALING MINERAL DEFICIENCIES IN MAJOR LEGUMES

12.3.1 Phosphorous (Pi) Deficiency

Phosphorus (P) is an important component of biological molecules like DNA, RNA, amino acid, proteins, phospholipids, and other secondary metabolites. It plays a crucial role in diverse metabolic and biosynthetic processes like photosynthesis, enzyme activity and others which are essential for proper growth and development of plants (Pang *et al.*, 2018). In legumes, also P drives in the energy required for the formation of nodules (Sa and Israel, 1991; Tang *et al.*, 2001; Dokwal *et al.*, 2021). Although Pi is abundant in soil, but its biological availability to plants is limited when it forms complexes with other soil minerals and compounds, particularly in acidic soil. A deficiency in P can affect nodulation by limiting processes like initiation, reduced number, size, and activity of nodules, affecting N acquisition and metabolism, ultimately hindering SNF (Tang *et al.*, 2001; Isidra-Arellano *et al.*, 2020). Since legumes are an important source of plant protein, their capacity to

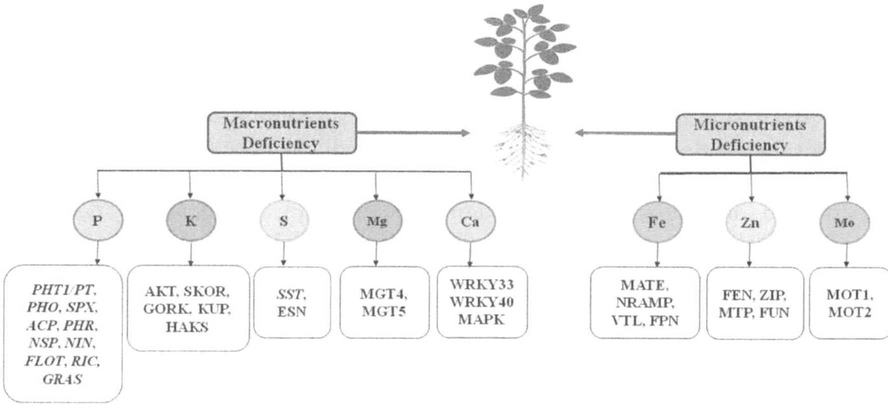

FIGURE 12.1 List of genes showing differential expression during respective mineral deficiency in legumes as identified though varied genomic analysis.

PHT/PHO – phosphate transporter; SPX-SPX – domain-related protein; ACP – acyl carrier protein; PHR – phosphate starvation response; NSP – nodulation signalling pathway; NIN – nodule inception; FLOT – flotillin; RIC – rhizobially induced CLE peptides; AKT – serine threonine kinases; SKOR – stelar K+ outward rectifier; GORK – gated outwardly rectifying K+ channel; KUP-K+ – transporter; HAKS – high-affinity potassium transporter; SST – sulphate transporter gene; ESN/FUN – fixation under nitrate; MGT – magnesium transporter; MAPK – mitogen-activated protein kinase; MATE – multidrug and toxic compound extrusion; NRAMP – natural resistance-associated macrophage protein; VTL – vacuolar ion transporter like; FPN/FEN – ferroportin; ZIP – Zrt/Irt-like protein; MTP – metal ion transport protein; MOT – molybdenum transporter; GRAS and WRKY– gene family transcription factors.

metabolise N during Pi deficiency is critical for agriculture. Transcriptome analysis of legumes during Pi starvation provides better understanding of the mechanisms adapted during stress by providing valuable insights into Pi deficiency-responsive genes (as listed in Table 12.1). Many such studies have been reported in various leguminous plants during Pi deficiency conditions like soybean, common bean, chickpea and medicago (Hernandez *et al.*, 2009; Nasr-Esfahani *et al.*, 2017; Liese *et al.*, 2017; Zhang *et al.*, 2017). P transporters like PHT1/PT help in acquisition of Pi, whose regulation is essential for Pi homeostasis especially during P starvation conditions (Nussaume *et al.*, 2011; Sulieman and Trans, 2015; Gu *et al.*, 2016). Genes with P transport activities have been recognised in *G. max* like *GmPT5* transporting P from roots to nodules and *GmPT7* transporting P from nodule cortex to the fixation zone (Qin *et al.*, 2012; Chen *et al.*, 2019). In *M. trunctula* Pi deficiency has resulted in downregulation of genes related to proteins like nicotinamine-synthase like and leghemoglobin thereby disturbing cellular distribution of iron and lowers nitrogenase activity in nodules, thereby reducing SNF efficiency (Liese *et al.*, 2017). Similarly in medicago during low Pi stress genes *MtVPT2* and *MtVPT3* maintains stable Pi levels in cytosol of the nodules thereby regulating homeostatic level of P (Liu *et al.*, 2023). Another genome-wide-associated mapping-based study in chickpea during Pi starvation, identified genes and quantitative trait loci associated with root structure and characteristics along with Pi concentration responsive genes in shoots, encoding dynamin related protein 1C, NAD kinase 2, fasciclin like arabinogalactan protein, 4-Coumarate CoA ligase 1 and ABC1 family protein (Thudi *et al.*, 2021). Similarly, Pi deficient conditions in soybean identified *GmACP1*, an acid phosphatases-encoding gene as significant candidate gene (Zhang *et al.*, 2017). Genes involved in Pi transport and homeostasis like *SPX5* and *8*, *PHO1.1* and *1.2* are essential for nodulation as reported in soybean and medicago, namely *GmSPX5*, *GmSPX8*, *MtPHO1.1* and *MtPHO1.2* (Xing *et al.*, 2022). The regulatory role of miRNA during Pi deficiency has also been explored in *M. truncatula* by an enrichment analysis for miRNA-mRNA, which identified miRNA target genes related to metabolic activities and hormone signal transduction regulating the Pi starvation response. Further study identified novel miR27 that regulates transcription factors like SPX domain for Pi transport and novel-miR3 targeting sulphoquinovosyl transferase (SQD2) regulated sulpholipid synthesis and miR159 and miR169j/k mediating glutathione S-transferase (GST) for metabolising glutathione (Li *et al.*, 2022). Responses of legume root growth during Pi starvation has been reviewed elsewhere (Chen *et al.*, 2023). The review discusses strategies adopted by legumes to overcome Pi deficiency through gene regulation and modulating specific root traits.

12.3.2 Potassium Deficiency

Potassium is an important mineral nutrient particularly for plants facing abiotic stress conditions. It plays vital role in managing osmotic and turgor pressure as well as ion homeostasis (Kanai *et al.*, 2007; Karnik *et al.*, 2017; Ragel *et al.*, 2019). In legume K transport is critical especially during root development and nodulation. In *M. truncatula* for example, severe decline of K within infected cells has been observed (Fedorova *et al.*, 2021). This causes mis-localisation of key K channel proteins primarily encoded by *MtAKT1* and *MtSKOR/GORK* (Fedorova *et al.*, 2021). Role of K is also reported in supporting N fixation by ATP enhancement (Mengel *et al.*, 1974). It also affects C input and N output in plants by mediating the transport of carbohydrate from legume plants to root nodules (Cakmak *et al.*, 1994). Effect of K deficiency in root development of *A. hypogaea* was studied, which showed the role of miRNA in mediating root development (Li *et al.*, 2021). The study identified upregulated miRNA, namely miR390, miR156, ARF2, NAC4, and AFB3, and certain downregulated miRNA, such as SPL10, miR393, miR160. Various genes for K transportation are reported in legumes, as in *Lotus japonicus*, a K transporter gene *LjKUP* expressed during later stages of nodule development (Desbrosses *et al.*, 2004); *GmHAK5* was reported in *G. max* (Liu *et al.*, 2022b); and high-affinity K ion transporter (*HAK*) genes were identified in common bean genome (Khan *et al.*, 2023).

12.3.3 Sulphur

Among macronutrients sulphur (S) is the fourth essential nutrient for plants, following N, P and K. It is a significant component of amino acids like methionine and cysteine, coenzymes and cofactors, as well as secondary metabolites as reviewed elsewhere (Davidian and Kopriva, 2010). S deficiency also affects SNF and legume plants growing on deficient S conditions have low nitrogenase activity along with less nodulation and associated variations in nodule metabolism, as observed in white clover (*Trifolium repens)* (Anderson *et al.*, 1950; Varin *et al.*, 2010). Importance of sulphate transporter gene (*SST1*) in *L. japonicus* in SNF has been reported (Krusell *et al.*, 2005). There are sulphate transporters to facilitate absorption and transport of sulphate, for example *SST1* in *L. japonicus* and *MtESN2* in *M. truncatula* (Kalloniati *et al.*, 2015; Li *et al.*, 2020b). Another gene *MtSULTR4;1* has been identified in *M. truncatula* under S deficiency particularly at flowering stage, where it is required for remobilising sulphate in leaves to maintain active metabolism (Zuber *et al.*, 2013).

12.3.4 Calcium

Calcium (Ca) is significant for plant nutrition from signal transduction to cell wall component as well as intracellular secondary messengers. In legumes Ca is also required from pre-infection to nodule development (Capoen *et al.*, 2011; Ehrhardt *et al.*, 1996; Sieberer *et al.*, 2009). C deficiency is mainly observed in acidic soil, where it disrupts nodule formation and growth (Watkin *et al.*, 1997; Reeve *et al.*, 1993; Redondo-nieto *et al.*, 2003). It plays significant role in early symbiosis during signalling of Nod factor (D'Haeze and Holsters, 2002; Kanamori *et al.*, 2006). Effect on nitrogenase activity in lupin during Ca deficiency has been reported (Andreev *et al.*, 2001). Its deficiency at reproductive stages in plants may causes embryo disruption and abnormal development of pods. Similar work in peanut identified various DEGs related to plant defence responses and antioxidant activities as well as genes for normal photosynthesis and signal transduction like WRKY33 and WRKY40, mitogen-activated protein kinase signalling pathway (MAPK) (Tang *et al.*, 2023). Effect of Ca in legume root nodules has been reviewed elsewhere (Zartdinova and Nikitin, 2023).

12.3.5 Iron

Iron (Fe) is yet another micronutrient that forms important component of enzymes like ferroproteins, cytochrome, hydrogenase, leghemoglobin and nitrogenase, thus it is vital for SNF (O'Hara *et al.*, 1988; Rodriguez-Hass *et al.*, 2013; Brear *et al.*, 2013; Gonzalez-Guerrero *et al.*, 2014; Ohki *et al.*, 2022). Fe deficiency impairs nodule initiation as well as SNF efficiency as reported in *L. usangustifolius* whereas in other legumes like soyabean, peanut, common bean it affects nodule maturation, its number, size and N fixation efficiencies (O' Hara *et al.*, 1988; Terry *et al.*, 1991; Slatni *et al.*, 2007). Modifications like enhanced proton secretion and ferric reductase activity are adopted by legumes to enhance Fe uptake from the soil (Terry *et al.*, 1991). Fe transport within the symplastic and apoplastic route of infected cell (Day and Smith, 2021) requires specific transporters like multi drug and toxic compound extrusion involved in Fe transportation and homeostasis like *LjMATE1* (Takanashi *et al.*, 2013) and *MtMATE67* (Kryvoruchko *et al.*, 2018). In *M. truncatula* various Fe transporters have been identified like *MtNRAMP1*, *MtVTL4* and *MtVTL8* (Tejada-Jimenez *et al.*, 2015). Similarly, in soybean *GmVTL1* plays important role in Fe transportation to nodules (Brear *et al.*, 2020), its homologous gene *Ferroportin 2* is reported in medicago *MtFPN2* (Escudero *et al.*, 2020).

12.3.6 Magnesium

Magnesium cations are required as cofactor for maximum enzymes involved in cellular biosynthesis and metabolism particularly photosynthesis (Chen *et al.*, 2018; Li *et al.*, 2020a). It is also involved in nitrogenase activity during SNF (Seefeldt *et al.*, 2009). Like other minerals, Mg transporters have been reported in legumes like GmMGT4 and GmMGT5 whose disruptions affects vividly like inhibiting nodule growth, increasing callose accumulation, decreases sugar content, and also reducing N export rates, thereby emphasising the effective role of Mg in symbiosis (Cao *et al.*, 2022). Hence its deficiency causes reduction in growth and yield, especially visible as yellowing and chlorosis in leaf (Farhat *et al.*, 2016). Effect of Mg deficiency in Turkish bean was observed through genome wide associated studies (GWAS), the study identified six DArTseq markers (Baloch *et al.*, 2022).

12.3.7 Molybdenum

Molybdenum is another important trace element regulating various physiological and biochemical phenomena in legumes during symbiotic nitrogen fixation (Banerjee *et al.*, 2022). The importance of Mo in legume growth and nitrogen assimilation has been reviewed elsewhere (Bursakov *et al.*, 2023). The limitation of Mo affects normal growth and development of legume crops particularly due to lack of nitrogen (Rehman *et al.*, 2008). Requirement of Mo is higher in Rhizobium as compared to legume plant (Alam *et al.*, 2015) and the efficiency of rhizobium complex increases in presence of molybdenum (Zhou *et al.*, 2023b). Mo is also a crucial co-factor for Fe-Mo nitrogenase (Kaiser *et al.*, 2005; Hille *et al.*, 2011). It is required by SNF enzymes like nitrogenase, nitrate reductase, xanthine dehydrogenase for electron transfer (Mendel and Hansch, 2002). In the deficiency of Mo legumes adopts other regulatory methods for Mo homeostasis, three different pathways for molybdate oxyanion transport for example *MOT1* and *MOT2*. Mo transporter genes *MtMOT1.2* and *MtMOT1.3* have been listed in *M. truncatula* (Tejada-Jimenez *et al.*, 2017; Gil-Diez *et al.*, 2019) which plays a significant role in synthesising functional nitrogenase.

12.3.8 Zinc

In legumes, besides all basal requirements, zinc (Zn) is particularly required for root nodule formation, as it is involved in homo-citrate synthase that synthesises homocitrate, which is a precursor for the Fe-Mo cofactor of nitrogenase. Most of the rhizobacteria cannot synthesise homocitrate, for which they are dependent on the host plant, and it is produced through the *FEN1* gene (Nouwen *et al.*, 2017). Many Zn transporters have been identified in legumes like *GmZIP1* in *G. max*, *MtZIP6* and *MtMTP2* in *M. truncatula*; these facilitate roles such as Zn uptake and efflux (Moreau *et al.*, 2002; Abreu *et al.*, 2017). Genomic studies in *M. truncatula* identified F-bZIP transcription factors during Zn deficiency (Liao *et al.*, 2022). Zn is also involved as the intracellular secondary messenger in root nodules, converting soil nutrients into signals, thereby linking environmental changes to N homeostasis, involving a transcription factor regulator like fixation under nitrate (FUN) (Lin *et al.*, 2024).

12.4 METAGENOMICS OF ORPHAN LEGUMES

Owing to their resilient characteristics, orphan crops serve as a potent treasure trove for identifying traits useful during extreme growing conditions. These traits can be introduced into crops for combating extreme climatic conditions. Advanced genomics studies are now applied to these underutilised orphan plants. Transcriptomic analysis of leaves and roots of white lupin (*Lupinus albus*) during Pi deficiency identified 1342 and 904 DEGs. In leaves most of the identified genes were related to metabolism, such as amino acid and carbohydrate metabolism, photosynthesis, biosynthesis of tetrapyrrole and flavonoid; DEGs identified in roots were related to signalling and transportation as well as metabolism and biosynthesis. Other genes related to P_i transportation, such as phosphate transporter 1, and others, such as pyrophosphatase, phospholipase, purple acid phosphatase, and monogalactosyl diacyl glycerol synthase, were also differentially expressed (O'Rourke *et al.*, 2013). *L. albus* is an orphan crop exhibiting tolerance to nutrient deficiency, especially Pi deficiency. A few putative long non-coding RNA (lncRNAs) were identified in different stages of root development in *L. albus* during Pi deficiency (Aslam *et al.*, 2022). Nearly 2028 lncRNAs were identified as intergenic (1564), and natural antisense intergenic transcript (NAT) (464) lncRNAs among these 13 putative lncRNAs showed a specific expression pattern (Aslam *et al.*, 2022). Further work revealed 51 *LaGRAS* genes analysed through bioinformatics, which identified the distribution of these genes in 10 distinct phylogenetic clades. The role of *LaGRAS* gene in nutrient acquisition was reported and identified *LaGRAS38* and *LaGRAS39* as potential candidate genes during Pi deficiency (Aslam *et al.*, 2023). Another work in pigeon pea (*C. cajan*) genome during Pi starvation identified four gene families, namely *PHT1*, *fasciclin-like arabinogalactan protein* (*FLA*), *phosphoethanolamine/phosphocholine phosphatase (PECP)*, and *glutamate decarboxylase* (*GAD*), involved in Pi uptake, root architecture, and exudation of organic acid (Liu *et al.*, 2022a). Similar works were reported in *Stylosanthes spp*, an important forage legume and cultivars grown in hydroponic conditions. Genes related to Pi uptake and Pi homeostasis were identified like Pi transporters, MATE, and PAP, as well as transcription factors like *WRKY*, *phosphate starvation response 1* (*PHR1*) and SYG1/PHO81/XPR1 (SPX) domains (Chen *et al.*, 2021). Hairy root transformation studies of *sgEXB1*, a β expansin gene upregulated in *Styloxanthes* under Pi deficient condition, were performed that identified increase in root growth and Pi concentration under low Pi conditions (Wang *et al.*, 2023). Another work reported purple acid phosphatase (PAP) gene family in *Styloxanthes*, identifying *SgPAP7*, as a key gene in the adaptation to low Pi conditions (Luo *et al.*, 2024). Despite some preliminary data, orphan legumes are still deprived of omic-based studies. Enhanced inputs into the molecular characterisation of these orphan legumes will identify potential candidate genes and markers for further exploration towards stress tolerance and human health benefits.

This chapter sums the metagenomic works up taken in legumes under varied mineral deficiency (Table 12.1). Although the amount of information is uneven, wherein major of the works are performed in Pi deficiency conditions and comparatively, lesser information are known in micronutrient deficiency. However, the chapter attempts to cover recent works reported in metagenomic analysis of legumes under

TABLE 12.1
List of Differentially Expressed Genes and Regulatory RNAs as Identified in Major and Minor Legumes as Identified through the Metagenomics Approach

Mineral Deficiency	Name of plant	Name of differentially expressing/responsive and regulatory genes	Function	Citation
Phosphorus (Pi)		PHT1, PT	Pi Transporter gene	
	Glycine max	GmPT5, GmPT7		Qin *et al.* (2012) and Chen *et al.* (2019)
		Pi transporters (PTs), SYG1/PHO81/XPR1 domain *(SPXs), purple acid phosphatases (PAPs)*		Xue *et al.* (2018)
		GmACP1	Acid phosphatase gene	Zhang *et al.* (2017)
		GmSPX8, GmSPX5	Phosphorus signalling and homeostasis	Xing *et al.* (2022)
	Medicago truncatula	MtVPT2, MtVPT3	Pi Vacuolar Transporter gene	Liu *et al.* (2023)
		MtPHO1.1, and *MtPHO1.2*	Pi exporter gene	Chen *et al.* (2023)
		MtPHO2A, B and C	miR399 binding sites	Huertas *et al.* (2023)
		SPX domain	miR27 regulated	Li *et al.* (2022)
		Sulphoquinosyl transferase (SQD2)	miR3 targeted	
		Glutathione S transferase (GST)	miR169j/k and miR159 mediated	
	Phaseolus vulgaris	PvPHR-Like 7 (PvPHR-L7)	Phosphate starvation response transcription factor	Singh *et al.* (2025)
		PvNSP2, PvNIN, and *PvFLOT2, PvRIC1* and *PvRIC2*	autoregulation of nodule numbers	Isidra-Arellano *et al.* (2018)
	Stylosanthes guianensis	*SgPT2-SgPT5*	Pi uptake and transport	An *et al.* (2023)
		PHR1, WRKY, SYG1/PHO81/XPR1 (SPX) domains	Transcription factors	Chen *et al.* (2021)
		SgPAP7	Purple acid phosphatases	Luo *et al.* (2024)
		SgEXB1	B expansin gene	Wang *et al.* (2023)
	Cajanus cajan	*Phosphate transporter 1 (PHT1)*, Phosphoethanolamine/Phosphocholine phosphatases (PECP), fasciclin-like arabinogalactan protein (FLA), glutamate decarboxylase (GAD), Phosphate transporter	Phospholipid recycling; root morphological remodelling; regulation of organic acid exudation	Liu *et al.* (2022a)

	Lupinus albus	*LaGRAS38, LaGRAS39*	Transcription factors	Aslam *et al*. (2022)
Potassium	*M. truncatula*	*MtAKT1, MtSKOR/GORK*	K channel proteins	Federova *et al*. (2021)
	Lotus japonicus	*LjKUP*	K transporter protein	Desbrosses *et al*. (2004)
	G. max	*GmHAK5*	High affinity potassium ion transporter genes	Liu *et al*. (2022b)
	P. vulgaris	*HAK*		Khan *et al*. (2023)
	Arachis hypogea	*SPL10, NAC4, ARF2, AFB3,* miR160, miR393, miR156, miR390,	Regulatory miRNA	Li *et al*. (2021)
Sulphur	*M. truncatula*	*MtESN2, MtSULTR4, 1*	Sulphate transport	Li *et al*. (2020b) and Zuber *et al*. (2013)
	L. japonicus	*SST1*		Li *et al*. (2020b)
Calcium	*A. hypogea*	*WRKY33, WRKY40, MAPK*	Regulatory and signalling activities	Tang *et al*. (2023)
Iron	*L. japonicus*	*LjMATE 1*	MutiDrug and Toxic compound Extrusion genes	Takanashi *et al*. (2013)
	M. truncatula	*MtMATE 67*		Kryvoruchko *et al*. (2018)
		MtNRAMP 1	Iron transporter genes	Tejada-Jimenez *et al*. (2015)
		MtVTL 4, MtVTL 8		
		MtFPN2		Escudero *et al*. (2020)
	G. max	*GmVTL1*		Brear *et al*. (2020)
Magnesium	*G. max*	*GmMGT4, GmMGT5*	Magnesium transport	Cao *et al*. (2022)
	Turkish bean	DArTseq marker	Biological markers for genetic diversity	Baloch *et al*. (2022)
Molybdenum	*M. truncatula*	*MtMOT1.2, MtMOT1.3*	Molybdenum transport	Gil-Diez *et al*. (2019) and Tejada-Jimenez *et al*. (2017)
Zinc	*G. max*	*GmZIP 1*	Transcription factors controlling Zinc and iron transport	Moreau *et al*. (2002)
	M. truncatula	*MtZIP 6, F-bZIP*		Abreu *et al*. (2017) and Liao *et al*. (2022)
		MtMTP 2	Metal transport protein	Abreu *et al*. (2017)
		FUN	Transcription factor regulating Fixation under nitrate	Lin *et al*. (2024)

mineral deficiency to the best. Some sections have been framed to discuss the orphan legumes, highlighting their importance and lack of information. In the wake of promoting sustainable agriculture and nutritional security, along with popular legumes lesser known, understudied legumes should also be considered in the mainstream. Therefore, major refocusing on legume genetics is the need of the hour, including genetics, genomics and biology of untapped, treasure trove legumes.

REFERENCES

Abreu, I., Saéz, A., Castro-Rodríguez, R., Escudero, V., Rodríguez-Haas, B., Senovilla, M., Larue, C., Grolimund, D., Tejada-Jiménez, M., Imperial, J. and González-Guerrero, M. (2017). *Medicago truncatula* Zinc-Iron Permease6 provides zinc to rhizobia-infected nodule cells. *Plant, Cell & Environment*, *40*(11), 2706–2719. https://doi.org/10.1111/pce.13035

Ajmera, I., Hodgman, T. C. and Lu, C. (2019). An integrative systems perspective on plant phosphate research. *Genes*, *10*(2), 139. https://doi.org/10.3390/genes10020139

Alam, F., Bhuiyan, M., Alam, S. S., Waghmode, T. R., Kim, P. J. and Lee, Y. B. (2015). Effect of *Rhizobium* sp: BARIRGm901 inoculation on nodulation, nitrogen fixation and yield of soybean (*Glycine max*) genotypes in gray terrace soil. *Bioscience Biotechnology Biochemistry.*, *79*, 1660–1668.

An, N., Huang, J., Xue, Y., Liu, P., Liu, G., Zhu, S. and Chen, Z. (2023). Characterization of phosphate transporter genes and the function of SgPT1 involved in phosphate uptake in *Stylosanthes guianensis. Plant Physiology and Biochemistry*, *194*, 731–741.

Anderson, A. J. and Spencer, D. (1950). Sulphur in nitrogen metabolism of legumes and non-legumes. *Australian Journal of Scientific Research Series B*, *3*, 431–449.

Andreev, I. M., Andreeva, I. N., Dubrovo, P. N., Krylova, V. V., Kozharinova, C. M. and Izmailov, S. F. (2001). Calcium status of yellow lupin symbiosomes as a potential regulator of their nitrogenase activity: The role of the peribacteroid membrane. *Russian Journal of Plant Physiology*, *48*, 308–317.

Aslam, M. M., Fritschi, F. B., Di, Z., Wang, G., Li, H., Lam, H. M., Waseem, M., Weifeng, X. and Zhang, J. (2023). Overexpression of LaGRAS enhances phosphorus acquisition via increased root growth of phosphorus-deficient white lupin. *Physiology Plant*, *175*(4), e13962. https://doi.org/10.1111/ppl.13962. PMID: 37343119.

Aslam, M. M., Waseem, M., Weifeng, X. and Qamar, M. T. U. (2022, April 30). Identification and expression analysis of phosphate transporter (PHT) gene family in Lupinus albus cluster root under phosphorus stress. *International Journal of Biological Macromolecules*, *205*, 772–781.

Baloch, F. S., Nadeem, M. A., Sönmez, F., Habyarimana, E., Mustafa, Z., Karaköy, T., Cömertpay, G., Alsaleh, A., Çiftçi, V., Sun, S., Chung, G. and Chung, Y. S. (2022 May 2). Magnesium: A forgotten element: Phenotypic variation and genome wide association study in Turkish common bean germplasm. *Frontiers in Genetics*, *13*, 848663.

Banerjee, P. and Nath, R. (2022). Prospects of molybdenum fertilization in grain legumes: A review. *Journal of Plant Nutrition*, *45*, 1425–1440.

Bechtaoui, N., Rabiu, M. K., Raklami, A., Oufdou, K., Hafidi, M. and Jemo, M. (2021). Phosphate-dependent regulation of growth and stress management in plants. *Frontiers in Plant Science*, *12*, 679916. https://doi.org/10.3389/fpls.2021.679916

Brear, E. M., Bedon, F., Gavrin, A., Kryvoruchko, I. S., Torres-Jerez, I., Udvardi, M. K., Day, D. A. and Smith, P. M. C. (2020). GmVTL1a is an iron transporter on the symbiosome membrane of soybean with an important role in nitrogen fixation. *The New Phytologist*, *228*(2), 667–681. https://doi.org/10.1111/nph.16734

Brear, E. M., Day, D. A. and Smith, P. M. (2013). Iron: An essential micronutrient for the legume-rhizobium symbiosis. *Frontiers in Plant Science*, *4*, 359.

Bursakov, S. A., Kroupin, P. Y., Karlov, G. I. and Divashuk, M. G. (2023). Tracing the element: The molecular bases of molybdenum homeostasis in legumes. *Agronomy*, *13*(9), 2300.

Cakmak, I., Hengeler, C. and Marschner, H. (1994). Partitioning of shoot and root dry-matter and carbohydrates in bean-plants suffering from phosphorus, potassium and magnesium-deficiency. *Journal of Experimental Botany*, *45*, 1245–1250.

Cao, H. R., Peng, W. T., Nie, M. M., Bai, S., Chen, C. Q., Liu, Q., Guo, Z. L., Liao, H. and Chen, Z. C. (2022). Carbon-nitrogen trading in symbiotic nodules depends on magnesium import. *Current Biology: CB*, *32*(20), 4337–4349.e5. https://doi.org/10.1016/j.cub.2022.08.019

Capoen, W., Sun, J., Wysham, D., Otegui, M. S., Venkateshwaran, M., Hirsch, S., Miwa, H., Downie, J. A., Morris, R. J., Ané, J. M. and Oldroyd, G. E. (2011). Nuclear membranes control symbiotic calcium signaling of legumes. *Proceedings of the National Academy of Sciences of the United States of America*, *108*(34), 14348–14353. https://doi.org/10.1073/pnas.1107912108

Chen, L., Qin, L., Zhou, L., Li, X., Chen, Z., Sun, L., Wang, W., Lin, Z., Zhao, J., Yamaji, N., Ma, J. F., Gu, M., Xu, G. and Liao, H. (2019). A nodule-localized phosphate transporter GmPT7 plays an important role in enhancing symbiotic N_2 fixation and yield in soybean. *The New Phytologist*, *221*(4), 2013–2025. https://doi.org/10.1111/nph.15541

Chen, Z. C., Peng, W. T., Li, J. and Liao, H. (2018). Functional dissection and transport mechanism of magnesium in plants. *Seminars in Cell & Developmental Biology*, *74*, 142–152. https://doi.org/10.1016/j.semcdb.2017.08.005

Chen, Z. C., Song, J., Li, X., Arango, J., Cardoso, J. A., Rao, I., Schultze-Kraft, R., Peters, M., Mo, X. and Liu, G. (2021). Physiological responses and transcriptomic changes reveal the mechanisms underlying adaptation of *Stylosanthes guianensis* to phosphorus deficiency. *BMC Plant Biology*, *21*(1), 466. https://doi.org/10.1186/s12870-021-03249-2

Chen, Z. C., Wang, L., Cardoso, J. A., Zhu, S., Liu, G., Rao, I. M. and Lin, Y. (2023). Improving phosphorus acquisition efficiency through modification of root growth responses to phosphate starvation in legumes. *Frontiers Plant Science*, *14*, 1094157.

Chikoti, Y. F., Supriadi, Duangkhet, M., Chungopast, S., Tajima, S., Ma, J. F. and Nomura, M. (2020). Effect of ferritin on nitrogen fixation in *Lotus japonicus* nodules under various iron concentrations. *Journal of Plant Physiology*, *252*, 153247.

Chiurazzi, M., Frugis, G. and Navazio, L. (2025). Symbiotic nitrogen fixation: A launchpad for investigating old and new challenges. *Journal of Experimental Botany*, *76*(6), 1473–1477. https://doi.org/10.1093/jxb/erae510

Cullis, C. and Kunert, K. J. (2017 April 1). Unlocking the potential of orphan legumes. Journal of Experimental Botany, *68*(8):1895–1903.

Dagan, R., Dovrat, G., Masci, T. and Sheffer, E. (2023). Competition-induced downregulation of symbiotic nitrogen fixation. *The New Phytologist*, *240*(6), 2288–2297. https://doi.org/10.1111/nph.19322

Davidian, J. C. and Kopriva, S. (2010). Regulation of sulfate uptake and assimilation: The same or not the same? *Molecular Plant*, *3*, 314–325. https://doi.org/10.1093/mp/ssq001

Day, D. A. and Smith, P. M. C. (2021). Iron transport across symbiotic membranes of nitrogen-fixing legumes. *International Journal of Molecular Sciences*, *22*, 432.

Desbrosses, G., Kopka, C., Ott, T. and Udvardi, M. K. (2004). *Lotus japonicus* LjKUP is induced late during nodule development and encodes a potassium transporter of the plasma membrane. *Molecular Plant-Microbe Interactions: MPMI*, *17*(7), 789–797. https://doi.org/10.1094/MPMI.2004.17.7.789

D'Haeze, W. and Holsters, M. (2002). Nod factor structures, responses, and perception during initiation of nodule development. *Glycobiology*, *12*, 79R–105R.

Dhaliwal, S. S., Sharma, V., Shukla, A. K., Kaur, J., Verma, V., Kaur, M., Singh, P., Brestic, M., Gaber, A. and Hossain, A. (2022). Interactive effects of molybdenum, Zinc and Iron on the grain yield, quality, and nodulation of cowpea (*Vigna unguiculata* (L.) Walp.) in North-Western India. *Molecules (Basel, Switzerland)*, *27*(11), 3622.

Dokwal, D., Cocuron, J. C., Alonso, A. P. and Dickstein, R. (2022). Metabolite shift in *Medicago truncatula* occurs in phosphorus deprivation. *Journal of Experimental Botany*, *73*(7), 2093–2111.

Dokwal, D., Romsdahl, T. B., Kunz, D. A., Alonso, A. P. and Dickstein, R. (2021). Phosphorus deprivation affects composition and spatial distribution of membrane lipids in legume nodules. *Plant Physiology*, *185*(4), 1847–1859. https://doi.org/10.1093/plphys/kiaa115

Ehrhardt, D. W., Wais, R. and Long, S. R. (1996). Calcium spiking in plant root hairs responding to rhizobium nodulation signals. *Cell*, *85*, 673–81.

Escudero, V., Abreu, I., Tejada-Jiménez, M., Rosa-Núñez, E., Quintana, J., Prieto, R. I., Larue, C., Wen, J., Villanova, J., Mysore, K. S., Argüello, J. M., Castillo-Michel, H., Imperial, J. and González-Guerrero, M. (2020). *Medicago truncatula* Ferroportin2 mediates iron import into nodule symbiosomes. *The New Phytologist*, *228*(1), 194–209. https://doi.org/10.1111/nph.16642

Farhat, N., Elkhouni, A., Zorrig, W., Smaoui, A., Abdelly, C. and Rabhi, M. (2016). Effects of magnesium deficiency on photosynthesis and carbohydrate partitioning. *Acta Physiologiae Plant*, *38*, 145.

Fedorova, E. E., Coba de la Peña, T., Lara-Dampier, V., Trifonova, N. A., Kulikova, O., Pueyo, J. J. and Lucas, M. M. (2021). Potassium content diminishes in infected cells *of Medicago truncatula* nodules due to the mislocation of channels MtAKT1 and MtSKOR/GORK. *Journal of Experimental Botany*, *72*(4), 1336–1348. https://doi.org/10.1093/jxb/eraa508

Ferguson, B. J., Mens, C., Hastwell, A. H., Zhang, M., Su, H., Jones, C. H., Chu, X. and Gresshoff, P. M. (2019). Legume nodulation: The host controls the party. *Plant, Cell & Environment*, *42*(1), 41–51.

Gil-Diez, P., Tejada-Jiménez, M., León-Mediavilla, J., Wen, J., Mysore, K. S., Imperial, J. and González-Guerrero, M. (2019). MtMOT1.2 is responsible for molybdate supply to *Medicago truncatula* nodules. *Plant, Cell & Environment*, *42*(1), 310–320. https://doi.org/10.1111/pce.13388

Gonzalez-Guerrero, M., Matthiadis, A., Sáez, Á. and Long, T. A. (2014). Fixating on metals: New insights into the role of metals in nodulation and symbiotic nitrogen fixation. *Frontiers in Plant Science*, *5*, 45. https://doi.org/10.3389/fpls.2014.00045

Goyal, R. K., Mattoo, A. K. and Schmidt, M. A. (2021). Rhizobial-host interactions and symbiotic nitrogen fixation in legume crops toward agriculture sustainability. *Frontiers in Microbiology*, *12*, 669404. https://doi.org/10.3389/fmicb.2021.669404

Gu, B., Chen, Y., Xie, F., Murray, J. D. and Miller, A. J. (2022). Inorganic nitrogen transport and assimilation in pea (*Pisum sativum*). *Genes*, *13*(1), 158.

Gu, M., Chen, A., Sun, S. and Xu, G. (2016). Complex regulation of plant phosphate transporters and the gap between molecular mechanisms and practical application: What is missing? *Molecular Plant*, *9*(3), 396–416. https://doi.org/10.1016/j.molp.2015.12.012

Hernandez, G., Valdés-López, O., Ramírez, M., Goffard, N., Weiller, G., Aparicio-Fabre, R., Fuentes, S. I., Erban, A., Kopka, J., Udvardi, M. K. and Vance, C. P. (2009). Global changes in the transcript and metabolic profiles during symbiotic nitrogen fixation in phosphorus-stressed common bean plants. *Plant Physiology*, *151*(3), 1221–1238. https://doi.org/10.1104/pp.109.143842

Hille, R., Nishino, T. and Bittner, F. (2011). Molybdenum enzymes in higher organisms. *Coordination Chemistry Reviews, 255*, 1179–205.

Hinsinger, P. (2001). Bioavailability of soil inorganic P in the rhizosphere as affected by root-induced chemical changes: A review. *Plant and Soil, 237*, 173–195. https://doi.org/10.1023/A:1013351617532

Huertas, R., Torres-Jerez, I., Curtin, S. J., Scheible, W. and Udvardi, M. (2023). *Medicago truncatula PHO2* genes have distinct roles in phosphorus homeostasis and symbiotic nitrogen fixation. *Frontiers in Plant Science, 13*, 1211107.

Isidra-Arellano, M. C., Pozas-Rodríguez, E. A., Del Rocío Reyero-Saavedra, M., Arroyo-Canales, J., Ferrer-Orgaz, S., Del Socorro Sánchez-Correa, M., Cardenas, L., Covarrubias, A. A. and Valdés-López, O. (2020). Inhibition of legume nodulation by Pi deficiency is dependent on the autoregulation of nodulation (AON) pathway. *The Plant Journal for Cell and Molecular Biology, 103*(3), 1125–1139. https://doi.org/10.1111/tpj.14789

Isidra-Arellano, M. C., Reyero-Saavedra, M. D. R., Sánchez-Correa, M. D. S., Pingault, L., Sen, S., Joshi, T., Girard, L., Castro-Guerrero, N. A., Mendoza-Cozatl, D. G., Libault, M. and Valdés-López, O. (2018). Phosphate deficiency negatively affects early steps of the symbiosis between common bean and rhizobia. *Genes (Basel), 9*(10), 498.

Kaiser, B. N., Gridley, K. L., Ngaire Brady, J., Phillips, T. and Tyerman, S. D. (2005). The role of molybdenum in agricultural plant production. *Annals of Botany, 96*(5), 745–754. https://doi.org/10.1093/aob/mci226

Kalloniati, C., Krompas, P., Karalias, G., Udvardi, M. K., Rennenberg, H., Herschbach, C. and Flemetakis, E. (2015). Nitrogen-fixing nodules are an important source of reduced sulfur, which triggers global changes in sulfur metabolism in *Lotus japonicus. The Plant Cell, 27*(9), 2384–2400. https://doi.org/10.1105/tpc.15.00108

Kanai, S., Ohkura, K., Adu-Gyamfi, J. J., Mohapatra, P. K., Nguyen, N. T., Saneoka, H. and Fujita, K. (2007). Depression of sink activity precedes the inhibition of biomass production in tomato plants subjected to potassium deficiency stress. *Journal of Experimental Botany, 58*(11), 2917–2928. https://doi.org/10.1093/jxb/erm149

Kanamori, N., Madsen, L. H., Radutoiu, S., Frantescu, M., Quistgaard, E. M., Miwa, H., Downie, J. A., James, E. K., Felle, H. H., Haaning, L. L., Jensen, T. H., Sato, S., Nakamura, Y., Tabata, S., Sandal, N. and Stougaard, J. (2006). A nucleoporin is required for induction of Ca2+ spiking in legume nodule development and essential for rhizobial and fungal symbiosis. *Proceedings of the National Academy of Sciences of the United States of America, 103*(2), 359–364. https://doi.org/10.1073/pnas.0508883103

Karnik, R., Waghmare, S., Zhang, B., Larson, E., Lefoulon, C., Gonzalez, W. and Blatt, M. R. (2017). Commandeering channel voltage sensors for secretion, cell turgor, and volume control. *Trends in Plant Science, 22*(1), 81–95. https://doi.org/10.1016/j.tplants.2016.10.006

Khan, A., Shah, Z., Ali, S., Ahmad, N., Iqbal, M., Ullah, A. and Ayub, F. (2023). Genome wide identification, structural characterization and phylogenetic analysis of High-Affinity potassium (HAK) ion transporters in common bean (*Phaseolus vulgaris* L.). *BMC Genomic Data, 24*(1), 66. https://doi.org/10.1186/s12863-023-01163-0

Kosa, A., Preininger, É. and Böddi, B. (2015). Nitrogen deficiency hinders etioplast development in stems of dark-grown pea (*Pisum sativum*) shoot cultures. *Physiologia Plantarum, 155*(3), 330–337.

Krusell, L., Krause, K., Ott, T., Desbrosses, G., Krämer, U., Sato, S., Nakamura, Y., Tabata, S., James, E. K., Sandal, N., Stougaard, J., Kawaguchi, M., Miyamoto, A., Suganuma, N. and Udvardi, M. K. (2005 May). The sulfate transporter SST1 is crucial for symbiotic nitrogen fixation in Lotus japonicus root nodules. *Plant Cell, 17*(5), 1625–1636.

Kryvoruchko, I. S., Routray, P., Sinharoy, S., Torres-Jerez, I., Tejada-Jiménez, M., Finney, L. A., Nakashima, J., Pislariu, C. I., Benedito, V. A., González-Guerrero, M., Roberts, D. M. and Udvardi, M. K. (2018). An iron-activated citrate transporter, MtMATE67, is required for symbiotic nitrogen fixation. *Plant Physiology*, *176*(3), 2315–2329. https://doi.org/10.1104/pp.17.01538

Leng, J., Xu, R., Liu, Y., Jiang, T., Hu, H., Ding, Z. and Dai, S. (2025). Genome-wide analysis of GmMYB S20 transcription factors reveals their critical role in soybean nodulation. *Plants (Basel, Switzerland)*, *14*(14), 2240. https://doi.org/10.3390/plants14142240

Li, J., Yokosho, K., Liu, S., Cao, H. R., Yamaji, N., Zhu, X. G., Liao, H., Ma, J. F. and Chen, Z. C. (2020a). Diel magnesium fluctuations in chloroplasts contribute to photosynthesis in rice. *Nature Plants*, *6*(7), 848–859. https://doi.org/10.1038/s41477-020-0686-3

Li, L., Li, Q., Davis, K. E., Patterson, C., Oo, S., Liu, W., Liu, J., Wang, G., Fontana, J. E., Thornburg, T. E., Pratt, I. S., Li, F., Zhang, Z., Zhou, Y., Pan, X. and Zhang, B. (2021). Response of root growth and development to nitrogen and potassium deficiency as well as microRNA-mediated mechanism in peanut (*Arachis hypogaea* L.). *Frontiers in Plant Science*, *12*, 695234.

Li, Q., Gao, Y. and Yang, A. (2020b). Sulfur homeostasis in plants. *International Journal of Molecular Sciences*, *21*, 8926.

Li, Z., Tong, Z., He, F., Li, X. and Sun, J. (2022). Integrated mRNA and microRNA expression analysis of root response to phosphate deficiency in *Medicago sativa*. *Frontiers in Plant Science*, *13*, 989048.

Liao, F., Lilay, G. H., Castro, P. H., Azevedo, H. and Assunção, A. G. L. (2022). Regulation of the Zinc deficiency response in the legume model *Medicago truncatula*. *Frontiers in Plant Science*, *13*, 916168.

Liese, R., Schulze, J. and Cabeza, R. A. (2017). Nitrate application or P deficiency induce a decline in *Medicago truncatula* N2-fixation by similar changes in the nodule transcriptome. *Scientific Reports*, *7*, 46264.

Lin, J., Bjørk, P. K., Kolte, M. V., Poulsen, E., Dedic, E., Drace, T., Andersen, S. U., Nadzieja, M., Liu, H., Castillo-Michel, H., Escudero, V., González-Guerrero, M., Boesen, T., Pedersen, J. S., Stougaard, J., Andersen, K. R. and Reid, D. (2024). Zinc mediates control of nitrogen fixation via transcription factor filamentation. *Nature*, *631*(8019), 164–169. https://doi.org/10.1038/s41586-024-07607-6

Liu, A., Contador, C. A., Fan, K. and Lam, H. M. (2018). Interaction and regulation of carbon, nitrogen, and phosphorus metabolisms in root nodules of legumes. *Frontiers in Plant Science*, *9*, 1860.

Liu, C., Tai, Y., Luo, J., Wu, Y., Zhao, X., Dong, R., Ding, X., Zhao, S., Luo, L., Liu, P. and Liu, G. (2022a). Integrated multi-omics analysis provides insights into genome evolution and phosphorus deficiency adaptation in pigeonpea (*Cajanus cajan*), *Horticulture Research*, *9*, 107.

Liu, J., Liu, J., Cui, M., Chen, X., Liu, J., Chen, J., Chen, A. and Xu, G. (2022b). Investigate the effect of potassium on nodule Symbiosis and uncover an HAK/KUP/KT member, GmHAK5, strongly responsive to root nodulation in soybean. *Journal of Plant Biology*, *65*, 459–471.

Liu, J., Yang, R., Yan, J., Li, C., Lin, X., Lin, L., Cao, Y., Xu, T., Li, J., Yuan, Y., Wen, J., Mysore, K. S. and Luan, S. (2023). VPT-like genes modulate Rhizobium-legume symbiosis and phosphorus adaptation. *The Plant Journal for Cell and Molecular Biology*, *116*(1), 112–127. https://doi.org/10.1111/tpj.16363

Luo, J., Chen, Z., Huang, R., Wu, Y., Liu, C., Cai, Z., Dong, R., Arango, J., Rao, I. M., Schultze-Kraft, R., Liu, G. and Liu, P. (2024). Multi-omics analysis reveals the roles of purple acid phosphatases in organic phosphorus utilization by the tropical legume *Stylosanthes guianensis*. *The Plant Journal*, *117*(3), 729–746.

Ma, Y., Xiao, C., Liu, J. and Ren, G. (2024). Nutrient-dependent regulation of symbiotic nitrogen fixation in legumes. *Horticulture Research*, *12*(3), uhae321. https://doi.org/10.1093/hr/uhae321

Mendel, R. R. and Hansch, R. (2002). Molybdoenzymes and molybdenum cofactor in plants. *Journal of Experimental Botany*, *53*, 1689–1698.

Mengel, K., Haghparast, M. and Koch, K. (1974). The effect of potassium on the fixation of molecular nitrogen by root nodules of *Vicia faba*. *Plant Physiology*, *54*, 535–538.

Moreau, S., Thomson, R. M., Kaiser, B. N., Trevaskis, B., Guerinot, M. L., Udvardi, M. K., Puppo, A. and Day, D. A. (2002). GmZIP1 encodes a symbiosis-specific zinc transporter in soybean. *The Journal of Biological Chemistry*, *277*(7), 4738–4746. https://doi.org/10.1074/jbc.M106754200

Murray, J. D., Liu, C. W., Chen, Y. and Miller, A. J. (2017). Nitrogen sensing in legumes. *Journal of Experimental Botany*, *68*(8), 1919–1926.

Nasr-Esfahani, M., Inoue, K., Chu, H. D., Nguyen, K. H., Van-Ha, C., Watanabe, Y., Burritt, D. J., Herrera-Estrella, L., Mochida, K. and Tran, L. P. (2017). Comparative transcriptome analysis of nodules of two Mesorhizobium-chickpea associations with differential symbiotic efficiency under phosphate deficiency. *The Plant Journal*, *91*(5), 911–926.

Nouwen, N., Arrighi, J. F., Cartieaux, F., Chaintreuil, C., Gully, D., Klopp, C. and Giraud, E. (2017 March 27). The role of rhizobial (NifV) and plant (FEN1) homocitrate synthases in Aeschynomene/photosynthetic Bradyrhizobium symbiosis. *Science Reports*, *7*(1), 448.

Nussaume, L., Kanno, S., Javot, H., Marin, E., Pochon, N., Ayadi, A., Nakanishi, T. M. and Thibaud, M. C. (2011). Phosphate import in plants: Focus on the PHT1 transporters. *Frontiers in Plant Science*, *2*, 83. https://doi.org/10.3389/fpls.2011.00083

O'Hara, G. W., Dilworth, M. J., Boonkerd, N. and Parkpian, P. (1988). Iron-deficiency specifically limits nodule development in peanut inoculated with *Bradyrhizobium sp. The New Phytologist*, *108*(1), 51–57. https://doi.org/10.1111/j.1469-8137.1988.tb00203.x

Ohki, Y., Munakata, K., Matsuoka, Y., Hara, R., Kachi, M., Uchida, K., Tada, M., Cramer, R. E., Sameera, W. M. C., Takayama, T., Sakai, Y., Kuriyama, S., Nishibayashi, Y. and Tanifuji, K. (2022). Nitrogen reduction by the Fe sites of synthetic [Mo_3S_4Fe] cubes. *Nature*, *607*(7917), 86–90. https://doi.org/10.1038/s41586-022-04848-1

O'Rourke, J. A., Yang, S. S., Miller, S. S., Bucciarelli, B., Liu, J., Rydeen, A., Bozsoki, Z., Uhde-Stone, C., Tu, Z. J., Allan, D., Gronwald, J. W. and Vance, C. P. (2013). An RNA-Seq transcriptome analysis of orthophosphate-deficient white lupin reveals novel insights into phosphorus acclimation in plants. *Plant Physiology*, *161*(2), 705–724. https://doi.org/10.1104/pp.112.209254

Pang, J., Ryan, M. H., Lambers, H. and Siddique, K. H. (2018). Phosphorus acquisition and utilisation in crop legumes under global change. *Current Opinion in Plant Biology*, *45*(Pt B), 248–254. https://doi.org/10.1016/j.pbi.2018.05.012

Plaza-Bonilla, D., Nolot, J., Raffaillac, D. and Justes, E. (2017). Innovative cropping systems to reduce N inputs and maintain wheat yields by inserting grain legumes and cover crops in southwestern France. *European Journal of Agronomy*, *82*, 331–341. https://doi.org/10.1016/j.eja.2016.05.010

Popoola, J. O., Aworunse, O. S., Ojuederie, O. B., Adewale, B. D., Ajani, O. C., Oyatomi, O. A., Eruemulor, D. I., Adegboyega, T. T. and Obembe, O. O. (2022 May 17). The exploitation of orphan legumes for food, income, and nutrition security in sub-saharan Africa. *Frontiers in Plant Science*, *13*, 782140.

Qin, L., Zhao, J., Tian, J., Chen, L., Sun, Z., Guo, Y., Lu, X., Gu, M., Xu, G. and Liao, H. (2012). The high-affinity phosphate transporter GmPT5 regulates phosphate transport to nodules and nodulation in soybean. *Plant Physiology*, *159*(4), 1634–1643. https://doi.org/10.1104/pp.112.199786

Ragel, P., Raddatz, N., Leidi, E. O., Quintero, F. J. and Pardo, J. M. (2019). Regulation of K^+ nutrition in plants. *Frontiers in Plant Science*, *10*, 281. https://doi.org/10.3389/fpls.2019.00281

Rahmat, Z., Sohail, M. N., Perrine-Walker, F. and Kaiser, B. N. (2023). Balancing nitrate acquisition strategies in symbiotic legumes. *Planta*, *258*(1), 12.

Redondo-nieto, M., Wilmot, A. R., El-Hamdaoui, A., Bonilla, I. and Bolanos, L. (2003). Relationship between boron and calcium in the N2-fixing legume– rhizobia symbiosis. *Plant Cell & Environment*, *26*, 1905–1915.

Reeve, W. G., Tiwari, R. P., Dilworth, M. J. and Glenn, A. R. (1993). Calcium affects the growth and survival of rhizobium-meliloti. *Soil Biology and Biochemistry*, *25*, 581–586.

Rodriguez-Haas, B., Finney, L., Vogt, S., González-Melendi, P., Imperial, J. and González-Guerrero, M. (2013). Iron distribution through the developmental stages of Medicago truncatula nodules. *Metallomics: Integrated Biometal Science*, *5*(9), 1247–1253. https://doi.org/10.1039/c3mt00060e

Roy, S., Liu, W., Nandety, R. S., Crook, A., Mysore, K. S., Pislariu, C. I., Frugoli, J., Dickstein, R. and Udvardi, M. K. (2020). Celebrating 20 years of genetic discoveries in legume nodulation and symbiotic nitrogen fixation. *The Plant Cell*, *32*(1), 15–41. https://doi.org/10.1105/tpc.19.00279

Sa, T. M. and Israel, D. W. (1991). Energy status and functioning of phosphorus-deficient soybean nodules. *Plant Physiology*, *97*, 928–935, 112.

Seefeldt, L. C., Hoffman, B. M. and Dean, D. R. (2009). Mechanism of Mo dependent nitrogenase. *Annual Review of Biochemistry*, *78*, 701–722.

Shohag, M. J. I., Máximo Salgado, E., Curtis Gluck, M. and Liu, G. (2024). Evaluating the impact of phosphorus and solid oxygen fertilization on snap bean (*Phaseolus vulgaris* L.): A two-year field study. *Plants (Basel, Switzerland)*, *13*(23), 3384. https://doi.org/10.3390/plants13233384

Shumilina, J., Soboleva, A., Abakumov, E., Shtark, O. Y., Zhukov, V. A. and Frolov, A. (2023). Signaling in legume-rhizobia symbiosis. *International Journal of Molecular Sciences*, *24*(24), 17397.

Sieberer, B. J., Chabaud, M., Timmers, A. C., Monin, A., Fournier, J. and Barker, D. G. (2009). A nuclear-targeted cameleon demonstrates intranuclear Ca2+ spiking in *Medicago truncatula* root hairs in response to rhizobial nodulation factors. *Plant Physiology*, *151*(3), 1197–1206. https://doi.org/10.1104/pp.109.142851

Singh, J., Mendoza-Soto, A. B., Tiwari, M., Acevedo-Sandoval, T. T., Formey, D., Ané, J. M., Isidra-Arellano, M. C. and Valdés-López, O. (2025). Phosphate deficiency reduces nodule formation through a phosphate starvation response like protein in *Phaseolus vulgaris*. *Plant and Cell Physiology*, pcaf069.

Sinha, R., Bala, M., Prabha, P., Ranjan, A., Chahota, R. K., Sharma, T. R., Singh, A. K. (2021 December). Identification and validation of reference genes for qRT-PCR based studies in horse gram (*Macrotyloma uniflorum*). *Physiology and Molecular Biology of Plants*, *27*(12), 2859–2873.

Sinha, R., Pal, A. K. and Singh, A. K. (2018). Physiological, biochemical and molecular responses of lentil (Lens culinaris Medik.) genotypes under drought stress. *Indian Journal of Plant Physiology*, *23*(4).

Sinha, R., Sharma, T. R. and Singh, A. K. (2019 January). Validation of reference genes for qRT-PCR data normalisation in lentil (*Lens culinaris*) under leaf developmental stages and abiotic stresses. *Physiology and Molecular Biology of Plants*, *25*(1), 123–134.

Slatni, T., Krouma, A., Aydi, S., Chaiffi, C., Gouia, H. and Abdelly, C. (2007). Growth, nitrogen fixation and ammonium assimilation in common bean (*Phaseolus vulgaris* L) subjected to iron deficiency. *Plant Soil*, *312*(1), 49–57. https://doi.org/10.1007/s11104-007-9481-4

Sol, S., Valkov, V. T., Rogato, A., Noguero, M., Gargiulo, L., Mele, G., Lacombe, B. and Chiurazzi, M. (2019). Disruption of the Lotus japonicus transporter LjNPF2.9 increases shoot biomass and nitrate content without affecting symbiotic performances. *BMC Plant Biology*, *19*(1), 380.

Speck, J. J., James, E. K., Sugawara, M., Sadowsky, M. J. and Gyaneshwar, P. (2019). An alkane sulfonate monooxygenase is required for symbiotic nitrogen fixation by *Bradyrhizobium diazoefficiens* (syn. *Bradyrhizobium japonicum*) USDA110^{T}. *Applied and Environmental Microbiology*, *85*(24), e01552-19. https://doi.org/10.1128/AEM.01552-19

Sulieman, S. and Tran, L. S. (2015). Phosphorus homeostasis in legume nodules as an adaptive strategy to phosphorus deficiency. *The Plant Science*, *239*, 36–43, 117.

Sun, X., Zhang, H., Yang, Z., Xing, X., Fu, Z., Li, X., Kong, Y., Li, W., Du, H. and Zhang, C. (2024). Overexpression of GmPAP4 enhances symbiotic nitrogen fixation and seed yield in soybean under phosphorus-deficient condition. *International Journal of Molecular Sciences*, *25*(7), 3649. https://doi.org/10.3390/ijms25073649

Takanashi, K., Yokosho, K., Saeki, K., Sugiyama, A., Sato, S., Tabata, S., Ma, J. F. and Yazaki, K. (2013). LjMATE1: A citrate transporter responsible for iron supply to the nodule infection zone of *Lotus japonicus*. *Plant & Cell Physiology*, *54*(4), 585–594. https://doi.org/10.1093/pcp/pct019

Tang, C., Hinsinger, P., Drevon, J. J. and Jaillard, B. (2001). Phosphorus deficiency impairs early nodule functioning and enhances proton release in roots of *Medicago truncatula* L. *Annals of Botany*, *88*(1), 131–138.

Tang, K., Li, L., Zhang, B., Zhang, W., Zeng, N., Zhang, H., Liu, D. and Luo, Z. (2023 July 27). Gene co-expression network analysis identifies hub genes associated with different tolerance under calcium deficiency in two peanut cultivars. *BMC Genomics*, *24*(1), 421.

Tejada-Jimenez, M., Castro-Rodríguez, R., Kryvoruchko, I., Lucas, M. M., Udvardi, M., Imperial, J. and González-Guerrero, M. (2015). *Medicago truncatula* natural resistance-associated macrophage Protein1 is required for iron uptake by rhizobia-infected nodule cells. *Plant Physiology*, *168*(1), 258–272. https://doi.org/10.1104/pp.114.254672

Tejada-Jimenez, M., Gil-Díez, P., León-Mediavilla, J., Wen, J., Mysore, K. S., Imperial, J. and González-Guerrero, M. (2017). *Medicago truncatula* Molybdate Transporter type 1 (MtMOT1.3) is a plasma membrane molybdenum transporter required for nitrogenase activity in root nodules under molybdenum deficiency. *The New Phytologist*, *216*(4), 1223–1235. https://doi.org/10.1111/nph.14739

Terry, R. E., Soerensen, K. U., Jolley, V. and Brown, J. C. (1991). The role of active *Bradyrhizobium japonicum* in iron stress response of soybeans. *Plant Soil*, *130*(1), 225–230.

Thudi, M., Chen, Y., Pang, J., Kalavikatte, D., Bajaj, P., Roorkiwal, M., Chitikineni, A., Ryan, M. H., Lambers, H., Siddique, K. H. M. and Varshney, R. K. (2021). Novel genes and genetic loci associated with root morphological traits, phosphorus-acquisition efficiency and phosphorus-use efficiency in Chickpea. *Frontiers in Plant Science*, *12*, 636973.

Valentine, A. J., Kleinert, A. and Benedito, V. A. (2017). Adaptive strategies for nitrogen metabolism in phosphate deficient legume nodules. *Plant Science: An International Journal of Experimental Plant Biology*, *256*, 46–52.

Vance, C. P., Uhde-Stone, C. and Allan, D. (2003). Phosphorus acquisition and use: Critical adaptations by plants for securing a nonrenewable resource. *New Phytologist*, *157*, 423–447. https://doi.org/10.1046/j.1469-8137.2003.00695.x

Varin, S., Cliquet, J. B., Personeni, E., Avice, J. C. and Lemauviel-Lavenant, S. (2010). How does sulphur availability modify N acquisition of white clover (*Trifolium repens* L.)? *Journal of Experimental Botany*, *61*(1), 225–234. https://doi.org/10.1093/jxb/erp303

Wang, L., Wang, L., Zhou, Y. and Duanmu, D. (2017). Use of CRISPR/Cas9 for symbiotic nitrogen fixation research in legumes. *Progress in Molecular Biology and Translational Science*, *149*, 187–213. https://doi.org/10.1016/bs.pmbts.2017.03.010

Wang, L., Wang, W., Miao, Y., Peters, M., Schultze-Kraft, R., Liu, G. and Chen, Z. (2023). Development of transgenic composite *Stylosanthes* plants to study root growth regulated by a β-expansin gene, SgEXPB1, under phosphorus deficiency. *Plant Cell Reports*, *42*(3), 575–585. https://doi.org/10.1007/s00299-023-02978-y

Watkin, E. L. J., O'Hara, G. W. and Glenn, A. R. (1997). Calcium and acid stress interact to affect the growth of *Rhizobium leguminosarum biovartrifolii*. *Soil Biology Biochemistry*, *29*, 1427–1232.

Xing, X., Du, H., Yang, Z., Li, X., Kong, Y., Li, W. and Zhang, C. (2022). GmSPX8, a nodule-localized regulator confers nodule development and nitrogen fixation under phosphorus starvation in soybean. *BMC Plant Biology*, *22*(1), 161. https://doi.org/10.1186/s12870-022-03556-2

Xu, Y., Gao, Q., Xue, L., Zhang, J. and Wang, C. (2025). Optimized nitrogen fertilizer management enhances soybean (*Glycine max* (L.) Merril.) yield and nitrogen use efficiency by promoting symbiotic nitrogen fixation capacity. *Frontiers in Plant Science*, *16*, 1604251. https://doi.org/10.3389/fpls.2025.1604251

Xue, Y., Zhuang, Q., Zhu, S., Xiao, B., Liang, C., Liao, H. and Tian, J. (2018). Genome wide transcriptome analysis reveals complex regulatory mechanisms underlying phosphate homeostasis in soybean nodules. *International Journal of Molecular Science*, *19*(10), 2924.

Yang, J., Lan, L., Jin, Y., Yu, N., Wang, D. and Wang, E. (2022). Mechanisms underlying legume-rhizobium symbioses. *Journal of Integrative Plant Biology*, *64*(2), 244–267.

Yazaki, W., Shimasaki, T., Aoki, Y., Masuda, S., Shibata, A., Suda, W., Shirasu, K., Yazaki, K. and Sugiyama, A. (2021). Nitrogen deficiency-induced bacterial community shifts in soybean roots. *Microbes and Environments*, *36*(3), ME21004.

Zartdinova, R. and Nikitin, A. (2023 December). Calcium in the life cycle of legume root nodules. *Indian Journal of Microbiology*, *63*(4), 410–420.

Zhang, H., Chu, S. and Zhang, D. (2017). Transcriptome dataset of soybean (*Glycine max*) grown under phosphorus-deficient and-sufficient conditions. *Data*, *2*, 17.

Zhang, X., Wu, J. and Kong, Z. (2024). Cellular basis of legume-rhizobium symbiosis. *Plant Communications*, *5*(11), 101045. https://doi.org/10.1016/j.xplc.2024.101045

Zhong, Y., Tian, J., Li, X. and Liao, H. (2023). Cooperative interactions between nitrogen fixation and phosphorus nutrition in legumes. *The New Phytologist*, *237*(3), 734–745. https://doi.org/10.1111/nph.18593

Zhou, H., Liu, Y., Mu, B., Wang, F., Feng, N. and Zheng, D. (2023a). Nitrogen limitation affects carbon and nitrogen metabolism in mung bean (*Vigna radiata* L.). *Journal of Plant Physiology*, *290*, 154105.

Zhou, J., Sun, X., Chen, C. and Chen, J. (2023b). The effect of molybdenum fertilizer on the growth of grass–legume mixtures related to symbiotic rhizobium. *Agronomy*, *13*, 495.

Zhu, Y. G., Peng, J., Chen, C., Xiong, C., Li, S., Ge, A., Wang, E. and Liesack, W. (2023). Harnessing biological nitrogen fixation in plant leaves. *Trends in Plant Science*, *28*(12), 1391–1405.

Zuber, H., Poignavent, G., Le Signor, C., Aimé, D., Vieren, E., Tadla, C., Lugan, R., Belghazi, M., Labas, V., Santoni, A. L., Wipf, D., Buitink, J., Avice, J. C., Salon, C. and Gallardo, K. (2013). Legume adaptation to sulfur deficiency revealed by comparing nutrient allocation and seed traits in *Medicago truncatula*. *The Plant Journal*, *76*(6), 982–996.

13 Metagenome–Metabolome Crosstalk

Exploring the Functional Microbiome of Legumes under Field Conditions

Nikhil Kumar Singh, Kreeti Kajal, and Vijai Pal Bhadana

13.1 INTRODUCTION: IMPORTANCE OF LEGUMES IN SUSTAINABLE AGRICULTURE

Legumes are central to sustainable agriculture due to their capacity to fix atmospheric nitrogen via symbiosis with rhizobia such as *Rhizobium* and *Bradyrhizobium*. This process reduces dependence on synthetic fertilisers, improves soil fertility, and enables productivity in nutrient-poor soils. As major sources of plant protein and oil, legumes contribute significantly to global food and nutritional security. In crop rotations, they enhance soil structure, water-use efficiency, and carbon sequestration, strengthening climate-resilience. Legumes also actively shape the rhizosphere by releasing amino acids, flavonoids, and organic acids that structure microbial communities involved in nutrient cycling, pathogen suppression, and plant growth (Zhalnina et al. 2018).

13.1.1 Limits of Single-Omics and the Need for Integrative Rhizosphere Analysis

The legume rhizosphere is a dynamic interface where roots interact with nitrogen-fixing symbionts, plant growth–promoting rhizobacteria, and arbuscular mycorrhizal fungi that collectively regulate nutrient acquisition, stress tolerance, and disease resistance (Moreau et al. 2015). While metagenomics reveals the taxonomic composition and functional potential of these communities, gene presence alone does not reflect metabolic activity (Aransiola, Atta, and Maddela 2024). Metabolomics complements this by profiling signalling molecules and metabolites involved in plant–microbe communication (Jacoby et al. 2020), but cannot independently resolve plant versus microbial origins. As a result, single-omics approaches often

DOI: 10.1201/9781003733324-14

provide incomplete or ambiguous insights, particularly under variable field conditions. Integrating metagenomic and metabolomic data addresses these limitations by directly linking microbial identity and functional capacity with metabolite dynamics, enabling mechanistic interpretation of rhizosphere processes in realistic agricultural settings (Morton et al. 2019).

13.2 FUNCTIONAL ROLES OF RHIZOSPHERE MICROBIOTA IN LEGUMES

The legume rhizosphere hosts complex microbial communities that regulate nutrient acquisition, stress tolerance, and disease resistance, making them central to sustainable legume production. Among these, nitrogen-fixing rhizobia form the most influential symbiosis. Members of *Rhizobium*, *Bradyrhizobium*, *Mesorhizobium*, and *Sinorhizobium* initiate nodulation through flavonoid-induced activation of nod genes and Nod factor signalling, leading to the formation of specialised root nodules. Within nodules, bacteroids reduce atmospheric nitrogen to ammonia via the nitrogenase complex under strict oxygen control (Udvardi and Poole 2013). This biological nitrogen fixation (BNF) markedly lowers fertiliser demand and contributes substantially to global nitrogen inputs (Giller et al. 2021). Genomic, transcriptomic, and multi-omics studies show that symbiotic efficiency is shaped by microbial genomic plasticity, host genotype, and adaptive processes such as horizontal gene transfer, redox regulation, and carbon allocation, enabling nitrogenase activity under variable field conditions (Ledermann, Schulte, and Poole 2021).

Beyond rhizobia, legumes associate with diverse free-living plant growth–promoting rhizobacteria (PGPR) and arbuscular mycorrhizal fungi that enhance nutrient availability and stress resilience. PGPR from genera such as *Pseudomonas*, *Bacillus*, *Azospirillum*, and *Burkholderia* contribute to phosphorus solubilisation, non-symbiotic nitrogen fixation, siderophore-mediated micronutrient acquisition, and phytohormone production, thereby promoting root development and nodulation(Backer et al. 2018). Under abiotic stress, ACC deaminase–producing microbes reduce ethylene accumulation, alleviating growth inhibition during drought, salinity, and heat stress (Orozco-Mosqueda, Glick, and Santoyo 2020). Arbuscular mycorrhizal fungi, such as *Rhizophagus irregularis*, further enhance phosphorus and water uptake while modulating host immune responses (Q. Wang et al. 2020). Field-based and metagenomic studies show that these microbial assemblages vary with soil type, season, and host genotype (de Vries et al. 2020), highlighting the potential of targeted inoculants and microbiome engineering for climate-resilient legume production.

Rhizosphere microbes also influence legumes through the production of bioactive metabolites that regulate growth, defence, and signalling. PGPR-derived indole-3-acetic acid, volatile organic compounds such as acetoin and 2,3-butanediol, and siderophores directly shape root architecture, nutrient acquisition, and systemic resistance responses (Fincheira et al. 2021). Additional secondary metabolites, including lipopeptides, phenazines, and polyketides, suppress soil-borne pathogens and contribute to disease control (Jha et al. 2025; Cadot et al. 2021). Metabolomic and stable-isotope studies in legume rhizospheres link microbial metabolite production with host flavonoid profiles and stress-related osmoprotectants, highlighting metabolites as key mediators of plant–microbe communication (Fan et al. 2025).

Collectively, legume-associated microbes drive nutrient cycling, carbon turnover, and stress adaptation through coordinated biochemical and ecological interactions. In addition to nitrogen fixation, rhizosphere microbes mediate phosphorus, sulphur, and potassium transformations, organic matter decomposition, and soil aggregation via extracellular polymers, improving soil structure and water retention (Costa, Raaijmakers, and Kuramae 2018). Stress-adapted microbial communities further enhance plant resilience by producing antioxidants, osmolytes, and ion-regulating metabolites, including in saline environments dominated by halotolerant taxa (Tian et al. 2025; X. Gao et al. 2024). Advances in multi-omics now enable reconstruction of these interaction networks, providing a foundation for microbiome-informed breeding, targeted inoculants, and biofertilisers to support low-input, climate-resilient legume agriculture.

13.3 PROFILING THE RHIZOSPHERE: METAGENOMICS AND METABOLOMICS APPROACHES

13.3.1 Sampling Strategies and Field Study Design

Effective rhizosphere omics depends on rigorous and standardised sampling. The rhizosphere – soil tightly adhering to roots and influenced by root exudates – must be clearly separated from bulk soil by gently uprooting plants, removing loosely attached soil, and collecting root-adhering soil via shaking, brushing, or buffer-assisted extraction (Tkacz and Poole 2021). Rhizosphere communities are spatially heterogeneous; therefore, field studies require multiple biological replicates and randomised plot designs (Trivedi et al. 2022). Sampling depth, root developmental stage, and distance from the crown or root tip should be standardised to minimise confounding effects. Plant phenology strongly influences microbial assembly, with early growth stages favouring fast-growing bacteria and later stages such as flowering or nodulation enriching for symbionts and secondary metabolism–associated microbes (Pascale et al. 2019). Time-series sampling across growth stages or seasons can therefore capture successional shifts in nutrient cycling and microbial functions. Comprehensive metadata, including soil properties, microclimate, management history, and plant genotype, should accompany each sample, and adoption of MIxS standards improves cross-study comparability (Thompson et al. 2017). Optimised extraction methods are then required to address soil heterogeneity and inhibitory compounds prior to downstream metagenomic and metabolomic analyses.

13.3.2 Metagenomic Workflows: DNA Extraction and Sequencing

DNA extraction from rhizosphere soils is challenging due to inhibitory humic compounds, clays, and polysaccharides; optimised CTAB–PVP protocols and kits such as DNeasy PowerSoil generally yield DNA suitable for sequencing (Schenk et al. 2023). Sequencing strategies depend on study objectives: marker-gene amplicon sequencing (16S rRNA, ITS) is cost-effective for large comparative studies, whereas whole-metagenome shotgun sequencing enables functional profiling and, when combined with long-read technologies, reconstruction of near-complete genomes and plasmids from complex communities (Zhou, Liu, and Yang 2022). Sequencing reads are quality filtered and assembled using tools such as Trimmomatic, fastp, and MEGAHIT, followed by binning with algorithms including MetaBAT2 to generate metagenome-assembled genomes. Genome quality is assessed with CheckM, and functional annotation is performed using databases such as KEGG, eggNOG, CAZy, and Pfam. Taxonomic assignment with GTDB-Tk or Kraken2 links these functions to specific microbial lineages. Functional profiling commonly targets genes related to nitrogen fixation, phosphorus solubilisation, stress response, and secondary metabolite biosynthesis and can be quantified using HUMAnN3 (Beghini et al. 2021). Metatranscriptomics complements metagenomics by identifying actively expressed genes under field conditions, revealing developmental and environmental regulation of nitrogen fixation, stress tolerance, and root signalling pathways (Rilling et al. 2018). Integration of metagenomic, metatranscriptomic, and metabolomic data links microbial identity, gene activity, and metabolite exchange, enabling mechanistic insight into nutrient cycling, stress adaptation, and legume performance.

13.3.3 Metabolomic Workflows: Root Exudate and Rhizosphere Metabolite Extraction

Root exudates can be collected under controlled conditions using hydroponic or semi-hydroponic systems, which enable continuous sampling of low-abundance compounds, though they may not fully mimic natural soils (Canarini et al. 2019). Field-oriented methods such as rhizoboxes or soil-core extractions provide more realistic insights into legume–soil interactions (Vranova et al. 2013). Rhizosphere soil is typically recovered by shaking or sonication of root-adhering soil in solvent mixtures, allowing extraction of both plant- and microbe-derived metabolites. Solvent selection is tailored to chemical class: polar metabolites are extracted with methanol–water, lipids with chloroform–methanol, and volatiles with solvents such as ethyl acetate (Fiehn 2002; Patti, Yanes, and Siuzdak 2012). Internal standards and preconcentration methods such as solid-phase extraction improve quantification, while rapid freezing and cryogenic grinding minimise post-harvest metabolite degradation. These standardised workflows underpin reproducible metabolomic profiling across controlled and field settings.

13.3.4 Analytical Platforms

A full procedural guide is beyond the scope of this chapter; instead, this section summarises the principal platforms used in rhizosphere metabolomics. Rhizosphere metabolomics primarily relies on GC–MS, LC–MS, and NMR spectroscopy, which offer complementary metabolite coverage (Pang et al. 2021). GC–MS provides high reproducibility for volatile and derivatised polar metabolites (Lisec et al. 2006), while LC–MS offers broad detection of semi-polar secondary metabolites including flavonoids, alkaloids, and phytohormones central to legume–microbe interactions. NMR delivers quantitative and structurally informative data for abundant compounds, though with lower sensitivity than MS-based platforms (Kim, Choi, and Verpoorte 2011). Advances in high-resolution MS, including Orbitrap and QTOF platforms, enhance untargeted discovery and isomer discrimination. Data processing uses tools such as XCMS, MS-DIAL, and MZmine, with annotation supported by GNPS, METLIN, HMDB, and KEGG Compound under MSI guidelines (Sumner et al. 2007). Multivariate analyses (e.g. PCA, PLS-DA) and univariate statistics are then used to identify metabolites associated with microbial activity or plant physiological states.

13.3.5 Linking Metabolites to Function and Microbial Origin

A central challenge in rhizosphere metabolomics is distinguishing plant-metabolites from microbe-derived metabolites. Stable isotope labelling (e.g. ^{13}C or ^{15}N tracers) helps track nutrient flow from root exudates to microbial metabolism (Rummel et al. 2021). Recent spatial metabolomics approaches using imaging MS or MALDI–MS map metabolite localisation along the root surface and nodules, revealing zones of microbial activity and metabolite exchange (Etalo et al. 2018). Integration with metagenomic and transcriptomic data through correlation or co-occurrence networks (e.g. mmvec, sPLS, sCCA) enables linking taxa with specific metabolites (Bauermeister et al. 2022; Morton et al. 2019). In legume systems, metabolomics has illuminated the dynamic biochemical dialogue that underpins symbiosis. Flavonoids such as daidzein and genistein act as signalling molecules that induce *nod* gene expression in *Rhizobium* (Subramanian, Stacey, and Yu 2007), while organic acids and amino acids in exudates regulate microbial colonisation and competition (Venturi and Keel 2016). Under drought or salinity stress, increased secretion of osmolytes (e.g. proline, trehalose) and antioxidants reflects adaptive plant–microbe coordination (Ma, Dias, and Freitas 2020). Such metabolite patterns not only reflect plant stress physiology but also identify microbial groups contributing to resilience.

13.4 INTEGRATION STRATEGIES: LINKING METAGENOME AND METABOLOME DATA

As discussed in earlier sections, integrating metagenomic and metabolomic data is essential for mechanistic interpretation of how microbial communities and small molecules jointly influence legume performance. Because microbiome datasets are compositional and sparse, while metabolomic intensities require scaling, each omic layer is normalised independently using clr or ILR transformations for microbial abundances and log or quantile scaling for metabolites (Mangnier et al. 2025). Initial integration commonly evaluates global concordance between datasets using Mantel or Procrustes tests, while multivariate approaches such as canonical correspondence analysis (CCA) or partial least squares (PLS) project both omics into shared latent spaces to assess covariance (Abdi 2010). Feature-level integration, as outlined above, focuses on resolving specific microbe–metabolite associations. Correlation-based networks are widely used but must account for compositional bias; SparCC and SPIEC-EASI are commonly applied to infer robust or conditional associations, including bipartite taxon–metabolite networks (Friedman and Alm 2012). Such approaches have revealed metabolite-linked microbial modules in legume rhizospheres, including associations between flavonoids, coumarins, and nitrogen-fixing taxa under crop diversification (Qiao et al. 2024). Supervised integration frameworks incorporate phenotypic outcomes. Random Forest models combine microbial and metabolite features to classify treatments or predict yield while ranking influential variables. DIABLO, implemented in mixOmics, identifies sparse feature sets that jointly discriminate groups and maximise covariance across omics layers. Unsupervised latent-factor models such as MOFA+ partition shared and modality-specific sources of variation (Argelaguet et al. 2018), while sparse CCA/PLS and redundancy analysis (RDA) support targeted correlation and regression analyses. Comparative benchmarks indicate that method choice depends on whether the objective is global similarity, prediction, or feature-level inference.

Several tools facilitate exploration of integrated data. MixOmics (R package) provides visualisation for integration: multi-omics PCA/PLS plots, clustered image maps (CIM), network plots of correlated features, etc. (Mangnier et al. 2025). For network visualisation, Cytoscape (open-source) is commonly used to draw microbial/metabolite association graphs and apply graph algorithms (Demchak et al. 2014). Metabolite-specific tools like MetaboAnalyst (webserver) offer integrated analysis modules (including joint pathway analysis) and interactive plots (heatmaps, correlation networks) for metabolome versus microbiome data (Pang et al. 2024). Combined microbiome–metabolome plots (e.g. co-inertia or Procrustes ordination of two datasets) can also be produced with packages like vegan (Procrustes) or custom scripts. Other visual tools include OmicsNet or Cytoscape plugins (e.g. MetScape) for mapping data onto metabolic networks (J. Gao et al. 2010). In the tobacco example, Jia et al. used RDA plots and correlation heatmaps to link leaf bacteria with metabolites (Jia et al. 2025). In general, user-friendly open-source platforms predominate (R, Python, Cytoscape, QIIME2, MZmine), while proprietary software (e.g. Ingenuity Pathway Analysis) are noted but not detailed here due to scope.

13.5 CASE STUDIES FROM FIELD CONDITIONS

Legume–microbiome interactions play a central role in nitrogen fixation, nutrient cycling, and crop performance, yet early field studies relying on single-omic approaches provided limited insight into coordinated plant–soil–microbe processes. Recent field-based multi-omics studies now demonstrate how integrating microbial composition, metabolite profiles, and plant gene expression enables mechanistic interpretation of legume productivity and stress resilience.

13.5.1 Peanut Intercropping System

A long-term peanut intercropping experiment in China showed how multi-omics can resolve the full cascade from root signalling to biological nitrogen fixation (Qiao et al. 2024). Intercropping peanuts with maize and rotation with oilseed rape increased yield and nitrogen uptake relative to monoculture. Integrated metabolomic and transcriptomic analyses revealed enhanced production of flavonoid- and coumarin-derived signals in peanut roots, accompanied by induction of corresponding biosynthetic genes. These chemical shifts coincided with enrichment of rhizosphere taxa carrying nitrogen fixation genes, linking root exudation, microbial recruitment, and enhanced nitrogen fixation in a manner not detectable through agronomic or microbiological measurements alone.

13.5.2 Soybean Rhizosphere Feedback

A field study in soybean revealed how plants reshape their own rhizosphere through coordinated chemical and biological feedbacks (Yamazaki et al. 2021). Multi-omics integration showed that localised changes in soil chemistry around roots were associated with enrichment of microbial taxa involved in nutrient cycling, alongside upregulation of plant genes related to nutrient uptake. This feedback loop-root-mediated soil modification, microbiome restructuring, and altered plant gene expression revealed system-level interactions obscured by single-omic analyses.

13.5.3 Fava Bean Stress Tolerance in Heterogeneous Soil

In a heterogeneous field environment near an e-waste site, adjacent fava bean plants exhibited contrasting growth phenotypes driven by micro-scale variation in heavy-metal exposure (L. Wang et al. 2025). Multi-omics profiling revealed that stress-tolerant plants hosted microbial communities associated with detoxification and accumulated defence-related metabolites, whereas stressed plants lacked these protective interactions. This plant–microbe–soil feedback would not have been resolved using soil chemistry or microbiome surveys alone.

Collectively, these case studies demonstrate that multi-omics reveals coordinated plant–microbe–metabolite interactions that single-omic approaches cannot capture. Across systems, secondary metabolites particularly flavonoids and related compounds emerge as key mediators linking root signalling, microbial function, and field performance, providing a mechanistic basis for designing resilient legume–microbiome partnerships.

13.6 CHALLENGES AND LIMITATIONS OF FIELD MULTI-OMICS IN LEGUME MICROBIOME STUDIES

Applying multi-omics to legume microbiomes under field conditions faces several constraints. Major challenges include distinguishing plant- from microbe-derived metabolites, large annotation gaps in both genes and metabolites, strong environmental variability, and difficulties in preserving high-quality samples. These factors can obscure biological signals and complicate data interpretation, but can be partially mitigated through careful experimental design and emerging analytical tools.

13.6.1 Metabolite Source Attribution (Plant versus Microbial)

In the legume rhizosphere, both roots and microbes release overlapping classes of metabolites, including sugars, amino acids, organic acids, and phenolics, making source assignment difficult in untargeted metabolomics datasets (Mashabela et al.

2022). As a result, rhizosphere profiles often reflect mixtures of plant exudates, microbial products, and transformed soil organic matter. Stable isotope labelling approaches, such as $^{13}CO_2$ pulse-labelling, enable tracing of plant-derived carbon into metabolites and microbial biomass and can be combined with DNA/RNA stable isotope probing to identify active consumers (Pétriacq et al. 2017). Spatial metabolomics methods, including mass spectrometry imaging, further resolve metabolite localisation at the root–soil interface (Vanni et al. 2022). Despite these advances, definitive source attribution remains challenging and typically requires integrated experimental designs and specialised instrumentation.

13.6.2 Annotation Gaps in Metagenomes and Metabolomes

A large fraction of features in multi-omics datasets remains unannotated. In metagenomes, 40–60% of predicted genes lack functional assignments and are classified as hypothetical, limiting inference of microbiome capabilities and interaction pathways. Although community efforts and AI-based tools increasingly cluster unknown proteins and predict function, experimental validation remains slow. Metabolomics faces a comparable "dark matter" problem: untargeted LC–MS or GC–MS typically detects hundreds to thousands of features, yet 70–80% remain unidentified in field rhizosphere studies (Singh et al. 2022). These signals may represent novel metabolites or known compounds lacking reference spectra. Molecular networking, in silico fragmentation, and expanding spectral libraries are improving annotation rates, but most interpretations still rely on a limited annotated subset, requiring cautious biological inference.

13.6.3 Environmental and Temporal Variability in Field Trials

Field-grown legumes experience strong spatial and temporal heterogeneity that can obscure treatment effects in multi-omics datasets. Soil texture, moisture, pH, nutrient hotspots, and microbial biogeography often vary over short distances, producing divergent microbiome and metabolite profiles among replicate plots. Temporal drivers such as diurnal cycles, rainfall events, heatwaves, and drought further alter root exudation and microbial activity, sometimes generating variability comparable to experimental treatments (Jeanne, D'Astous-Pagé, and Hogue 2022). Mitigation requires rigorous experimental design, including replicated and randomised plots, blocking along soil gradients, standardised sampling across growth stages, and parallel recording of soil and weather variables. Mixed-effect or hierarchical models then allow variance attributable to environment and treatment to be partitioned, enabling robust biological inference despite unavoidable field variability.

13.6.4 Limited Field-Deployable Pipelines (Sample Preservation Challenges)

High-quality field multi-omics requires rapid stabilisation of DNA, RNA, proteins, and metabolites, which is often difficult at remote sites. Delays in freezing or suboptimal storage can alter microbial composition and accelerate degradation of labile metabolites, producing artefactual profiles; even short delays can significantly shift metabolomes (De Spiegeleer et al. 2020). RNA-stabilising buffers preserve nucleic acids at ambient temperature but are frequently incompatible with metabolomics and may bias community structure. Drying methods or filter cards stabilise DNA but do not preserve metabolites. Portable extraction kits and handheld sequencers allow limited on-site analysis, yet comprehensive multi-omics still requires laboratory processing, ideally following rapid snap-freezing in liquid nitrogen or on dry ice. Standardised handling and minimal time to stabilisation therefore remain essential, and logistical constraints continue to limit field multi-omics in legumes.

13.7 EMERGING SOLUTIONS AND FUTURE OUTLOOK

New analytical, computational, and synthetic-biology tools are rapidly expanding the scope of field-based legume microbiome research. These innovations address major limitations of traditional multi-omics especially metabolite source attribution, real-time monitoring, and the design of beneficial microbial communities. Stable isotope probing (SIP) using $^{13}CO_2$ or $^{15}N_2$ enables direct tracing of plant-derived carbon and nitrogen into microbial taxa, resolving ambiguities inherent to static omics datasets. Field SIP studies have quantified biological nitrogen fixation and cross-feeding in legume–cereal systems, improving attribution of nutrient fluxes (Picone et al. 2025). Emerging fluxomics platforms, including microfluidic rhizosphere chips coupled with localised mass spectrometry, further enable real-time mapping of exudate gradients and nutrient exchange at the root–soil interface (Aufrecht et al. 2022).

Field-deployable technologies now support in situ microbiome characterisation. Portable nanopore sequencers and miniaturised mass spectrometers generate taxonomic and metabolite profiles comparable to laboratory workflows, enabling rapid and spatially extensive sampling (Bloemen et al. 2023). In parallel, AI and machine learning approaches increasingly integrate soil, environmental, and multi-omics data to predict microbial taxa or consortia associated with stress tolerance and yield improvement, supporting decision-making for soil health management and inoculant selection (Pace et al. 2025)

Advances in metabolic modeling and synthetic biology are enabling the assembly of microbial consortia tailored to plant biochemistry. Genome-scale metabolic reconstructions have been used to design minimal consortia enriched for functions such as iron acquisition, exopolysaccharide production, and auxin biosynthesis. AI-guided frameworks predict compatibility and stability among strains, improving the success of top-down simplification or bottom-up SynCom construction. These rationally designed communities aim to overcome the variability of broad-spectrum inoculants by matching microbial traits with host exudation profiles and soil conditions.

REFERENCES

Abdi, H. 2010. Partial Least Squares Regression and Projection on Latent Structure Regression (PLS Regression): PLS Regression. *Wiley Interdisciplinary Reviews. Computational Statistics* 2, no. 1 (January): 97–106.

Aransiola, S.A., H.I. Atta, and N.R. Maddela. 2024. *Soil Microbiome in Green Technology Sustainability*. Springer Nature.

Argelaguet, R., B. Velten, D. Arnol, et al. 2018. Multi-Omics Factor Analysis-A Framework for Unsupervised Integration of Multi-Omics Data Sets. *Molecular Systems Biology* 14, no. 6: e8124.

Aufrecht, J., M. Khalid, C.L. Walton, K. Tate, J.F. Cahill, and S.T. Retterer. 2022. Hotspots of Root-Exuded Amino Acids Are Created within a Rhizosphere-on-a-Chip. *Lab on a Chip* 22, no. 5 (March 1): 954–963.

Backer, R., J.S. Rokem, G. Ilangumaran, J. Lamont, D. Praslickova, E. Ricci, S. Subramanian, and D.L. Smith. 2018. Plant Growth-Promoting Rhizobacteria: Context, Mechanisms of Action, and Roadmap to Commercialization of Biostimulants for Sustainable Agriculture. *Frontiers in Plant Science* 9 (October 23): 1473.

Bauermeister, A., H. Mannochio-Russo, L.V. Costa-Lotufo, A.K. Jarmusch, and P.C. Dorrestein. 2022. Mass Spectrometry-Based Metabolomics in Microbiome Investigations. *Nature Reviews. Microbiology* 20, no. 3 (March): 143–160.

Beghini, F., L.J. McIver, A. Blanco-Míguez, L. Dubois, F. Asnicar, S. Maharjan, A. Mailyan, et al. 2021. Integrating Taxonomic, Functional, and Strain-Level Profiling of Diverse Microbial Communities with bioBakery 3. *eLife* 10, no. e65088 (May 4). http://dx.doi.org/10.7554/eLife.65088.

Bloemen, B., M. Gand, K. Vanneste, K. Marchal, N.H.C. Roosens, and S.C.J. De Keersmaecker. 2023. Development of a Portable on-Site Applicable Metagenomic Data Generation Workflow for Enhanced Pathogen and Antimicrobial Resistance Surveillance. *Scientific Reports* 13, no. 1 (November 11): 19656.

Cadot, S., V. Gfeller, L. Hu, N. Singh, A. Sánchez-Vallet, G. Glauser, D. Croll, M. Erb, M.G.A. van der Heijden, and K. Schlaeppi. 2021. Soil Composition and Plant Genotype Determine Benzoxazinoid-Mediated Plant-Soil Feedbacks in Cereals. *Plant, Cell & Environment* 44, no. 12 (December): 3502–3514.

Canarini, A., C. Kaiser, A. Merchant, A. Richter, and W. Wanek. 2019. Root Exudation of Primary Metabolites: Mechanisms and Their Roles in Plant Responses to Environmental Stimuli. *Frontiers in Plant Science* 10 (February 21): 157.

Costa, O.Y.A., J.M. Raaijmakers, and E.E. Kuramae. 2018. Microbial Extracellular Polymeric Substances: Ecological Function and Impact on Soil Aggregation. *Frontiers in Microbiology* 9 (July 23): 1636.

Demchak, B., T. Hull, M. Reich, T. Liefeld, M. Smoot, T. Ideker, and J.P. Mesirov. 2014. Cytoscape: The Network Visualization Tool for GenomeSpace Workflows. *F1000 Research* 3 (July 1): 151.

De Spiegeleer, M., M. De Graeve, S. Huysman, A. Vanderbeke, L. Van Meulebroek, and L. Vanhaecke. 2020. Impact of Storage Conditions on the Human Stool Metabolome and Lipidome: Preserving the Most Accurate Fingerprint. *Analytica Chimica Acta* 1108 (April 29): 79–88.

de Vries, F.T., R.I. Griffiths, C.G. Knight, O. Nicolitch, and A. Williams. 2020. Harnessing Rhizosphere Microbiomes for Drought-Resilient Crop Production. *Science (New York, N.Y.)* 368, no. 6488 (April 17): 270–274.

Etalo, D.W., C. Díez-Simón, R.C.H. de Vos, and R.D. Hall. 2018. Laser Ablation Electrospray Ionization-Mass Spectrometry Imaging (LAESI-MS) for Spatially Resolved Plant Metabolomics. *Methods in Molecular Biology (Clifton, N.J.)* 1778: 253–267.

Fan, H., Q. Wang, J. Bai, Y. Chen, C. Yang, D. Hrynsphan, T. Savitskaya, Z. Wang, and J. Chen. 2025. Review on the Metabolic Synergistic Mechanisms in Fungal-Bacterial Co-Culture Systems for VOCs Biodegradation: From a Microscopic Perspective. *Re/views in Environmental Science and Bio/technology* 24, no. 3 (September): 733–752.

Fiehn, O. 2002. Metabolomics – the Link between Genotypes and Phenotypes. *Plant Molecular Biology* 48, no. 1–2 (January): 155–171.

Fincheira, P., A. Quiroz, G. Tortella, M.C. Diez, and O. Rubilar. 2021. Current Advances in Plant-Microbe Communication via Volatile Organic Compounds as an Innovative Strategy to Improve Plant Growth. *Microbiological Research* 247, no. 126726 (June): 126726.

Friedman, J., and E.J. Alm. 2012. Inferring Correlation Networks from Genomic Survey Data. *PLoS Computational Biology* 8, no. 9 (September 20): e1002687.

Gao, J., V.G. Tarcea, A. Karnovsky, B.R. Mirel, T.E. Weymouth, C.W. Beecher, J.D. Cavalcoli, et al. 2010. Metscape: A Cytoscape Plug-in for Visualizing and Interpreting Metabolomic Data in the Context of Human Metabolic Networks. *Bioinformatics (Oxford, England)* 26, no. 7 (April 1): 971–973.

Gao, X., S. Wang, W. Kong, G. Li, L. Zhang, and X. Yin. 2024. Floristic Changes and Environmental Drivers of Soil Fungi and Archaea in Different Salt-Tolerant Plant Communities in the Intertidal Habitat of Coastal Wetlands. *Environmental Geochemistry and Health* 46, no. 5 (April 9): 167.

Giller, K.E., R. Hijbeek, J.A. Andersson, and J. Sumberg. 2021. Regenerative Agriculture: An Agronomic Perspective. *Outlook on Agriculture* 50, no. 1 (March): 13–25.

Jacoby, R.P., L. Chen, M. Schwier, A. Koprivova, and S. Kopriva. 2020. Recent Advances in the Role of Plant Metabolites in Shaping the Root Microbiome. *F1000 Research* 9 (February 26): 151.

Jeanne, T., J. D'Astous-Pagé, and R. Hogue. 2022. Spatial, Temporal and Technical Variability in the Diversity of Prokaryotes and Fungi in Agricultural Soils. *Frontiers in Soil Science* 2 (August 4). https://doi.org/10.3389/fsoil.2022.945888.

Jha, M., A. Nautiyal, K. Pant, and N. Kumar. 2025. Current Status of Antimicrobial Peptides Databases and Computational Tools for Optimization. *Environment Conservation Journal* 26, no. 1 (February 5): 281–292.

Jia, Y., J. Wang, X. Lin, T. Liang, H. Dai, B. Wu, M. Yang, Y. Zhang, and R. Li. 2025. Integrated Metabolomics and Metagenomics Reveal Plant-Microbe Interactions Driving Aroma Differentiation in Flue-Cured Tobacco Leaves. *Frontiers in Plant Science* 16 (June 3): 1588888.

Kim, H.K., Y.H. Choi, and R. Verpoorte. 2011. NMR-Based Plant Metabolomics: Where Do We Stand, Where Do We Go? *Trends in Biotechnology* 29, no. 6 (June): 267–275.

Ledermann, R., C.C.M. Schulte, and P.S. Poole. 2021. How Rhizobia Adapt to the Nodule Environment. *Journal of Bacteriology* 203, no. 12 (May 20): e0053920.

Lisec, J., N. Schauer, J. Kopka, L. Willmitzer, and A.R. Fernie. 2006. Gas Chromatography Mass Spectrometry-Based Metabolite Profiling in Plants. *Nature Protocols* 1, no. 1: 387–396.

Ma, Y., M.C. Dias, and H. Freitas. 2020. Drought and Salinity Stress Responses and Microbe-Induced Tolerance in Plants. *Frontiers in Plant Science* 11 (November 13): 591911.

Mangnier, L., A. Bodein, M. Mariaz, A. Mathieu, M.-P. Scott-Boyer, N. Vashist, M.S. Bramble, and A. Droit. 2025. A Systematic Benchmark of Integrative Strategies for Microbiome-Metabolome Data. *Communications Biology* 8, no. 1 (July 25): 1100.

Mashabela, M.D., L.A. Piater, I.A. Dubery, F. Tugizimana, and M.I. Mhlongo. 2022. Rhizosphere Tripartite Interactions and PGPR-Mediated Metabolic Reprogramming towards ISR and Plant Priming: A Metabolomics Review. *Biology* 11, no. 3 (February 22): 346.

Moreau, D., B. Pivato, D. Bru, H. Busset, F. Deau, C. Faivre, A. Matejicek, F. Strbik, L. Philippot, and C. Mougel. 2015. Plant Traits Related to Nitrogen Uptake Influence Plant-Microbe Competition. *Ecology* 96, no. 8 (August): 2300–2310.

Morton, J.T., A.A. Aksenov, L.F. Nothias, J.R. Foulds, R.A. Quinn, M.H. Badri, T.L. Swenson, et al. 2019. Learning Representations of Microbe-Metabolite Interactions. *Nature Methods* 16, no. 12 (December): 1306–1314.

Orozco-Mosqueda, M.D.C., B.R. Glick, and G. Santoyo. 2020. ACC Deaminase in Plant Growth-Promoting Bacteria (PGPB): An Efficient Mechanism to Counter Salt Stress in Crops. *Microbiological Research* 235, no. 126439 (May): 126439.

Pace, R., V. Schiano Di Cola, M.M. Monti, A. Affinito, S. Cuomo, F. Loreto, and M. Ruocco. 2025. Artificial Intelligence in Soil Microbiome Analysis: A Potential Application in Predicting and Enhancing Soil Health: A Review. *Discover Applied Sciences* 7, no. 2 (January 16). https://doi.org/10.1007/s42452-024-06381-4.

Pang, Z., J. Chen, T. Wang, C. Gao, Z. Li, L. Guo, J. Xu, and Y. Cheng. 2021. Linking Plant Secondary Metabolites and Plant Microbiomes: A Review. *Frontiers in Plant Science* 12 (March 2): 621276.

Pang, Z., Y. Lu, G. Zhou, F. Hui, L. Xu, C. Viau, A.F. Spigelman, et al. 2024. Metabo Analyst 6.0: Towards a Unified Platform for Metabolomics Data Processing, Analysis and Interpretation. *Nucleic Acids Research* 52, no. W1 (July 5): W398–W406.

Pascale, A., S. Proietti, I.S. Pantelides, and I.A. Stringlis. 2019. Modulation of the Root Microbiome by Plant Molecules: The Basis for Targeted Disease Suppression and Plant Growth Promotion. *Frontiers in Plant Science* 10: 1741.

Patti, G.J., O. Yanes, and G. Siuzdak. 2012. Innovation: Metabolomics: The Apogee of the Omics Trilogy. *Nature Reviews. Molecular Cell Biology* 13, no. 4 (March 22): 263–269.

Pétriacq, P., A. Williams, A. Cotton, A.E. McFarlane, S.A. Rolfe, and J. Ton. 2017. Metabolite Profiling of Non-Sterile Rhizosphere Soil. *The Plant Journal: For Cell and Molecular Biology* 92, no. 1 (October): 147–162.

Picone, R., G. Pietramellara, G. Guggenberger, S.I. Pathan, and N. Gentsch. 2025. The Contemporary Plant-Soil Feedback in Legume-Cereal Intercropping Systems: A Review of Carbon, Nutrient, and Microbial Dynamics. *Plant and Soil.*

Qiao, M., R. Sun, Z. Wang, K. Dumack, X. Xie, C. Dai, E. Wang, et al. 2024. Legume Rhizodeposition Promotes Nitrogen Fixation by Soil Microbiota under Crop Diversification. *Nature Communications* 15, no. 1 (April 4): 2924.

Rilling, J.I., J.J. Acuña, M.J. Sadowsky, and M.A. Jorquera. 2018. Putative Nitrogen-Fixing Bacteria Associated with the Rhizosphere and Root Endosphere of Wheat Plants Grown in an Andisol from Southern Chile. *Frontiers in Microbiology* 9 (November 20): 2710.

Rummel, P.S., R. Well, B. Pfeiffer, K. Dittert, S. Floßmann, and J. Pausch. 2021. Nitrate Uptake and Carbon Exudation – Do Plant Roots Stimulate or Inhibit Denitrification? *Plant and Soil* 459, no. 1–2 (February): 217–233.

Schenk, J.J., L.E. Becklund, S.J. Carey, and P.P. Fabre. 2023. What Is the "Modified" CTAB Protocol? Characterizing Modifications to the CTAB DNA Extraction Protocol. *Applications in Plant Sciences* 11, no. 3 (May): e11517.

Singh, N.K., S.M. Tralamazza, L.N. Abraham, G. Glauser, and D. Croll. 2022. Genome-Wide Association Mapping Reveals Genes Underlying Population-Level Metabolome Diversity in a Fungal Crop Pathogen. *BMC Biology* 20, no. 1 (October 8): 224.

Subramanian, S., G. Stacey, and O. Yu. 2007. Distinct, Crucial Roles of Flavonoids during Legume Nodulation. *Trends in Plant Science* 12, no. 7 (July): 282–285.

Sumner, L.W., A. Amberg, D. Barrett, M.H. Beale, R. Beger, C.A. Daykin, T.W.-M. Fan, et al. 2007. Proposed Minimum Reporting Standards for Chemical Analysis Chemical Analysis Working Group (CAWG) Metabolomics Standards Initiative (MSI). *Metabolomics: Official Journal of the Metabolomic Society* 3, no. 3 (September): 211–221.

Thompson, L.R., J.G. Sanders, D. McDonald, A. Amir, J. Ladau, K.J. Locey, R.J. Prill, et al. 2017. A Communal Catalogue Reveals Earth's Multiscale Microbial Diversity. *Nature* 551, no. 7681 (November 23): 457–463.

Tian, J., Y. Qing, M. Zang, X. Long, G. Liu, and Y. Ma. 2025. Genomic Insights into Phenol Degradation by Halophilic Bacteria and Their Potential Application in Saline Soil Remediation. *Scientific Reports* 15, no. 1 (September 24): 32740.

Tkacz, A., and P. Poole. 2021. The Plant Microbiome: The Dark and Dirty Secrets of Plant Growth. *Plants, People, Planet* 3, no. 2 (March): 124–129.

Trivedi, P., B.D. Batista, K.E. Bazany, and B.K. Singh. 2022. Plant-Microbiome Interactions under a Changing World: Responses, Consequences and Perspectives. *The New Phytologist* 234, no. 6 (June): 1951–1959.

Udvardi, M., and P.S. Poole. 2013. Transport and Metabolism in Legume-Rhizobia Symbioses. *Annual Review of Plant Biology* 64, no. 1 (March 1): 781–805.

Vanni, I., M. Croce, L. Pastorino, et al. 2025. Identification of Prognostic Liquid Biopsy Biomarkers in Patients with Cutaneous Squamous Cell Carcinoma Treated with Cemiplimab. *Journal of Translational Medicine* 23, no. 1: 965.

Venturi, V., and C. Keel. 2016. Signaling in the Rhizosphere. *Trends in Plant Science* 21, no. 3 (March): 187–198.

Vranova, V., K. Rejsek, K.R. Skene, D. Janous, and P. Formanek. 2013. Methods of Collection of Plant Root Exudates in Relation to Plant Metabolism and Purpose: A Review. *Journal of Plant Nutrition and Soil Science* 176, no. 2 (April): 175–199.

Wang, L., Y. Wu, Z.-B. Zhao, and T. Jia. 2025. Small-Scale Heterogeneity of Soil Properties in Farmland Affected Fava Beans Growth through Rhizosphere Differential Metabolites and Microorganisms. *Environmental Microbiome* 20, no. 1 (May 1): 45.

Wang, Q., M. Ma, X. Jiang, D. Guan, D. Wei, F. Cao, Y. Kang, C. Chu, S. Wu, and J. Li. 2020. Influence of 37 Years of Nitrogen and Phosphorus Fertilization on Composition of Rhizosphere Arbuscular Mycorrhizal Fungi Communities in Black Soil of Northeast China. *Frontiers in Microbiology* 11 (September 8): 539669.

Yamazaki, S., H. Mardani-Korrani, R. Kaida, K. Ochiai, M. Kobayashi, A.J. Nagano, Y. Fujii, A. Sugiyama, and Y. Aoki. 2021. Field Multi-Omics Analysis Reveals a Close Association between Bacterial Communities and Mineral Properties in the Soybean Rhizosphere. *Scientific Reports* 11, no. 1 (April 23): 8878.

Zhalnina, K., K.B. Louie, Z. Hao, N. Mansoori, U.N. da Rocha, S. Shi, H. Cho, et al. 2018. Dynamic Root Exudate Chemistry and Microbial Substrate Preferences Drive Patterns in Rhizosphere Microbial Community Assembly. *Nature Microbiology* 3, no. 4 (April): 470–480.

Zhou, Y., M. Liu, and J. Yang. 2022. Recovering Metagenome-Assembled Genomes from Shotgun Metagenomic Sequencing Data: Methods, Applications, Challenges, and Opportunities. *Microbiological Research* 260, no. 127023 (July): 127023.

14 Dynamics of the Legume Rhizosphere Microbiota

Sakshi Suman, Aarti Yadav, Gopal Chowdhary, and Yamini Agrawal

14.1 INTRODUCTION

The rhizosphere is the narrow zone of soil surrounding plant roots, its root secretions, and associated micro-organisms (Berendsen et al., 2012). It forms dynamic and complex ecological niche where interactions occur between microbial communities, soil, and plant roots (Mendes et al., 2013). This microenvironment plays a crucial role in the healthy growth of plants, hormone production, soil health, and nutrient cycling (nitrogen, phosphorus solubilisation), pathogen suppression, and helps in stress mitigation (Yu et al., 2025; Trivedi et al., 2021) (Figure 14.1). The roots use their innate immune system to detect and regulate the colonisations; with time it also evolved to evade host immunity (Grundy et al., 2023). Legumes (Fabaceae) are one of the essential crops that are ecological and economic pillars of sustainable agriculture as they fix the atmospheric nitrogen into bioavailable forms. It occupies a unique position due to its special ability to establish endosymbiotic relationships with nitrogen-fixing rhizobia and arbuscular mycorrhizal fungi (AMF), which helps in thriving in nutrient-poor soils, therefore improving the structure and functionality of the rhizosphere microbiota (Wang et al., 2021). The rhizosphere microbiome primarily consists of rhizobial and non-rhizobial microbial groups like bacteria (*Rhizobium, Bradyrhizobium, Sinorhizobium, Mesorhizobium*), actinomycetes (*Frankia*), protists, and fungi (*Glomus, Rhizophagus, Funneliformis*) that inhabit soil and root surfaces (Agyekum et al., 2025). The incorporation of legumes in intercropping systems or crop rotations helps to improve soil structure, increase microbial diversity, and suppress weeds (Sharma et al., 2025).

Rhizobia live inside the special root cells and form the structure called "nodules," which enter through the thread-like structure called infection threads (Yang et al., 2022). Diazotrophs fix atmospheric N_2, making nitrogen available for plant uptake with the help of the nif gene cluster. The soil microbiota, having different soil properties, play crucial roles and strongly influence the nodule microbiome. The plant genotype not only shapes it but also determines its compatibility with the rhizobia symbiont (Wagner et al., 2016). Inside the nodules, it creates a unique environment, characterised by high single taxon bacterial density, low oxygen concentrations, an ample supply of photosynthates, carbon, nitrogen, and metals such as iron and molybdenum (Ruttelib et al., 2019).

 DOI: 10.1201/9781003733324-15

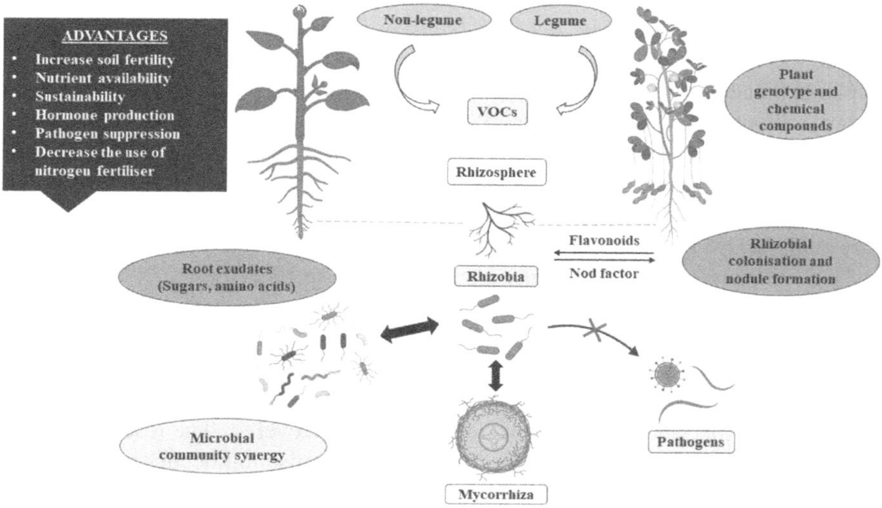

FIGURE 14.1 Overview of the dynamics of rhizosphere microbiome interactions (Modified from Trivedi et al., 2021; Checcucci and Marchetti, 2020)

Understanding of rhizosphere microbiota dynamics has revolutionised due to the advancement in molecular biology and next-generation sequencing (NGS), which uncovers the microbial interactions, networks and their regulatory mechanisms. This understanding helps in sustainable agricultural practices such as soil health management, microbiome engineering, and biofertiliser formulation. This chapter aims to explore the legume rhizosphere microbiota composition, factors influencing the dynamics, their functional roles, and their agronomic application and ecological significance.

14.2 COMPOSITION OF THE LEGUME RHIZOSPHERE MICROBIOTA AND FACTORS INFLUENCING RHIZOSPHERE MICROBIAL DYNAMICS

The rhizosphere microbiota plays a vital role against both biotic and abiotic stress in nutrient acquisition and enhances plant resistance (Wang et al., 2022; Berendsen et al., 2012). These microbes colonise mostly all parts of plants, either as epiphytes or as endophytes (Brown et al., 2020). Additionally, the microbial composition varies from one to another, which includes the phyllosphere, rhizosphere, and endosphere (Wagner et al., 2016). However, previous studies suggested a conserved rhizosphere microbiota structure across various plant hosts, including *Arabidopsis thaliana*, *Medicago truncatula*, *Oryza sativa*, *Populus*, *Hordeum vulgare*, and *Triticum aestivum*. The microbiota enhances nutrient bioavailability through mineral solubilisation and siderophore production. *Pseudomonas*, *Bacillus*, and *Rhizobium* spp. releases phosphatases that solubilise inorganic phosphates and releases some organic acids (Khan et al., 2009). Further, *Acidithiobacillus*, *Thiobacilli,* and *Desulfovibrio* etc. reduce the sulphur compounds to sulphate or sulphide forms which are accessible to plants (Chaudhary et al., 2023). Siderophore producing microbes chelate Fe^{3+} and improve its uptake under deficiency (Adal, 2023). Beneficial rhizobacteria provoke systemic resistance in plants and strengthen immunity. Jasmonic acid (JA) and ethylene (ET) signalling pathways trigger induced systemic resistance (ISR) while systemic acquired resistance (SAR) is mediated by salicylic acid (Pieterse et al., 2014). Roy et al. (2025) used *Bacillus cereus* and *Bacillus safensis* as PGPR to induce ISR in *Lens culinaris* against *Alternaria.*

Zgadzaj et al. (2016) found a mutant *Lotus japonicus* that was defective in nodulation, compared to wild-type plants and detected a shift in microbiome community composition. This finding suggests that core symbiosis genetic pathways can influence interactions with a broader range of microbes beyond the specific symbiosis partners in which they were originally discovered. The researchers have proposed a two-step model for how microbiome colonisation occurs in roots. First, root exudates that comprise common sugars like glucose, sucrose, fructose, secondary metabolites like phenolics, and amino acids like glutamate, tryptophan etc., alter the community composition in the rhizosphere, followed by plant genetic factors that regulate and determine entry inside the root endosphere (Brown et al., 2020). Brown et al. (2020) studied the *M. truncatula* microbiome and demonstrated that both plant genetics and soil origin significantly shape distinct compartments. Surprisingly, *Ensifer* rhizobia (formerly *Sinorhizobium*) were found not just in nodules but also abundant throughout plant tissues and still retained their symbiosis plasmids (pSymA and pSymB), although their role beyond nodules remains uncertain. Several studies have shown that plant host genotypes influence the composition and structural variations of plant-associated microbiota. The symbiotic interaction is highly specific between *Cicer arientinum* genotypes (G_L) and *Mesorhizobium* strains genotypes (G_R). The microbiome in the rhizosphere changes across different developmental stages, and specific bacteria are actively present that promote plant growth during these phases.

Sahil et al. (2025) studied rhizosphere microbial communities at vegetative, reproductive, and maturation stages and used a multiomics meta-analysis approach, combining genetic and genomic data from eight legume species. The species that were chosen are agronomically important legumes. They are *M. truncatula, Phaseolus vulgaris, C. arietinum, Vigna ssubterranea, Glycine max, Arachis hypogaea, Trifolium pratense, and G. soja.* It revealed that the number of active bacterial genes was more abundant at the reproductive stage as compared to the vegetative stage. It recognises 10 key marker bacteria whose presence correlates positively with plant growth traits and is involved in pathways like biotin metabolism, carbon fixation, and production of secondary metabolites. The 10 key marker bacteria identified in the study, each associated with a specific developmental stage of legume growth, includes as: (i) Vegetative stage microbial communities: *Lysobacter soli, L. dokdonensis, Stenotrophomonas rhizophila,* aids in indole-3-acetic acid (IAA) production, phosphate solubilisation, and siderophore production. (ii) Reproductive stage microbial communities: *Dyadobacter fermentans, Dyella japonica, D.*

koreensis, Sphingobium yanoikuyae, Nordella oligomobilis, helpful in biotin synthesis and metabolism of nitrogen and cobalt. (iii) Maturation stage microbial communities: *Pseudoduganella guangdongensis, Nocardioides iriomotensis,* benefits in nutrient cycling.

There are various factors that influence the microbiota composition environmentally and anthropogenically. Optimal pH is critical for survival and function of microbiota. Acidic or alkaline soils hinder growth and nodulation which reduces the biological nitrogen fixation (Huang, 2024). High temperature denatures the enzyme systems, while low temperature slows the metabolism and nodule formation (Zhao et al., 2024). For the soil moisture, both drought and waterlogging negatively affect the viability as it reduces root infection, while other limits impairing of nitrogenase gene and oxygen supply, respectively (De Silva et al., 2025). Rotation of the crops or intercropping enriches the soil nitrogen and maintains a microbial balance and diversity, while continuous monocropping depletes soil nutrients and reduces microbiota activity as well as diversity and increases the pathogenic taxa (Huang, 2024). The conversion of natural vegetation to agricultural land decreases microbial diversity and alters N-and P-cycling (Yang et al., 2025). Genetic factors also play a significant role, and host-strain compatibility is the foremost factor to be considered. It determines nodulation efficiency, and only the genetically compatible strain forms effective nodules (Yang et al., 2022). The agrochemical exerts selective pressure on rhizosphere microbial assemblages. Herbicides such as glyphosate and insecticides disrupt microbial enzymatic activity and reduce the efficiency in biological nitrogen fixation (Kepler et al., 2020). Overuse of NPK fertilisers decreases the diversity from symbiotic to free-living forms (Geisseler and Scow, 2014). Contamination with heavy metal like Cd, Pb, or Cu reduces the survival of microbiota (Jach et al., 2022).

In crop rotation systems, soil microbes play direct and indirect roles in soil structure formation, nutrient cycles like carbon and nitrogen cycles, and energy flow within the ecosystem (Schleuss et al., 2019). You et al. (2024) explained how the common agricultural practice, like legume–vegetable rotation, leads to dynamic shifts in soil microbial communities that regulate nutrient cycling and greenhouse gas (GHG) emissions. During rotation, there is a reduction in soil organic matter (SOM), total phosphorus (TP), and soil organic carbon (SOC), which increases the richness and diversity of soil bacteria, particularly among taxa like Proteobacteria, Chloroflexi, and Gemmatimonadetes, potentially hosting CAZy enzyme genes, including carbohydrate esterases (CE), auxiliary activities (AA), and glycoside hydrolases (GH) etc. Functional genes involved in methane and nitrogen metabolism vary remarkably across rotation practices, reflecting microbial community restructuring. Soybean (*G. max* L.) is a major protein legume and a global oil crop with significant environmental and economic benefits (Peoples et al., 2009). Agyekum et al. (2025) explained the partnership between soybeans, *Rhizobium*, and different soil management practices. The non-nodulating mutant of soybean, En1282, shows no nitrogen-fixing symbiosis but reports having a more diverse bacterial community, which might recruit other helpful microbes, but there is depletion in the number of *Pseudomonas*, which acts as beneficial. These changes show a shift in the dynamics of the rhizosphere and also lead to a lack of symbiosis.

14.3 SYMBIOTIC INTERACTIONS IN LEGUMES AND FUNCTIONAL ROLES OF RHIZOSPHERE MICROBIOTA

Approximately 65 million years ago, some major lineages of Caesalpinioideae and Papilionoid legume plants evolved and gained the ability to establish mutualistic associations with rhizobia, which they still possess today (Zhao et al., 2021). The symbiotic interaction begins with recognition and communication. In the rhizosphere, legume root exudates attract the rhizobia by producing specific types of flavonoids, such as in soybean: isoliquiritigenin, apigenin, prunetin, afrormosin; in *M. sativa*: methoxychalcone; in *P. vulgaris*: genistein, coumestrol; in *T. subterraneum*: 4′,7-dihydroxyflavone; in *Pisum sativum*: pisatin; in *Vicia sativa*: methoxychalcone, etc., thus activating the nodulation (nod) genes in cognate rhizobia (Liu and Murray, 2016) thus determining the host range (Liu and Murray, 2016). As a result, rhizobia synthesise lipo-chitooligosaccharide (LCOs) signalling molecules called Nod factors (Poole et al., 2018). The transcriptional activator NodD activates the expression of various rhizobial *nod* genes (*nodA*, *nodB*, *nodC*) to synthesise LCOs (Peck et al., 2006). On root hairs, these nod factors are recognised by Nod factor receptors (NFR1/NFR5), which ensures host specificity. Once it gets recognised, it triggers root hair deformation and curling. The root hair cell wall formed a tubular structure by invagination, called as infection threads through which rhizobia penetrate. Simultaneously, in the inner cortex, cytokinin signalling stimulates cell division, leading to the nodule primordium, and the infection thread grows towards these cortical cells (Yang et al., 2022; Poole et al., 2018). Inside these rhizobia are released, and differentiate into bacteroid, the active nitrogen-fixing form (Figure 14.2).

Earlier legume nodules were thought to contain only nitrogen-fixing bacteria. New sequencing technologies such as amplicon sequence variant (ASV) shows that non-rhizobia endophytes like *Pseudomonas*, *Bacillus*, *Flavobacterium*, etc., also occur frequently (Yu et al., 2025). Also, their abundance depends on soil microbiota, pH, and host genotype. These bacteria can be retained throughout the nodule lifecycle and influence nitrogen fixation and plant productivity. Yu et al. (2025) highlighted the contribution of *Pseudomonas* in nodule formation using fluorescent tags to track bacterial distribution within nodules and plant health. Most enter nodules with rhizobia, either by crack-entry through natural root openings or hitchhiking via rhizobial infection threads. For example, *Pseudomonas* sp. PLb11B was unable to nodulate in *L. burttii* on its own, or colonise ineffective nodules formed by *Rhizobium* sp. BW8–2, but colonised nodules alongside *Mesorhizobium* sp. Qb1E3–1 intracellularly (Crosbie et al., 2022). The functional roles of rhizosphere microbiota having nodule-associated roles involved the (i) Stress alleviation-Under moderate Cu^{2+} stress, *Pseudomonas* sp. Zy-2–1 from *Sphaerophyssa salsula* nodules alleviated growth reduction in *M. lupulina* (Kong et al., 2017), were found to exhibit aminocyclopropane-1-carboxylate (ACC) deaminase activity (Flores-Duarte et al., 2022). *Pseudomonas* sp. N4 and N8 from *Medicago* nodules tolerated heavy metals *in vitro* and significantly increased heavy metal accumulation in roots

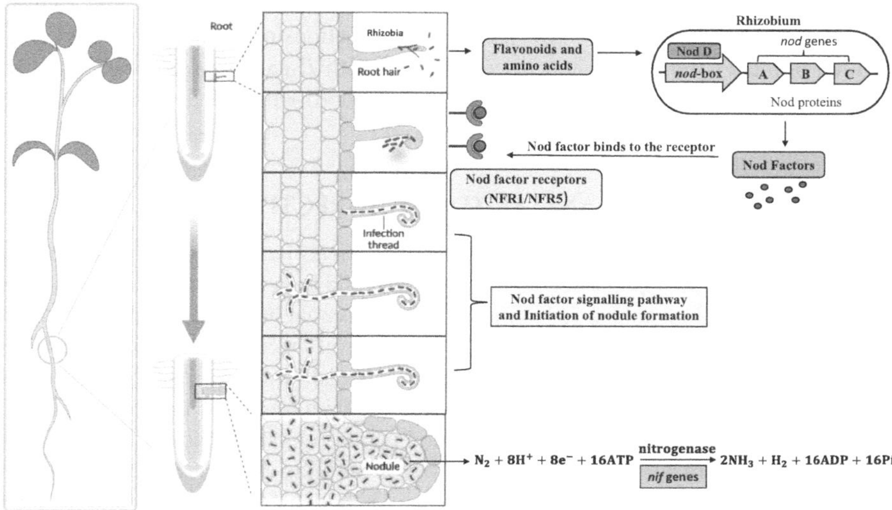

FIGURE 14.2 Schematic overview of the nodulation process and biological nitrogen fixation (Modified from Wang et al., 2018; Laranjo et al., 2014)

(Flores-Duarte et al., 2022). (ii) Phytohormone production-*in vitro* studies show *Pseudomonas* isolates stimulate root and nodule formation due to IAA (Kong et al., 2017). IAA-producing *Pseudomonas* sp. GR12–2 promoted root growth in canola seedlings and adventitious roots on mung bean cuttings, whereas the mutant strain lacking the IAA-producing gene (*ipdc* mutant) did not have these effects (Patten and Glick, 2002). (iii) Insoluble phosphate bioavailable-Phosphate-solubilising *Pseudomonas* have been shown to increase growth in *Arachis, Glycine,* and *Pisum,* etc., to release organic acids that lower pH, making phosphate more accessible to plants and improve nutrient uptake (Kumawat et al., 2019). (iv) Pathogen protection-by producing iron through siderophores (Miftakhov et al., 2022). *Pseudomonas* sp. DD201, isolated from *G. max*, has inhibited the oomycete pathogen *Phytophthora sojae* 01 (Zhao et al., 2018). Similarly, Rubina Noreen et al. (2018) reported that fluorescent *Pseudomonas* strains isolated from *Leucaena leucocephala* nodules suppressed fungal pathogens, including *Macrophomina phaseolina, Rhizoctonia solani, Fusarium solani,* and *F. oxysporum.* (v) Antimicrobial metabolites like proteolytic enzymes (Heywood and Lamont, 2020), and bacteriocins (Suleman et al., 2025). Proteases secreted by *P. protegens* CHA0 inhibited egg hatching and killed the root-knot nematode *Meloidogyne incognita* (Siddiqui et al., 2005). Tailocins, phage tail-like bacteriocins, inhibit competitors of the leaf pathogen *P. viridiflava* (Backman et al., 2024). Species with phenazine-producing genes, isolated from nodules of *L. corniculatus* and *M. sativa*, had antifungal activities (Knezevic et al., 2021). (vi) ISR is triggered by various molecules. *P. protegens* CHA0 produces a cyclic lipopeptide called orfamide that triggers ISR against *Cochliobolus miyabeanus* and reduces leaf lesions in rice (Ma et al., 2017). Using combinations of *P. capeferrum* WCS358 (which produces the siderophore pseudobactin) with *Pseudomonas* sp. RE8 (which can activate ISR but does not produce pseudobactin) provided stronger defence against *F. oxysporum* (de Boer et al., 2003).

14.4 APPLICATIONS AND FUTURE PERSPECTIVES

Microbial consortia are often more stable and efficient than single-strain inoculants, due to complementary plant-beneficial properties and better ecological compatibility. Synergistic microbial consortia having *Sinorhizobium meliloti* (the natural rhizobium symbiont of alfalfa) and various non-rhizobial soil bacteria with plant growth–promoting (PGP) traits (like IAA production, nitrogen fixation, phosphate solubilisation) have longer roots, more secondary roots, maintain native microbial diversity in soil, boost biological nitrogen fixation, and increase biomass yield (Cangioli et al., 2025). The identification of resilient, growth-promoting marker bacteria linked to developmental stages suggests new strategies for enhancing crop improvement and utilising these marker taxa to develop multi-species microbial inoculants, leading to sustainable yield increases in legume agriculture globally (Sahil et al., 2025). Synthetic microbial communities (SynComs) are simplified, controlled mixtures of microorganisms that are both functionally efficient and ecologically stable, and used to study and optimise microbial functions, such as nutrient cycling, pollutant degradation, or plant growth promotion (Wang et al., 2025). Integrating plant genetics with microbial symbiosis genes enables microbiome-assisted breeding and the development of responsive cultivars, which may enhance nutrient acquisition and resilience. Multi-omics integration may help in identifying keystone taxa that can create resilient agroecosystems. The advanced modelling and computational tools help to accelerate discoveries by predicting interactions, metabolic exchanges, and the design of customised SynComs formulations. Integrated data pipelines linked to the microbial function and plant physiology need to be developed and validated under controlled and field conditions. Machine learning and other metabolic modelling can create next-generation bioinoculants or region-specific customised inoculants that are ecologically compatible and resilient. It is expected in the future to establish legume-specific microbial reference databases and promotion of biofertilisers policy frameworks for proper quality control and biosafety that can support large scale adoption in sustainable agriculture.

14.5 CONCLUSION

The legume rhizosphere shows one of the most dynamic interfaces and interactions between plant roots and diverse microbial communities that drive ecological resilience. The symbiotic relationship between legumes and rhizobia has remained central for nitrogen fixation. *Pseudomonas* and other non-rhizobia endophytes open a new dimension and mechanisms to plant health, such as phytohormone production, bioavailability of insoluble phosphate, and pathogen suppression. Recent insights into host–microbe genetic interactions reveal it as a network of biological niches. The origin of the synthetic microbial consortia (SynComs) concept increased the efficiency of the rhizosphere as well as the plant growth. These advancements not only enhance biological nitrogen fixation or improve the soil microbiome but also promote sustainable crop productivity. Decoding the dynamics of the rhizosphere microbiota will be the key to achieve resilient agroecosystems, reducing the use of chemical fertilisers, and establishing a sustainable bio-based agricultural future.

REFERENCES

Adal, Yimam Mekonen. "Role of rhizosphere on the soil fertility and availability nutrient." *Journal of Global Ecology and Environment* 19, no. 3–4 (2023): 22–30. https://doi.org/10.56557/JOGEE/2023/v19i3-48492

Agyekum, Dominic V.A., Khondoker M.G. Dastogeer, and Shin Okazaki. "Deciphering the rhizosphere microbiota composition of nature farming soybean (Glycine max L.) with different nodulation phenotypes." *BMC Plant Biology* 25, no. 1 (2025): 520. https://doi.org/10.1186/s12870-025-06566-y

Backman, Talia, Sergio M. Latorre, Efthymia Symeonidi, Artur Muszyński, Ella Bleak, Lauren Eads, Paulina I. Martinez-Koury, et al. "A phage tail–like bacteriocin suppresses competitors in metapopulations of pathogenic bacteria." *Science* 384, no. 6701 (2024): eado0713. https://doi.org/10.1126/science.ado0713

Berendsen, Roeland L., Corné M.J. Pieterse, and Peter A.H.M. Bakker. "The rhizosphere microbiome and plant health." *Trends in Plant Science* 17, no. 8 (2012): 478–486. https://doi.org/10.1016/j.tplants.2012.04.001

Brown, Shawn P., Michael A. Grillo, Justin C. Podowski, and Katy D. Heath. "Soil origin and plant genotype structure distinct microbiome compartments in the model legume Medicago truncatula." *Microbiome* 8, no. 1 (2020): 139. https://doi.org/10.1186/s40168-020-00915-9

Cangioli, Lisa, Camilla Fagorzi, Francesca Vaccaro, Stefano Varriale, Maria Laura Amenta, Alessio Mengoni, Rosolino Ingraffia et al. "Soil bacteria and symbiotic rhizobia synergistically promote nitrogen fixation and biomass production of alfalfa (Medicago sativa L.) plants even under water shortage conditions without altering the native rhizospheric microbiota." *Applied Soil Ecology* 213 (2025): 106283. https://doi.org/10.1016/j.apsoil.2025.106283

Chaudhary, Suman, Satyavir Singh Sindhu, Rinku Dhanker, and Anju Kumari. "Microbes-mediated sulphur cycling in soil: Impact on soil fertility, crop production and environmental sustainability." *Microbiological Research* 271 (2023): 127340. https://doi.org/10.1016/j.micres.2023.127340

Checcucci, Alice, and Marta Marchetti. "The rhizosphere talk show: The rhizobia on stage." *Frontiers in Agronomy* 2 (2020): 591494. https://doi.org/10.3389/fagro.2020.591494

Crosbie, Duncan B., Maryam Mahmoudi, Viviane Radl, Andreas Brachmann, Michael Schloter, Eric Kemen, and Macarena Marín. "Microbiome profiling reveals that Pseudomonas antagonises parasitic nodule colonisation of cheater rhizobia in Lotus." *New Phytologist* 234, no. 1 (2022): 242–255. https://doi.org/10.1111/nph.17988

de Boer, Marjan, Peter Bom, Frodo Kindt, Joost J.B. Keurentjes, Ientse van der Sluis, L.C. Van Loon, and Peter A.H.M. Bakker. "Control of Fusarium wilt of radish by combining Pseudomonas putida strains that have different disease-suppressive mechanisms." *Phytopathology* 93, no. 5 (2003): 626–632. https://doi.org/10.1094/PHYTO.2003.93.5.626

De Silva, Sujani, Lithma Kariyawasam Hetti Gamage, and Vesh R. Thapa. "Impact of drought on soil microbial communities." *Microorganisms* 13, no. 7 (2025): 1625. https://doi.org/10.3390/microorganisms13071625

Flores-Duarte, Noris J., Sara Caballero-Delgado, Eloisa Pajuelo, Enrique Mateos-Naranjo, Susana Redondo-Gómez, Salvadora Navarro-Torre, and Ignacio D. Rodríguez-Llorente. "Enhanced legume growth and adaptation to degraded estuarine soils using Pseudomonas sp. nodule endophytes." *Frontiers in Microbiology* 13 (2022): 1005458. https://doi.org/10.3389/fmicb.2022.1005458

Geisseler, Daniel, and Kate M. Scow. "Long-term effects of mineral fertilizers on soil microorganisms: A review." *Soil Biology and Biochemistry* 75 (2014): 54–63. https://doi.org/10.1016/j.soilbio.2014.03.023

Grundy, Estelle B., Peter M. Gresshoff, Huanan Su, and Brett J. Ferguson. "Legumes regulate symbiosis with rhizobia via their innate immune system." *International Journal of Molecular Sciences* 24, no. 3 (2023): 2800. https://doi.org/10.3390/ijms24032800

Heywood, Astra, and Iain L. Lamont. "Cell envelope proteases and peptidases of Pseudomonas aeruginosa: Multiple roles, multiple mechanisms." *FEMS Microbiology Reviews* 44, no. 6 (2020): 857–873. https://doi.org/10.1093/femsre/fuaa036

Huang, Wenzhong. "Boosting soil health: The role of rhizobium in legume nitrogen fixation." *Molecular Soil Biology* 15 (2024). https://doi.org/10.5376/msb.2024.15.0014

Jach, Monika Elżbieta, Ewa Sajnaga, and Maria Ziaja. "Utilization of legume-nodule bacterial symbiosis in phytoremediation of heavy metal-contaminated soils." *Biology* 11, no. 5 (2022): 676. https://doi.org/10.3390/biology11050676

Kepler, Ryan M., Dietrich J. Epp Schmidt, Stephanie A. Yarwood, Michel A. Cavigelli, Krishna N. Reddy, Stephen O. Duke, Carl A. Bradley, Martin M. Williams Jr, Jeffrey S. Buyer, and Jude E. Maul. "Soil microbial communities in diverse agroecosystems exposed to the herbicide glyphosate." *Applied and Environmental Microbiology* 86, no. 5 (2020): e01744–19. https://doi.org/10.1128/AEM.01744-19

Khan, Mohammad Saghir, Almas Zaidi, and Parvaze A. Wani. "Role of phosphate solubilizing microorganisms in sustainable agriculture: A review." *Sustainable Agriculture* (2009): 551–570. https://doi.org/10.1007/978-90-481-2666-8_34

Knezevic, Magdalena, Tanja Berić, Aneta Buntić, Dušica Delić, Ivan Nikolić, Slaviša Stanković, and Olivera Stajković-Srbinović. "Potential of root nodule nonrhizobial endophytic bacteria for growth promotion of Lotus corniculatus L. and Dactylis glomerata L." *Journal of Applied Microbiology* 131, no. 6 (2021): 2929–2940. https://doi.org/10.1111/jam.15152

Kong, Zhaoyu, Zhenshan Deng, Bernard R. Glick, Gehong Wei, and Minxia Chou. "A nodule endophytic plant growth-promoting Pseudomonas and its effects on growth, nodulation and metal uptake in Medicago lupulina under copper stress." *Annals of Microbiology* 67, no. 1 (2017): 49–58. https://doi.org/10.1007/s13213-016-1235-1

Kumawat, K.C., Poonam Sharma, Asmita Sirari, Inderjeet Singh, B.S. Gill, Ummed Singh, and Krishna Saharan. "Synergism of Pseudomonas aeruginosa (LSE-2) nodule endophyte with Bradyrhizobium sp.(LSBR-3) for improving plant growth, nutrient acquisition and soil health in soybean." *World Journal of Microbiology and Biotechnology* 35, no. 3 (2019): 47. https://doi.org/10.1007/s11274-019-2622-0

Laranjo, Marta, Ana Alexandre, and Solange Oliveira. "Legume growth-promoting rhizobia: An overview on the Mesorhizobium genus." *Microbiological Research* 169, no. 1 (2014): 2–17. https://doi.org/10.1016/j.micres.2013.09.012

Liu, Cheng-Wu, and Jeremy D. Murray. "The role of flavonoids in nodulation host-range specificity: An update." *Plants* 5, no. 3 (2016): 33. https://doi.org/10.3390/plants5030033

Ma, Zongwang, Marc Ongena, and Monica Höfte. "The cyclic lipopeptide orfamide induces systemic resistance in rice to Cochliobolus miyabeanus but not to Magnaporthe oryzae." *Plant Cell Reports* 36, no. 11 (2017): 1731–1746. https://doi.org/10.1007/s00299-017-2187-z

Mendes, Rodrigo, Paolina Garbeva, and Jos M. Raaijmakers. "The rhizosphere microbiome: Significance of plant beneficial, plant pathogenic, and human pathogenic microorganisms." *FEMS Microbiology Reviews* 37, no. 5 (2013): 634–663. https://doi.org/10.1111/1574-6976.12028

Miftakhov, Aynur Kamilevich, Roderic Gilles Claret Diabankana, Mikhail Frolov, Marat Miratovich Yusupov, Shamil Zavdatovich Validov, and Daniel Mawuena Afordoanyi. "Persistence as a constituent of a biocontrol mechanism (Competition for nutrients and niches) in Pseudomonas putida PCL1760." *Microorganisms* 11, no. 1 (2022): 19. https://doi.org/10.3390/microorganisms11010019

Patten, Cheryl L., and Bernard R. Glick. "Role of Pseudomonas putida indoleacetic acid in development of the host plant root system." *Applied and Environmental Microbiology* 68, no. 8 (2002): 3795–3801. https://doi.org/10.1128/AEM.68.8.3795-3801.2002

Peck, Melicent C., Robert F. Fisher, and Sharon R. Long. "Diverse flavonoids stimulate NodD1 binding to nod gene promoters in Sinorhizobium meliloti." *Journal of Bacteriology* 188, no. 15 (2006): 5417–5427. https://doi.org/10.1128/jb.00376-06

Peoples, M.B., J. Brockwell, D.F. Herridge, I.J. Rochester, B.J.R. Alves, S. Urquiaga, R.M. Boddey et al. "The contributions of nitrogen-fixing crop legumes to the productivity of agricultural systems." *Symbiosis* 48, no. 1 (2009): 1–17. https://doi.org/10.1007/BF03179980

Pieterse, Corné M.J., Christos Zamioudis, Roeland L. Berendsen, David M. Weller, Saskia C.M. Van Wees, and Peter A.H.M. Bakker. "Induced systemic resistance by beneficial microbes." *Annual Review of Phytopathology* 52, no. 1 (2014): 347–375. https://doi.org/10.1146/annurev-phyto-082712-102340

Poole, Philip, Vinoy Ramachandran, and Jason Terpolilli. "Rhizobia: From saprophytes to endosymbionts." *Nature Reviews Microbiology* 16, no. 5 (2018): 291–303. https://doi.org/10.1038/nrmicro.2017.171

Roy, Tina, Anuradha Bandopadhyay, Sukanta Majumdar, Shariful Alam, and Nirmalendu Das. "Induced systemic resistance-mediated defense against alternaria blight disease in lentil by pesticide degrading plant growth-promoting rhizobacteria." *Current Microbiology* 82, no. 3 (2025): 109. https://doi.org/10.1007/s00284-025-04083-7

Rubina Noreen, Habiba, Afshan Rahman, Hafiza Asma Shafique, Viqar Sultana, and Syed Ehteshamul-Haque. "Biocontrol and plant growth promoting potential of endo-nodule fluorescent Pseudomonas and rhizobia associated with root nodules of Leucaena leucocephala." *International Journal of Biological Research* 6, no. 1 (2018): 15–26.

Ruttelib, Paul J., and Philip S. Poole. "Oxygen regulatory mechanisms of nitrogen fixation in rhizobia." *Advances in Microbial Physiology* 75 (2019): 325–389. https://doi.org/10.1016/bs.ampbs.2019.08.001

Sahil, Rishav, Vivek Pal, Arun S. Kharat, and Mukesh Jain. "A multi-omics meta-analysis of rhizosphere microbiome reveals growth-promoting marker bacteria at different stages of legume development." *Plant, Cell & Environment* (2025). https://doi.org/10.1111/pce.15429

Schleuss, Per-Marten, Meike Widdig, Anna Heintz-Buschart, Alexander Guhr, Sarah Martin, Kevin Kirkman, and Marie Spohn. "Stoichiometric controls of soil carbon and nitrogen cycling after long-term nitrogen and phosphorus addition in a mesic grassland in South Africa." *Soil Biology and Biochemistry* 135 (2019): 294–303. https://doi.org/10.1016/J.SOILBIO.2019.05.018

Sharma, Neetu, Rakesh Kumar, A.P. Singh, Rohit Sharma, Peeyush Sharma, Joy Samuel Mecarty, and Faraaz Farooq. "Legumes in cropping system for soil ecosystem improvement: A review." *Legume Research: An International Journal* 48, no. 1 (2025). https://doi.org/10.18805/LR-5289

Siddiqui, Imran Ali, Dieter Haas, and Stephan Heeb. "Extracellular protease of Pseudomonas fluorescens CHA0, a biocontrol factor with activity against the root-knot nematode Meloidogyne incognita." *Applied and Environmental Microbiology* 71, no. 9 (2005): 5646–5649. https://doi.org/10.1128/AEM.71.9.5646-5649.2005

Suleman, Muhammad, Allah Rakha Yaseen, Shahbaz Ahmed, Zoha Khan, Asma Irshad, Afsah Pervaiz, Hafiza Hiba Rahman, and Muteeba Azhar. "Pyocins and beyond: Exploring the world of bacteriocins in Pseudomonas aeruginosa." *Probiotics and Antimicrobial Proteins* 17, no. 1 (2025): 240–252. https://doi.org/10.1007/s12602-024-10322-3

Trivedi, Pankaj, Chakradhar Mattupalli, Kellye Eversole, and Jan E. Leach. "Enabling sustainable agriculture through understanding and enhancement of microbiomes." *New Phytologist* 230, no. 6 (2021): 2129–2147. https://doi.org/10.1111/nph.17319

Wagner, Maggie R., Derek S. Lundberg, Tijana G. Del Rio, Susannah G. Tringe, Jeffery L. Dangl, and Thomas Mitchell-Olds. "Host genotype and age shape the leaf and root microbiomes of a wild perennial plant." *Nature Communications* 7, no. 1 (2016): 12151. https://doi.org/10.1038/ncomms12151

Wang, Liyang, Zed Rengel, Kai Zhang, Kemo Jin, Yang Lyu, Lin Zhang, Lingyun Cheng, Fusuo Zhang, and Jianbo Shen. "Ensuring future food security and resource sustainability: Insights into the rhizosphere." *Iscience* 25, no. 4 (2022). https://doi.org/10.1016/j.isci.2022.104168

Wang, Qi, Jinge Liu, and Hongyan Zhu. "Genetic and molecular mechanisms underlying symbiotic specificity in legume-rhizobium interactions." *Frontiers in Plant Science* 9 (2018): 313. https://doi.org/10.3389/fpls.2018.00313

Wang, Xiaolin, Huan Feng, Yayu Wang, Mingxing Wang, Xingguang Xie, Huizhong Chang, Like Wang et al. "Mycorrhizal symbiosis modulates the rhizosphere microbiota to promote rhizobia–legume symbiosis." *Molecular Plant* 14, no. 3 (2021): 503–516. https://doi.org/10.1016/j.molp.2020.12.002

Wang, Zhihan, Shang Wang, Qing He, Xingsheng Yang, Bo Zhao, Haihan Zhang, and Ye Deng. "Ecological design of high-performance synthetic microbial communities: From theoretical foundations to functional optimization." *ISME Communications* 5, no. 1 (2025): ycaf133. https://doi.org/10.1093/ismeco/ycaf133

Yang, Haixin, Yuan Li, Jiasheng Li, Xinyue Zhao, KeJia Zhu, Wanrou Zhang, Wanqin Zhao et al. "Land-use changes alter the soil microbial community structure and key C, N, and S metabolic functional potentials in the Yuncheng Salt Lake wetland." *Catena* 260 (2025): 109479. https://doi.org/10.1016/j.catena.2025.109479

Yang, Jun, Liying Lan, Yue Jin, Nan Yu, Dong Wang, and Ertao Wang. "Mechanisms underlying legume–rhizobium symbioses." *Journal of Integrative Plant Biology* 64, no. 2 (2022): 244–267. https://doi.org/10.1111/jipb.13207

You, Xinxin, Sheng Wang, Linna Du, Yurong Chen, Ting Wang, and Xiaoxu Bo. "Metagenomics reveals the variations in functional metabolism associated with greenhouse gas emissions during legume-vegetable rotation process." *Ecotoxicology and Environmental Safety* 275 (2024): 116268. https://doi.org/10.1016/j.ecoenv.2024.116268

Yu, Yu-Hsiang, Duncan B. Crosbie, and Macarena Marín Arancibia. "Pseudomonas in the spotlight: Emerging roles in the nodule microbiome." *Trends in Plant Science* 30, no. 5 (2025): 461–470. https://doi.org/10.1016/j.tplants.2024.12.002

Zgadzaj, Rafal, Ruben Garrido-Oter, Dorthe Bodker Jensen, Anna Koprivova, Paul Schulze-Lefert, and Simona Radutoiu. "Root nodule symbiosis in Lotus japonicus drives the establishment of distinctive rhizosphere, root, and nodule bacterial communities." *Proceedings of the National Academy of Sciences* 113, no. 49 (2016): E7996–E8005. https://doi.org/10.1073/pnas.1616564113

Zhao, Jiayi, Xuan Xie, Yuying Jiang, Jiaxin Li, Qi Fu, Yingbo Qiu, Xianheng Fu et al. "Effects of simulated warming on soil microbial community diversity and composition across diverse ecosystems." *Science of the Total Environment* 911 (2024): 168793. https://doi.org/10.1016/j.scitotenv.2023.168793

Zhao, Long Fei, Ya Jun Xu, and Xin He Lai. "Antagonistic endophytic bacteria associated with nodules of soybean (Glycine max L.) and plant growth-promoting properties." *Brazilian Journal of Microbiology* 49, no. 2 (2018): 269–278. https://doi.org/10.1016/j.bjm.2017.06.007

Zhao, Yiyong, Rong Zhang, Kai-Wen Jiang, Ji Qi, Yi Hu, Jing Guo, Renbin Zhu et al. "Nuclear phylotranscriptomics and phylogenomics support numerous polyploidization events and hypotheses for the evolution of rhizobial nitrogen-fixing symbiosis in Fabaceae." *Molecular plant* 14, no. 5 (2021): 748–773. https://doi.org/10.1016/j.molp.2021.02.006

15 Challenges and Emerging Relevance of Metagenomics for Sustainable Legume Cultivation

Sangeeta Singh, Parichita Priyadarshini, and Anurag Batham

15.1 INTRODUCTION

Sustainable agriculture represents one of humanity's most pressing challenges: how to feed a growing global population while mitigating environmental degradation. In this context, legumes hold a unique and crucial position. They are foundational elements in both human nutrition (as a source of protein) and ecological health, primarily due to their remarkable ability to establish a symbiotic relationship with nitrogen-fixing bacteria (rhizobia), thereby reducing the need for synthetic nitrogen fertilisers whose production is energy-intensive and environmentally damaging (van der Heijden et al., 2008). The productivity and resilience of these vital crops are intrinsically linked not just to plant genetics but, more critically, to the unseen microbial communities residing in and around their roots, known as the rhizosphere. (Bulgarelli et al., 2015; Mendes et al., 2013). Traditionally, understanding these complex soil ecosystems has been limited by the constraint that over 99% of soil microbes are non-culturable in a laboratory setting. This "great plate anomaly" created a significant knowledge gap, hindering the development of microbiome-based solutions for agricultural improvement. The advent of metagenomics the study of genetic material recovered directly from environmental samples has fundamentally altered this landscape. This powerful approach bypasses the need for laboratory cultivation, allowing scientists to survey the collective genomes of entire microbial communities simultaneously. By sequencing the total DNA from a handful of soil, researchers can gain unprecedented insights into microbial diversity, community structure, and functional potential. This technological leap has profound implications for sustainable legume cultivation. Metagenomics offers the potential to move beyond the application of broad-spectrum chemical inputs towards precision microbiome management. It allows for the identification of novel genes responsible for enhanced nutrient acquisition, improved stress tolerance (to drought, salinity, and pests), and robust plant defence mechanisms

 DOI: 10.1201/9781003733324-16

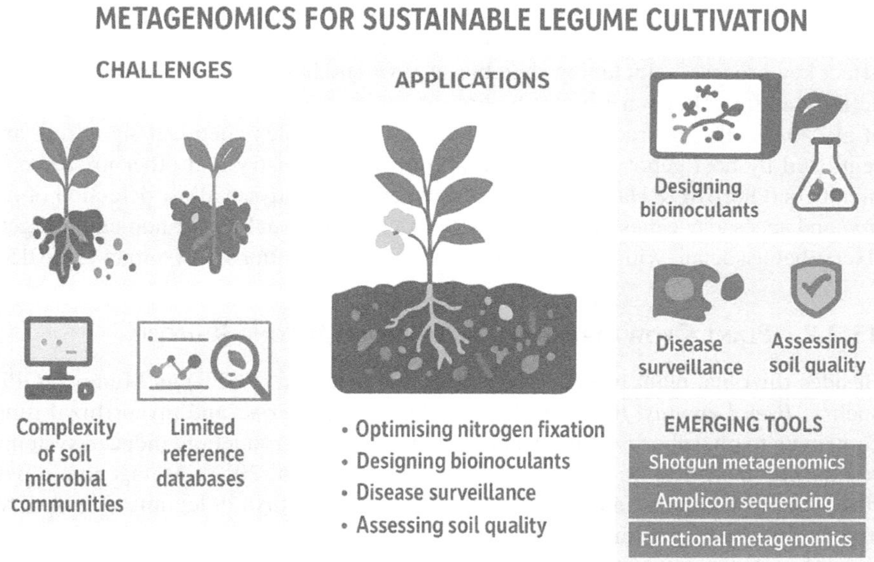

FIGURE 15.1 Metagenomics for sustainable legume cultivation

that support legume growth in marginal soils. However, the path from raw data to actionable agricultural practices is fraught with its own set of significant hurdles. The challenges inherent in metagenomics are substantial, encompassing the massive computational power required to analyse terabytes of sequence data, the complexity of assigning function to millions of novel genes (the "dark matter" of the microbial world), and the difficulty of translating intricate ecological data into practical, scalable farming solutions (Figure 15.1). This introduction will explore the dual narrative of metagenomics in legume cultivation: highlighting the revolutionary potential and emerging relevance of this technology for engineering a more sustainable agricultural future, while simultaneously addressing the critical technical, bioinformatic, and translational challenges that must be overcome to fully realise its promise.

15.2 BACKGROUND: WHY THE MICROBIOME MATTERS FOR LEGUMES

The microbiome is critical for legumes because it allows them to source their own nitrogen, resist environmental stress, and maximize nutrient uptake. Unlike most plants, legumes form an advanced, cooperative relationship with specific soil microbes to thrive in poor soils.

15.2.1 Biological Nitrogen Fixation and Symbiosis

The capacity of legumes to establish symbiotic relationships with rhizobia bacteria enables biological nitrogen fixation (BNF), reducing the dependence on chemical fertilisers while enhancing soil health. However, the efficiency of this symbiosis is significantly influenced by environmental factors, such as soil acidity, salinity, temperature, moisture content, light intensity, and nutrient availability. These factors affect key processes, including rhizobia survival, nodule formation, and nitrogenase activity, ultimately determining the growth and productivity of legumes (Yeremko et al., 2025). The interaction is also mediated by complex chemical signalling and regulated by host genotype, rhizobial strain, soil chemistry and other microbiome members (Oldroyd & Harrison, 2011). Metagenomic analyses allow detection of *nif*, *nod* and accessory genes across diverse taxa, and can reveal non-canonical nitrogen-fixers that associate with legumes under particular conditions (Peixoto et al., 2015).

15.2.2 Plant Growth–Promoting Microbes Beyond Rhizobia

Besides rhizobia, plant growth–promoting rhizobacteria (PGPR) and fungi (PGPF) such as *Pseudomonas*, *Bacillus*, *Azospirillum*, *Trichoderma* and mycorrhizal fungi contribute to phosphorus solubilisation, phytohormone production, induced systemic resistance, and stress mitigation (Bhattacharyya & Jha, 2012). Metagenomic data reveal the distribution and genetic capabilities of these taxa in legume rhizospheres and nodules, which is crucial for devising effective inoculant strategies. Recently, there has been interest in plant growth–promoting microorganisms (PGPMs), which seem to be a promising alternative for managing disease and improving soil health. By solubilising nutrients, enhancing soil structure, boosting water intake, fixing atmospheric nitrogen, and generating phytohormones, PGPMs promote plant growth. Additionally, a number of PGPMs exhibit biocontrol activity through competition, antagonism, or systemic resistance induction in host plants. Plant growth–promoting rhizobacteria (PGPR) and plant growth–promoting fungi (PGPF), which include a variety of *Trichoderma*, *Gliocladium*, and *Chaetomium* species as well as arbuscular mycorrhizal fungi (AMF), are the two groups of PGPMs (Marwa, 2025)

15.2.3 Soil Health, Nutrient Cycling, and Disease Suppression

The immediate area of soil that surrounds plant roots is known as the rhizosphere, and it is home to a sizeable population of microorganisms and animals, making its environment extremely complex. Root exudates from huge associations of soil microorganisms are used by plants to attract certain soil microorganisms. The first stage in protecting plants from soil-borne pathogenic organisms is the successful development of rhizospheric bacteria that promote plant growth. Microbial communities drive decomposition, carbon cycling, nutrient mineralisation, and antagonism against soil pathogens. Comprehending the distinct and dynamic ecology of the rhizosphere will facilitate better nutrient absorption, water use efficiency, and soil property modification, all of which will improve plant development, production, and the integration of disease management techniques (Jain et al., 2020). Understanding the gene repertoires that mediate these services for example, enzymes involved in organic matter turnover, siderophore biosynthesis, and secondary metabolite production is possible with functional metagenomics and comparative community profiling (Philippot et al., 2013).

15.3 METAGENOMIC APPROACHES: METHODS, STRENGTHS, AND LIMITATIONS

15.3.1 Sampling Design and Experimental Considerations

High-quality metagenomic inference begins with careful sampling design. Important considerations include: spatial replication (across plants and plots), temporal sampling (to capture dynamics), fractionation (rhizosphere versus bulk soil versus endosphere versus nodules), and controls (unplanted soil, non-legume reference). Soil physicochemical metadata (pH, moisture, organic C, nutrient profiles) must be recorded to interpret functional gene distributions.

15.3.2 Sequencing Strategies

The genetic diversity of the uncultured component of host-associated microbial communities can now be possible due to rapid development in DNA sequencing and biocomputing services. For instance, the most used technique for describing the variety of microbiota is amplicon sequencing. An alternate method for studying uncultured microbiota that gets beyond the drawbacks of amplicon sequencing is shotgun metagenomic sequencing (Sharpton, 2014)

Two common metagenomic strategies are used:

- **Amplicon sequencing (16S rRNA, ITS):** cost-effective for taxonomic surveys and community comparisons (e.g. QIIME2 pipelines), but limited for functional inference and resolution below species/strain levels (Table 15.1).
- **Shotgun metagenomics:** sequences total DNA enabling taxonomic profiling, gene content analysis, and recovery of metagenome-assembled genomes (MAGs). This approach is more informative but requires higher sequencing depth and computational resources (Quince et al., 2017).

15.3.3 Assembly, Binning, and MAG Recovery

Assembly and binning tools (e.g. MEGAHIT, MetaSPAdes, MetaBAT, MaxBin) enable reconstruction of MAGs that approximate genomes of uncultured taxa. MAGs

TABLE 15.1
Summary of Different Kinds of Bioinformatics Tools Used for Metagenomics

Step	Purpose	Tools/Methods	Output
Assembly	Build contigs from short reads	MEGAHIT, MetaSPAdes	Contigs
Binning	Cluster contigs into genome bins	MetaBAT2, CONCOCT, MaxBin	Draft genome bins
MAG Recovery	Refine bins into near-complete genomes	CheckM, BinSPreader	MAGs (Metagenome-Assembled Genomes)

allow gene context, plasmid detection, and metabolic reconstruction. However, assembly is hampered by microdiversity, low-abundance genomes, and repetitive elements typical of soil communities.

15.3.4 Functional Annotation and Pathway Inference

In metagenomic analysis, functional annotation and pathway inference are essential processes that enable researchers to go from unprocessed sequencing data to biological interpretation. Predicting the functions of genes found in metagenomic contigs or reads is known as functional annotation, and it usually utilises homology searches against carefully selected databases such as KEGG, COG, Pfam, or eggNOG. In order to find conserved protein families, this stage may also involve domain and motif analysis. More recently, machine learning techniques such as DeepFRI have been used to predict functions for new genes that have minimal resemblance to known references. Using tools such as KEGG or MetaCyc, route inference maps identified genes to metabolic and regulatory pathways, integrating these activities into higher-order biological processes. Because microbial communities are made up of several species, pathways must be reconstructed at the community level rather than for a single genome, making pathway inference in metagenomics more difficult. This procedure is automated by tools such as HUMAnN, MEGAN, MG-RAST, and MetaErg, allowing routes and functional profiles to be visualised. By connecting genetics to phenotype and exposing the aggregate metabolic capacity of microbial communities, these methods collectively offer insights into microbial ecology, human health, biotechnology, and environmental monitoring (De Filippo et al., 2012; Dong & Strous, 2019; Maranga et al., 2023). Limitations arise from large fractions of uncharacterised proteins and limited representation of soil microbes in reference databases; many sequences are annotated as "hypothetical proteins," reducing interpretability.

15.3.5 Functional Metagenomics and Activity-Based Screening

Functional metagenomics constructs expression libraries of environmental DNA screened in heterologous hosts for desired activities (e.g. cellulases, phosphatases, salt-tolerant enzymes). This approach can discover novel genes not detectable by homology alone, but requires suitable expression hosts and screening platforms. Functional metagenomics combined with activity-based screening provides a powerful strategy to discover novel genes and enzymes directly from environmental samples without the need for cultivation. In this approach, DNA extracted from microbial communities is cloned into suitable vectors and expressed in a host organism, allowing researchers to identify genes based on their expressed activity rather than sequence similarity alone. Activity-based screening then detects functional traits – such as substrate degradation, antibiotic resistance, or metabolite production – using assays like colorimetric, fluorometric, or selective growth tests. This direct link between genotype and phenotype enables the identification of previously unknown enzymes and bioactive compounds, expanding our understanding of microbial diversity and offering applications in biotechnology, medicine, and ecology (Handelsman, 2004; Uchiyama & Miyazaki, 2009).

TABLE 15.2 Emerging Tools and Techniques

Approach	Purpose	Example Tools
Shotgun Metagenomics	Community composition & function	MG-RAST, MetaPhlAn, HUMAnN
Amplicon Sequencing (16S/ITS)	Taxonomic profiling	QIIME2, Mothur
Metagenome-Assembled Genomes (MAGs)	Genome recovery from complex samples	MetaBAT, MaxBin, CheckM
Functional Metagenomics	Discovery of novel genes/ enzymes	pCC1FOS libraries, functional screening
Integrative Omics	Linking microbial function to phenotype	KEGG, STRING, Cytoscape
AI-Based Predictive Modelling	Predicting microbiome–plant interactions	Random forest, deep learning models

15.3.6 Integrating Metatranscriptomics, Metaproteomics, and Metabolomics

Metatranscriptomics (RNA), metaproteomics, and metabolomics provide activity and phenotype layers, indicating which genes are expressed and which metabolites are produced in situ. Integrative multi-omics increases confidence in functional assignments and helps move from potential to realised function (Franzosa et al., 2015).

15.4 APPLICATIONS IN SUSTAINABLE LEGUME CULTIVATION

15.4.1 Optimising Biological Nitrogen Fixation

Nitrogen fixation enhancement in sustainable legume cultivation is a major application of metagenomics, as it enables the identification and optimisation of microbial symbionts that improve plant productivity while reducing dependence on synthetic fertilisers. Legumes form symbiotic associations with rhizobia, which convert atmospheric nitrogen (N_2) into ammonia through the nitrogenase enzyme complex, supplying plants with bioavailable nitrogen. Metagenomic approaches allow researchers to profile the diversity of rhizobial strains in soils and root nodules, uncovering novel nitrogen-fixing bacteria beyond the well-studied *Rhizobium* and *Bradyrhizobium*. Functional annotation of metagenomic data highlights key genes involved in nodulation (e.g. *nod*, *nif*, *fix* genes) and regulatory pathways that control nitrogen fixation efficiency. Pathway inference further reveals how microbial communities collectively contribute to nitrogen cycling, including interactions with phosphate-solubilisers and mycorrhizal fungi that enhance nutrient uptake. Activity-based screening of metagenomic libraries has also led to the discovery of enzymes and regulatory proteins that improve nitrogenase activity under stress conditions such as drought or salinity. These insights are being applied to design biofertilizers and microbial consortia tailored to specific legume crops (soybean, chickpea, lentil), improving yields while minimising environmental impacts of chemical fertilisers. Metagenomics profiles can identify native rhizobial strains with superior *nif/nod* gene complements in local soils, guide strain selection for inoculants, and reveal accessory microbiome members that enhance nodulation efficiency. For marginal soils, metagenomic discovery of alternative diazotrophs provides new options when classic rhizobia perform poorly. Thus, metagenomics-driven nitrogen fixation enhancement represents a cornerstone of climate-smart and sustainable agriculture (Nwachukwu & Babalola, 2022; Mus et al., 2016; Hungria & Mendes, 2015).

15.4.2 Designing Bioinoculants and Microbial Consortia

By combining taxonomic abundance data, MAGs, and functional gene presence/absence, researchers can rationally design multispecies consortia that supply complementary services (nitrogen fixation, phosphate solubilisation, drought tolerance). Field-relevant consortia should be derived from local soils to ensure adaptation and persistence (Compant et al., 2019). Designing bioinoculants and microbial consortia is an advanced strategy in sustainable agriculture that leverages metagenomic insights to harness beneficial microbes for improved crop productivity and resilience. Bioinoculants are formulations containing selected microorganisms such as rhizobia, mycorrhizae, phosphate-solubilising bacteria, and plant growth–promoting rhizobacteria (PGPR), which are applied to seeds, soil, or plants to enhance nutrient uptake, nodulation, and stress tolerance. Moving beyond single-strain inoculants, microbial consortia combine multiple compatible microbes with complementary functions for example, nitrogen-fixing rhizobia paired with phosphate-solubilisers and biocontrol agents such as *Trichoderma* or *Pseudomonas* to deliver synergistic benefits. The design process involves microbiome profiling through metagenomics, functional annotation of microbial genes (e.g. *nif* for nitrogen fixation, *pho* for phosphorus metabolism), and careful selection of strains based on ecological compatibility, host specificity, and resilience to abiotic stresses such as drought or salinity. These consortia are then tested for stability and efficacy under laboratory and field conditions, ensuring they improve yield while reducing reliance on chemical fertilisers and pesticides. By integrating ecological principles with genomic data, bioinoculants and microbial consortia provide tailored, climate-smart solutions for crops such as soybean, chickpea, and lentil, contributing to sustainable legume cultivation and long-term soil health (Nwachukwu & Babalola, 2022; Vessey, 2003).

15.4.3 Disease Surveillance and Biocontrol

Metagenomics reveals shifts in microbial community composition preceding disease outbreaks and can identify biocontrol candidate taxa and antimicrobial biosynthetic gene clusters. Early detection of pathogen DNA and enrichment of antagonists offers a path towards microbiome-based disease management. This provides real-time epidemiological insights, allowing proactive interventions before outbreaks occur. Biocontrol, on the other hand, harnesses beneficial microbes to suppress plant pathogens through mechanisms such as competition for nutrients, antibiosis via antimicrobial compound production, parasitism of harmful fungi, and induction of systemic resistance in plants(Villavicencio-Vásquez et al., 2025).

15.4.4 Soil Quality Assessment and Carbon Sequestration

Functional gene profiles associated with carbon decomposition and stabilisation inform practices to increase soil organic carbon while maintaining nutrient cycling. Metagenomic indicators can serve as biological metrics for soil health monitoring under legume-based rotations. Together, soil quality assessment provides the baseline for identifying soils with high sequestration potential, while carbon sequestration transforms soils into climate allies, improving fertility, restoring degraded lands, and reducing greenhouse gas emissions(Lal, 2004).

15.4.5 Microbiome-Assisted Breeding

Emerging evidence suggests that plant genotypes can shape root microbiomes; incorporating microbiome profiles as phenotypes in breeding programs (selecting for genotypes that recruit beneficial microbes) may improve crop resilience (Berg & Smalla, 2009).

15.5 CASE STUDIES

Note: the following case studies synthesise findings reported in the literature to illustrate metagenomic applications across legumes. Citations link to representative primary studies and reviews.

15.5.1 Chickpea (*Cicer Arietinum*) Rhizosphere under Drought Stress

Several metagenomic and metatranscriptomic studies of chickpea rhizospheres under water limitation consistently report enrichment of Actinobacteria, Firmicutes, and genes for osmolyte biosynthesis and reactive oxygen species (ROS) detoxification (Vurukonda et al., 2018). Functional signatures include trehalose synthesis, betaine transporters, and compatible solute pathways that likely contribute to rhizosphere stability during drought. These observations have guided trials of drought-tolerant microbial consortia that combine desiccation-tolerant Bacillales with phosphate-solubilisers, showing yield maintenance in semi-arid trials.

15.5.2 Soybean (*Glycine Max*): Wild versus Cultivated Microbiomes

Comparative metagenomics between wild relatives and cultivated soybean varieties revealed that domestication and modern breeding can reduce root microbiome diversity and the abundance of beneficial diazotrophs, including certain Bradyrhizobium lineages and non-rhizobial nitrogen fixers (Bulgarelli et al., 2015; Peiffer et al., 2013). This "microbiome erosion" suggests unexploited microbial resources in wild gene pools; integrating wild-derived microbiome traits into breeding or inoculation programs may recover lost functions.

15.5.3 Groundnut (*Arachis Hypogaea*): Functional Metagenomics for Phosphate Solubilisation

Functional screening of groundnut rhizosphere metagenomic libraries has yielded novel phosphate-solubilising enzymes and siderophore biosynthesis genes. These enzymes, when tested in greenhouse inoculation trials, improved phosphorus uptake and early vigour in phosphorus-limited soils a key application for smallholder agriculture where P fertiliser access is limited (Bhattacharyya & Jha, 2012).

15.5.4 Common Bean (*Phaseolus Vulgaris*): Nodule Microbiome Complexity

Shotgun metagenomes of common bean nodules demonstrate that nodules harbour more than rhizobia, other bacteria (and occasionally fungi) are present, some carrying genes linked to stress tolerance and secondary metabolism. Understanding these nodule-associated communities opens opportunities to co-select nodule co-inhabitants that support BNF under stress (Oldroyd & Dixon, 2014).

15.5.5 Lentil (*Lens Culinaris*): Microbiome Taxa Correlated with Yield Stability

Field surveys combining amplicon and shotgun approaches across multiple sites have identified microbial taxa whose presence was correlated with yield stability across years and soils notably taxa involved in nitrification-denitrification balance and root exudate metabolism. These taxa are candidate biomarkers for site-specific management and inoculant formulation (Philippot et al., 2013).

15.5.6 *Medicago Truncatula*: Soil Origin and Genotype Effects

A common garden experiment using 16S rRNA amplicon and shotgun metagenomics revealed that both soil origin and plant genotype strongly shape the root microbiome of *Medicago truncatula*. Distinct microbial communities were associated with different soil types, and genotypic variation influenced the abundance of nitrogen-fixing and stress-adaptive taxa. This demonstrates that **host genetics interact with soil microbiomes**, suggesting breeding programs could integrate microbiome traits for improved nutrient acquisition and resilience (Brown P et al., 2020).

15.5.7 *Pea (Pisum Sativum)*: Rhizosphere Microbiome and Nutrient Cycling

Metagenomic surveys of pea rhizospheres have shown enrichment of **phosphate-solubilising bacteria and nitrifiers**, particularly under low-input farming systems. Functional annotation identified genes for organic acid secretion, siderophore biosynthesis, and denitrification pathways. These findings support the design of microbial consortia that enhance nutrient cycling, reduce fertiliser dependence, and stabilise yields in smallholder systems (Schaedel et al., 2021; Rajguru et al., 2024).

15.5.8 *Faba Bean (Vicia Faba)*: Symbiotic Efficiency and Stress Adaptation

Shotgun metagenomic analysis of faba bean nodules revealed **co-inhabiting non-rhizobial bacteria** carrying stress tolerance genes, including osmoprotectant biosynthesis and antioxidant pathways. These microbes complement rhizobia by enhancing nodulation efficiency under salinity and drought stress. Trials with inoculants combining rhizobia and stress-tolerant co-inhabitants have shown improved nitrogen fixation and biomass accumulation in marginal soils (Schaedel et al., 2021).

15.5.9 *Pigeon Pea (Cajanus Cajan)*: Metagenomics of Marginal Soils

Studies of pigeon pea rhizospheres in degraded soils identified microbial taxa enriched in **lignocellulose degradation and nitrogen fixation pathways**, including *Streptomyces* and *Burkholderia*. Functional metagenomics uncovered novel enzymes for cellulose breakdown, which improve organic matter turnover and soil fertility. These insights are guiding the development of bioinoculants for pigeon pea cultivation in semi-arid regions (Rajguru et al., 2024).

15.6 MAJOR CHALLENGES AND HOW THEY ARE BEING ADDRESSED

15.6.1 Biological and Ecological Complexity

Soil and rhizosphere communities are hyper-diverse and dynamic. Low-abundance, but functionally important taxa are often missed; strain-level variation can drastically alter function. Strategies to address this include deep sequencing, targeted enrichment (e.g. sequence capture), and single-cell genomics to resolve rare but important players.

15.6.2 Incomplete Reference Databases and Hypothetical Proteins

Many soil sequences lack close homologues in databases, limiting functional inference. Community initiatives to sequence cultured soil isolates, curate MAGs, and expand functional databases (e.g. OpenFLUX, IMG/M) are critical. Functional metagenomics and activity-based assays provide complementary routes to discover novel functions that homology-based annotation misses.

15.6.3 Computational and Analytical Bottlenecks

Large datasets require substantial compute and storage. Standardised pipelines, containerised workflows (e.g. Nextflow, Snakemake) and cloud platforms mitigate reproducibility and scaling issues. Training programs to build local bioinformatics capacity in regions where legumes are economically critical will accelerate uptake.

15.6.4 Translational Gaps: From Lab to Field

Many inoculants successful in greenhouse fail in field trials due to competition, environmental stress, or host specificity. Solutions include:

- Using **locally sourced strains** adapted to target soils
- Formulating **multispecies consortia** that provide functional redundancy
- Engineering **carrier and delivery systems** (granules, seed coatings) that protect microbes
- Co-designing trials with farmers to reflect realistic management practices.

15.6.5 Reproducibility, Standardisation, and Metadata

Lack of standardised sampling, sequencing, and metadata reporting hampers cross-study comparisons. Adoption of minimum information standards (MIxS) and public deposition of raw reads and rich metadata are necessary to build cumulative knowledge.

15.6.6 Regulatory, Ethical, and Ecological Concerns

Introducing engineered or non-native microbes raises biosafety and ecological risk questions. Regulatory frameworks must balance innovation with safeguards, using stepwise risk assessments, confined trials, and monitoring of persistence and horizontal gene transfer.

15.6.7 Temporal Dynamics and Seasonality

Soil and rhizosphere microbiomes change with **seasonal cycles, crop growth stages, and management practices**. A single sampling point may not capture these fluctuations, leading to incomplete functional insights. Longitudinal studies and time-series metagenomics are needed to understand microbial succession and its impact on legume productivity.

15.6.8 Multi-Omics Integration

Metagenomics alone provides genetic potential but not actual activity. Integrating **metatranscriptomics, metaproteomics, and metabolomics** is essential to link microbial genes to real-time functions. However, multi-omics integration faces challenges in data harmonisation, cross-platform variability, and computational complexity.

15.6.9 Host–Microbe Specificity

Legume–microbe interactions are highly specific, with certain rhizobial strains or endophytes performing well only in particular host genotypes. This specificity complicates inoculant design, as microbial consortia may not perform uniformly across different cultivars or local varieties. Breeding programs must consider **microbiome compatibility traits** alongside agronomic traits.

15.6.10 Climate Change Impacts

Rising temperatures, altered rainfall, and soil degradation affect microbial community stability and function. Bioinoculants designed under current conditions may fail under future climates. Predictive modeling and climate-resilient microbial selection are needed to ensure long-term sustainability.

15.6.11 Socio-economic and Adoption Barriers

Even when effective bioinoculants are developed, **farmer adoption** can be limited by cost, availability, lack of awareness, or scepticism about microbial products. Building trust through participatory trials, extension services, and demonstration plots is critical for scaling innovations.

15.6.12 Intellectual Property and Benefit Sharing

Microbial resources often originate from local soils and communities. Issues of **bioprospecting, intellectual property rights, and equitable benefit sharing** can arise, especially when microbes are commercialised globally. Clear frameworks for access and benefit-sharing (ABS) under the Nagoya Protocol are essential.

15.7 INTEGRATING METAGENOMICS WITH MULTI-OMICS AND MACHINE LEARNING

15.7.1 Systems-Level Understanding

Coupling metagenomics with metatranscriptomics, metaproteomics, and metabolomics creates a layered understanding: potential (DNA), expression (RNA), execution (proteins) and outcomes (metabolites). For legumes, this enables linking microbial community changes to nodulation gene expression, nitrogen fluxes, and root exudate chemistry.

15.7.2 Predictive Modelling with Machine Learning

Machine learning (ML) methods are being used to predict functional outcomes (e.g. BNF efficiency, disease risk) from complex omics and environmental data. Supervised models (random forests, gradient boosting) and deep learning architectures can learn non-linear relationships between microbiome features and crop phenotypes. Critical to success are large, well-annotated training datasets and careful model validation (e.g. across sites and years).

15.7.3 Network Analysis and Keystone Species

Network inference identifies co-occurrence modules and keystone taxa that disproportionately influence community structure. Targeting keystone taxa for manipulation (via inoculation or agronomic change) can produce outsized shifts in community function.

15.7.4 Spatial and Temporal Resolution

Multi-omics datasets often lack fine-scale resolution in space and time. Integrating metagenomics with **spatial transcriptomics** and **time-series sampling** can reveal how microbial communities and their functions shift across root zones (rhizoplane vs rhizosphere vs endosphere) and developmental stages. This helps identify critical windows when microbial interventions (e.g. inoculation) are most effective.

15.7.5 Integration with Environmental Metadata

Coupling omics data with **soil physicochemical parameters, climate variables, and agronomic practices** provides context for interpreting microbial functions. Machine learning models that incorporate both biological and environmental metadata can better predict outcomes such as nitrogen fixation efficiency, yield stability, or pathogen outbreaks across diverse agroecosystems.

15.7.6 Synthetic Biology and Functional Validation

Multi-omics and ML predictions often generate hypotheses about microbial functions. **Synthetic biology tools** (e.g. CRISPR editing, pathway reconstruction in model organisms) can validate these predictions by experimentally testing candidate genes or pathways. This closes the loop between computational inference and biological reality.

15.7.7 Precision Agriculture Applications

Integrating omics with ML enables **decision-support systems** for farmers. Predictive models can recommend site-specific inoculants, crop rotations, or soil amendments based on microbiome profiles. This moves metagenomics from research into actionable precision agriculture tools.

15.7.8 Data Sharing and Federated Learning

One challenge is that agricultural omics datasets are often fragmented across institutions and regions. **Federated learning approaches** allow ML models to be trained across distributed datasets without centralising sensitive data, improving generalisability while respecting data sovereignty.

15.8 PRACTICAL RECOMMENDATIONS FOR RESEARCHERS AND PRACTITIONERS

- **Design experiments with translation in mind:** include field trials early, collect rich metadata, and partner with agronomists/farmers
- **Prioritise local diversity:** focus on indigenous microbes adapted to the local ecosystem for inoculant discovery

- **Use multi-omics when possible:** add metatranscriptomics or metabolomics to distinguish potential versus active functions
- **Adopt standards:** follow MIxS and deposit raw data and metadata to public repositories for reproducibility
- **Test consortia, not single strains:** multispecies products are more likely to persist and provide redundancy
- **Iterative field testing:** use multi-site, multi-season evaluation to capture environmental variability
- **Risk assessment and monitoring:** implement ecological risk assessments for introduced microbes and monitor persistence and gene transfer
- **Capacity building:** invest in local bioinformatics and microbiology training to enable regionally relevant solutions.

15.9 FUTURE PROSPECTS AND EMERGING DIRECTIONS

- **Precision Microbiome Engineering:** CRISPR-based editing of microbial genomes and phage-based interventions could allow targeted modulation of soil communities, subject to ethical/regulatory constraints.
- **Microbiome-aware Breeding:** Incorporating microbiome recruitment traits into breeding pipelines to create varieties that recruit beneficial communities.
- **Digital Twins and Decision Support:** Combining sensor networks, remote sensing, soil metagenomics and ML to create farm-level decision support systems that recommend microbiome interventions.
- **Open Microbiome Repositories:** Crop-specific microbiome atlases (with MAGs, functional annotations, and isolate strains) that enable comparative analyses and accelerate inoculant discovery.
- **Affordable Diagnostics:** Point-of-care or field diagnostics using CRISPR or isothermal amplification to monitor key microbial biomarkers associated with soil health or disease risk.

15.10 CONCLUSION

Metagenomics is a transformative tool poised to revolutionise sustainable legume cultivation by providing an unprecedented, culture-independent view of the intricate plant–microbe interactions that drive soil health and crop productivity. By unlocking the vast genetic potential of previously uncultivable microorganisms, this technology enables the identification of novel genes and functions essential for nutrient cycling (especially nitrogen and phosphorus fixation), plant growth promotion, and disease resistance. However, significant challenges remain, primarily in data analysis because the sheer volume and complexity of metagenomic data require sophisticated and often computationally intense bioinformatics tools and expertise for accurate interpretation and functional assignment. The second challenge lies with translational Research as translating valuable genetic insights from the lab into practical, field-applicable agricultural solutions (e.g. effective biofertilisers and biopesticides) requires substantial collaboration among researchers, the biotechnology industry, and farmers. The third challenge is the experimental standards as a lack of standardised experimental design and data-sharing protocols across studies hinders comparative analyses and the development of universal models for managing agricultural microbiomes.

Despite these hurdles, the emerging relevance of metagenomics is undeniable. Continued technological advancements, such as next-generation sequencing and multi-omics integration, promise to deepen our understanding of legume agroecosystems. Overcoming current challenges will facilitate the development of targeted, ecologically sound strategies to enhance soil fertility and crop resilience, ultimately contributing to global food security and a more sustainable agricultural future.

REFERENCES

Berg, G., & Smalla, K. (2009). Plant species and soil type cooperatively shape the structure and function of microbial communities in the rhizosphere. *FEMS Microbiology Ecology*, 68(1), 1–13.

Bhattacharyya, P. N., & Jha, D. K. (2012). Plant growth-promoting rhizobacteria (PGPR): Emergence in agriculture. *World Journal of Microbiology and Biotechnology*, 28(4), 1327–1350.

Brown, S. P., Grillo, M. A., Podowski, J. C., & Heath, K. D. (2020). Soil origin and plant genotype structure distinct microbiome compartments in the model legume Medicago truncatula. *Microbiome*, 8(1), 139.

Bulgarelli, D., Schlaeppi, K., Spaepen, S., van Themaat, E. V. L., & Schulze-Lefert, P. (2015). Structure and functions of the bacterial microbiota of plants. *Annual Review of Plant Biology*, 66, 807–838.

Compant, S., Samad, A., Faist, H., & Sessitsch, A. (2019). A review on the plant microbiome: Ecology, functions, and emerging trends in microbial application. *Journal of Advanced Research*, 19, 29–37.

De Filippo, C., Ramazzotti, M., Fontana, P., & Cavalieri, D. (2012). Bioinformatic approaches for functional annotation and pathway inference in metagenomics data. *Briefings in Bioinformatics*, 13(6), 696–710.

Dong, X., & Strous, M. (2019). An integrated pipeline for annotation and visualization of metagenomic contigs. *Frontiers in Genetics*, 10, 999.

Franzosa, E. A., et al. (2015). Relating the metatranscriptome and metagenome of the human gut. *Proceedings of the National Academy of Sciences*, 111(22), E2329–E2338.

Handelsman, J. (2004). Metagenomics: Application of genomics to uncultured microorganisms. *Microbiology and Molecular Biology Reviews*, 68(4), 669–685.

Hungria, M., & Mendes, I. C. (2015). Nitrogen fixation with soybean: The perfect symbiosis? *Biological Nitrogen Fixation*, 1009–1024.

Jain, A., Chakraborty, J., & Das, S. (2020). Underlying mechanism of plant–microbe crosstalk in shaping microbial ecology of the rhizosphere. *Acta Physiologiae Plantarum*, 42(1), 8.

Lal, R. (2004). Soil carbon sequestration impacts on global climate change and food security. *Science*, 304(5677), 1623–1627.

Maranga, M., Szczerbiak, P., Bezshapkin, V., Gligorijevic, V., Chandler, C., Bonneau, R., . . . & Kosciolek, T. (2023). Comprehensive functional annotation of metagenomes and microbial genomes using a deep learning-based method. *MSystems*, 8(2), e01178–22.

Marwa, A. M. (2025). The utilization of microorganisms for controlling fungal root diseases in legume crops. *EC Microbiology*, 21, 1–3.

Mendes, R., Garbeva, P., & Raaijmakers, J. M. (2013). The rhizosphere microbiome: Significance of plant beneficial, plant pathogenic, and human pathogenic microorganisms. *FEMS Microbiology Reviews*, 37(5), 634–663.

Mus, F., Crook, M. B., Garcia, K., Garcia Costas, A., Geddes, B. A., Kouri, E. D., . . . & Peters, J. W. (2016). Symbiotic nitrogen fixation and the challenges to its extension to nonlegumes. *Applied and Environmental Microbiology*, 82(13), 3698–3710.

Nwachukwu, B. C., & Babalola, O. O. (2022). Metagenomics: A tool for exploring key microbiome with the potentials for improving sustainable agriculture. *Frontiers in Sustainable Food Systems*, 6, 886987.

Oldroyd, G. E. D., & Dixon, R. (2014). Biotechnological solutions to the nitrogen problem. *Current Opinion in Biotechnology*, 26, 19–24.

Oldroyd, G. E. D., & Harrison, M. J. (2011). Signalling in the legume symbiosis. *Current Opinion in Plant Biology*, 14(2), 1–9.

Peiffer, J. A., et al. (2013). Diversity and heritability of the maize rhizosphere microbiome under field conditions. *Proceedings of the National Academy of Sciences*, 110(16), 6548–6553.

Peixoto, R. S., Rosado, A. S., Leite, L. R. S., Rosado, A., & Bourneuf, R. (2015). Microbial diversity in the rhizosphere of plants. *Environmental Microbiology Reports*, 7(6), 110–120.

Philippot, L., Raaijmakers, J. M., Lemanceau, P., & van der Putten, W. H. (2013). Going back to the roots: The microbial ecology of the rhizosphere. *Nature Reviews Microbiology*, 11(11), 789–799.

Quince, C., et al. (2017). Shotgun metagenomics, from sampling to analysis. *Nature Biotechnology*, 35(9), 833–844.

Rajguru, B., Shri, M., & Bhatt, V. D. (2024). Exploring microbial diversity in the rhizosphere: A comprehensive review of metagenomic approaches and their applications. *3 Biotech*, 14(10), 224.

Schaedel, M., Hidrobo, G., & Grossman, J. (2021). From microns to meters: Exploring advances in legume microbiome diversity for agroecosystem benefits. *Frontiers in Sustainable Food Systems*, 5, 668195.

Sharpton, T. J. (2014). An introduction to the analysis of shotgun metagenomic data. *Frontiers in Plant Science*, 5, 209.

Uchiyama, T., & Miyazaki, K. (2009). Functional metagenomics for enzyme discovery: Challenges to efficient screening. *Current Opinion in Biotechnology*, 20(6), 616–622.

van der Heijden, M. G. A., Bardgett, R. D., & van Straalen, N. M. (2008). The unseen majority: Soil microbes as drivers of plant diversity and productivity in terrestrial ecosystems. *Ecology Letters*, 11(3), 296–310.

Vessey, J. K. (2003). Plant growth promoting rhizobacteria as biofertilizers. *Plant and Soil*, 255(2), 571–586.

Villavicencio-Vásquez, M., Espinoza-Lozano, F., Espinoza-Lozano, L., & Coronel-León, J. (2025). Biological control agents: Mechanisms of action, selection, formulation and challenges in agriculture. *Frontiers in Agronomy*, 7, 1578915.

Vurukonda, S. S. K. P., Vardharajula, S., Shrivastava, M., & SkZ, A. (2018). Enhancement of drought stress tolerance in crops by plant growth-promoting rhizobacteria. *Frontiers in Microbiology*, 9, 1147.

Yeremko, L., Czopek, K., Staniak, M., Marenych, M., & Hanhur, V. (2025). Role of environmental factors in legume-rhizobium symbiosis: A review. *Biomolecules*, *15*(1), 118.

16 Role of Key Microbiome with Potentials for Improving Sustainable Legume Cultivation

Seemon Giri, Zahra Parwez, Smaranika Pradha, Sourav Ranjan Parida, Subhransu Sekhar Behera, and Lopamudra Ray

16.1 INTRODUCTION

Legumes play a vital role in sustainable agriculture due to their nutritional importance and natural ability to improve soil fertility through biological nitrogen fixation. In recent years, the plant-associated microbiome has emerged as a key factor influencing legume growth, nutrient acquisition, stress tolerance, and disease resistance. Diverse microbial communities inhabiting the rhizosphere, endosphere, and phyllosphere contribute significantly to plant health through multiple beneficial interactions. Understanding the role of key microbiome members and their functional potential offers new opportunities for developing eco-friendly and climate-resilient strategies for sustainable legume cultivation. This chapter highlights the diversity, functional significance, and agricultural applications of beneficial microbiomes associated with legumes.

16.1.1 Legumes as Pivotal Components of the Agro-Ecosystem

Legumes, members of the Fabaceae family, occupy a fundamentally important position in global agriculture and food security. They are cultivated primarily as an affordable and critical source of protein, vitamins, and other essential nutrients necessary for feeding a rapidly expanding global population. Beyond their direct use in human and livestock diets, legumes contribute significantly to agro-ecosystem health when utilised as livestock silage, forage, or, notably, as green manure for soil enhancement (Smýkal et al. 2015). Despite their established role and immense production volumes worldwide, legume cultivation faces severe productivity hurdles. These limiting factors include intensifying biotic (pathogens, pests) and abiotic (drought, salinity, temperature extremes) stresses exacerbated by climate change, coupled with pervasive issues of soil degradation and historically low yields compared to major cereals. This confluence of challenges necessitates a rapid, sustainable, and environmentally sound solution to enhance legume growth and productivity. Microbial

DOI: 10.1201/9781003733324-17

technologies represent a cost-effective and environment-friendly approach to overcoming these constraints. These beneficial soil microbes, often administered as biofertilisers, function by mimicking and enhancing critical natural soil processes, thereby improving overall plant health and productivity in a manner conducive to climate-resilient farming (Abobatta, Hegab, and El-Hashash 2022).

16.1.2 The Legume Holobiont and Its Functional Compartments

The understanding of plant biology is increasingly moving beyond the individual organism to embrace the concept of the holobiont – the plant host together with its associated, complex community of microorganisms (the microbiome). In legumes, this integrated system confers substantial "functional plasticity," providing a collective selective advantage that allows the plant to more efficiently acquire resources, improve nutrient status, and enhance tolerance to both biological and environmental stresses (Hocine et al. 2020). The legume microbiome is functionally and taxonomically diversified, inhabiting several distinct microhabitats across the plant structure. The rhizosphere, the soil layer immediately influenced by root exudates, is a zone of intense microbial activity, competition, and primary interaction for many plant growth–promoting bacteria (PGPB) and arbuscular mycorrhizal fungi (AMF). The endosphere comprises microbes residing inside plant tissues, including the specialised rhizobia colonies within root nodules, which are responsible for systemic benefits such as nitrogen fixation (Vincze et al. 2024). Furthermore, above-ground microhabitats, such as the anthosphere (flowers) and phyllosphere (leaves), harbour their own communities, which contribute to the plant's broader immunity and overall ecological stability. The recognition of the legume–microbe association as a holobiont fundamentally redefines the unit of selection in agricultural management and crop breeding. This realisation indicates that successful strategies for developing climate-resilient legumes must not solely focus on host genetics but must optimise the host's genetic capacity to attract, select, and maintain a highly beneficial and resilient microbial community (Berlanga-Clavero et al. 2020).

16.2 KEY MICROBIOME PLAYERS IN THE LEGUME SOIL SYSTEM

16.2.1 Symbiotic Nitrogen Fixers: The Rhizobia

Rhizobia, including genera such as *Rhizobium* and *Bradyrhizobium*, are the preeminent group of symbiotic nitrogen fixers (SNF). Their interaction with legumes culminates in the formation of specialised root nodules where the enzyme nitrogenase catalyses the conversion of atmospheric dinitrogen (N_2) into biologically assimilable forms, primarily ammonia. This Biological Nitrogen Fixation (BNF) is the cornerstone of sustainable legume productivity (Masson-Boivin and Sachs 2018). The symbiotic relationship is tightly managed by the host. The initiation involves specific molecular signalling, as described previously, that must result in the expression of key nodulin genes. The maintenance of this symbiosis, however, is dynamic and subject to plant control. Evidence suggests that legumes actively regulate the symbiosis by imposing metabolic penalties or sanctions on rhizobia strains that exhibit suboptimal nitrogen-fixing activity. This host mechanism dictates that inoculant effectiveness relies critically on the strain's ability to maintain high nitrogenase performance within the host's rigorous metabolic and regulatory environment. Beyond their SNF capacity, rhizobia frequently express a suite of intrinsic Plant Growth-Promoting (PGP) traits. These include the synthesis of phytohormones (such as Indole-3-acetic acid or IAA), the ability to solubilise inorganic soil phosphorus, and mechanisms for protecting the host plant against pathogens (Kundan et al. 2015). These dual functions make rhizobia valuable assets, even aiding in novel applications such as the phytoremediation of heavy metal contaminated soils by promoting plant biomass accumulation under toxic conditions.

16.2.2 Plant Growth-Promoting Bacteria (PGPB)

PGPB encompass a broad functional and taxonomic range of bacteria, including prominent agricultural genera such as *Azotobacter*, *Azospirillum*, *Bacillus*, and *Pseudomonas*. These organisms enhance plant growth through diverse mechanisms operating both outside (rhizosphere) and inside (endosphere) the root tissue (Compant, Clément, and Sessitsch 2010). A key contribution of PGPB is non-symbiotic nitrogen fixation, conducted by free-living diazotrophs that supplement the nitrogen pool outside the root nodules. Equally crucial is their role in enhancing phosphorus acquisition. PGPB actively convert insoluble forms of soil phosphorus into soluble, plant-available forms, often through the secretion of organic acids, thereby maximising Phosphorus Use Efficiency (PUE), which is vital given the low natural availability of P in most soils (Dobbelaere, Vanderleyden, and Okon 2003). PGPB also provide significant biocontrol capabilities, notably through the production of siderophores. Siderophores are low-molecular-mass, high-affinity iron-binding proteins secreted by bacteria. They effectively chelate ferric ions (Fe^{3+}) in the rhizosphere, making this essential micronutrient unavailable to competing pathogenic fungi and bacteria. The PGPB strains, however, possess specific outer membrane receptors enabling them to utilise the ferric-siderophore complex, effectively starving pathogens while securing iron for themselves and the host plant. Furthermore, PGPB interactions often demonstrate synergistic effects with other key microbes. Root exudates stimulate both PGPB colonisation and AMF formation, accelerating beneficial tripartite interactions. Research confirms that controlled co-inoculation, such as with dual strains such as *Bacillus* sp. and *P. putida*, yields significantly enhanced growth parameters, including root/shoot ratio and fresh biomass (Khawula et al. 2025).

16.2.3 Beneficial Fungi: Mycorrhizal and Endophytic

Beneficial fungi are broadly categorised into two major functional groups: arbuscular mycorrhizal fungi and endophytic fungi. AMF establish highly beneficial mutualistic associations with legume roots, characterised by the formation of arbuscules within the root cortical cells and extensive hyphal networks that proliferate throughout the surrounding soil. These extra-radical hyphae dramatically extend the effective scavenging area of the root system, fundamentally improving nutrient, particularly phosphorus and water, acquisition (Farhaoui et al. 2025). AMF utilise a variety of PGP mechanisms that extend beyond nutrient scavenging. They significantly enhance plant tolerance to abiotic stresses (such as salinity and drought) through the regulation of phytohormones, the production of exopolysaccharides, and the accumulation of compatible osmolytes, such as proline and betaines, which protect cellular structures under stress. AMF are also implicated in mitigating specific soil constraints, such as aluminium toxicity in acidic soils, although the colonisation rate may be slightly reduced in the presence of high aluminium concentrations (Diagne et al. 2020). Endophytic fungi colonise plant tissues internally without causing discernible damage. They serve as potent biocontrol agents and biofertilisers, promoting crop improvement and soil enhancement. Their mechanisms for enhancing host fitness include nutrient facilitation, synthesis of key phytohormones, and the mitigation of biotic stress via antibiosis, lytic enzyme production, or the activation of Induced Systemic Resistance (ISR) in the host. While AMF focus heavily on Phosphorus Use Efficiency (PUE) and water relations, endophytic fungi often specialise in bolstering host immunity and elevating endogenous phytohormone levels, providing a comprehensive strategy for coping with various stresses (Yasmeen et al. 2024).

16.2.4 Other Functional Groups (Actinomycetes and Archaea)

While Rhizobia and AMF receive the most attention, other microbial functional groups, notably Actinomycetes and Archaea, play essential roles in maintaining the stability and productivity of the legume soil system. Actinomycetes are filamentous bacteria known for their contributions to soil mineralisation and functional diversity. Species such as *Streptomyces griseoincarnatus* RB7AG are famous for fixing nitrogen in non-leguminous plants, but as a broader group, they are vital for complex organic matter decomposition and mineral solubilisation, including phosphorus (Behera et al. 2023). Archaea, though less studied in plant–microbe interactions than bacteria and fungi, are critical participants in global biogeochemical cycles, particularly nitrogen transformation processes such as nitrification and immobilisation. These nitrogen transformations are highly sensitive to soil physicochemical properties and prevailing climatic conditions (Cabello, Roldan, and Moreno-Vivian 2004). A crucial contribution of legume cultivation, often mediated by these broad microbial communities, is its role in mitigating climate change through soil health dynamics. Legume inclusion enhances soil microbial biomass, diversity, and enzyme activity. The enhanced growth of labile organic carbon fractions, facilitated by legume production, structurally modifies the soil microbiome to significantly promote Carbon (C) sequestration. The Actinomycetes and Archaea, as key decomposers and nutrient cyclers, are indispensable intermediaries in driving these beneficial soil enzyme activities and overall ecosystem functions, including soil C storage and productivity. This illustrates that the full sustainability benefit of legumes is achieved not just through BNF, but through the synergistic enhancement of the entire microbial community structure (Yan et al. 2023).

16.3 FUNCTIONAL POTENTIALS FOR SUSTAINABLE LEGUME CULTIVATION

16.3.1 Enhancing Nutrient Acquisition (NUE and PUE)

The industrial production and extensive application of synthetic fertilisers pose significant environmental and economic problems. Nitrogen (N) fertiliser production is energy-intensive, and agricultural Nitrogen Use Efficiency (NUE) is remarkably low, with plants taking up only 30% to 50% of applied N. The resulting nutrient runoff contributes to

TABLE 16.1
Key Microbial Functional Mechanisms for Legume Cultivation

Microbial Group	Mechanism of Action	Functional Outcome	Sustainability Impact
Rhizobia	Symbiotic Nitrogen Fixation (Nod factor signalling)	Direct N input, enhanced yield	Reduced reliance on N fertilisers (NUE)
PGPB (*Bacillus, Pseudomonas*)	Phosphate Solubilisation (Organic acid secretion)	Increased P availability, improved P uptake	Enhanced Phosphorus Use Efficiency (PUE)
PGPB (*Pseudomonas* spp.)	Siderophore Production (Iron chelation)	Biocontrol against soil pathogens, improved Fe uptake	Biotic stress mitigation, improved micronutrient uptake
PGPB/Fungi	ACC Deaminase Activity	Lowers stress ethylene levels in roots	Abiotic stress tolerance (Drought, Salinity)
AMF (*Glomus* spp.)	Hyphal Network Extension and Phosphate Solubilisation	Expanded water/nutrient (P, Zn) scavenging zone	Enhanced water-use efficiency, improved PUE

widespread environmental issues, including the eutrophication of aquatic ecosystems (Mirzaee and Nafchi 2025). Microbial solutions provide a crucial natural pathway to maximise nutrient use efficiency. BNF by rhizobia directly delivers nitrogen to the plant, significantly reducing dependence on synthetic nitrogen fertilisers. Simultaneously, AMF and PGPB drive Phosphorus Use Efficiency (PUE). AMF extend the effective absorption zone, while PGPB solubilise mineral P, making these complementary strategies essential for efficient P uptake (Kumar et al. 2025). The co-occurrence of rhizobia and AMF creates a highly regulated tripartite symbiotic relationship. This interaction is characterised by complex contingent dynamics. Physiologically, AMF infection often exerts a positive influence on nodulation, typically increasing the number and mass of root nodules. This reciprocal regulation suggests a host mechanism designed to strategically balance N and P acquisition resources based on soil nutrient levels and the relative benefits provided by each symbiont. The maximisation of both NUE and PUE is achievable through these integrated microbial strategies (Kavadia et al. 2021).

16.3.2 Microbial Modulation of Plant Stress Physiology

Legumes frequently encounter biotic and abiotic stresses, necessitating robust physiological coping mechanisms. Microbes assist plants by upregulating the plant's native antioxidant systems, the accumulation of compatible osmolytes, enhanced photosynthetic capacity, and improved membrane stability, all of which contribute to the reduction of oxidative stress-induced damage. A profound mechanism observed involves the modulation of the host's innate immune system. Studies have demonstrated that purified Nod factors (NFs) from rhizobia can attenuate the host plant's immune response when the plant is simultaneously exposed to known Pathogen-Associated Molecular Patterns (PAMPs), such as the bacterial flagellin peptide flg22. This Nod factor-mediated suppression of host immunity is hypothesised to be a necessary precondition for successful rhizobia infection and nodule development, highlighting a delicate balance between defence and symbiosis (Gourion et al. 2015). Both AMF and PGPR utilise various, sometimes shared, mechanisms to combat abiotic challenges like salt stress, including the enhancement of water uptake and osmoregulation, leading to improved plant fitness in saline environments.

16.3.3 Phytohormone Synthesis and Growth Promotion

A direct mechanism through which root-associated microbes stimulate plant growth and stress tolerance is the synthesis and regulation of key phytohormones. Microbes produce and influence levels of Cytokinin's (CK), Gibberellic acid (GA), Indole-3-acetic acid (IAA), and Abscisic acid (ABA). The production of IAA, a crucial auxin, promotes the proliferation and elongation of the root system, which is vital for enhancing the acquisition of water and nutrients. CKs promote cell division, while ABA and salicylic acid (SA) are integral to managing water balance and systemic defence responses under stress conditions (Egamberdieva et al. 2017). Crucially, many Plant Growth-Promoting Rhizobacteria (PGPR) confer stress resistance through the activity of the enzyme 1-aminocyclopropane-1-carboxylate (ACC) deaminase. ACC is the immediate metabolic precursor of the stress hormone ethylene. Under abiotic stress conditions, such as drought or flooding, plants produce high levels of ethylene, which inhibits root growth and development. The microbial ACC deaminase cleaves ACC, thus effectively down-regulating the stress-ethylene level in the roots, allowing the plant to maintain growth and promoting resistance capacities under adverse conditions (Saleem et al. 2007).

16.3.4 Detoxification and Soil Health Dynamics

Microbial activity is fundamental to improving soil health and facilitating detoxification processes, which are critical for reclaiming marginal agricultural lands. The legume-rhizobium symbiosis is increasingly utilised in bio rhizoremediation to address heavy metal contamination. By fixing nitrogen and producing PGP factors (like phytohormones), rhizobia enhance overall legume biomass, thereby improving the plant's capacity for metal extraction or stabilisation. Co-augmentation using PGPB and AMF has been shown to significantly increase plant growth parameters and aid in the detoxification and exclusion of specific contaminants, such as sodium (Na^+) and nickel (Ni) (Mandal and Bhattacharyya 2012). Furthermore, beneficial microbes can mediate the chemical characteristics of the soil to mitigate toxicity. In acidic soils, highly exchangeable aluminium (Al) poses a severe threat to root morphology and crop yield. Certain PGPB strains have been observed to increase soil pH (e.g. from 4.0 to 6.0), which effectively chelates and detoxifies the aluminium, rendering it unavailable to the growing crop (Dutta and Bora 2019). At a macro level, the integration of legumes fundamentally enhances soil structure and fertility. Through symbiotic association with rhizobia, legumes provide a substantial annual nitrogen input, averaging between 5 and 19 kg/day per year, significantly reducing the demand for industrial N fertilisers, and mitigating the associated greenhouse gas emissions. The positive impact of legume diversification on microbial community structure and enzyme activities (e.g. nitrogen cycling) ultimately facilitates the growth of labile organic carbon fractions, which promotes long-term Carbon Sequestration in the soil (Kebede 2021).

16.4 TRANSLATIONAL MICROBIOME TECHNOLOGIES FOR SUSTAINABLE LEGUME SYSTEMS

16.4.1 Development and Formulation of High-Efficacy Bioinoculants

The successful translation of laboratory-identified microbial functions into applied agriculture hinges on the development of reliable, high-efficacy bioinoculants. The primary obstacles encountered in commercial application relate to the lack of consistent performance and poor persistence of inoculated strains in the complex soil matrix. Low field efficacy is often attributed to the inability of inoculant strains to compete effectively against robust native microbial populations or the potential transfer of beneficial genes to less effective resident strains via Horizontal Gene Transfer (Laskar et al. 2024). Consequently, the process of strain selection must prioritise "elite strains" that combine superior functional capacity (e.g. high nitrogen fixation efficiency) with inherent competitive advantages and resilience under local field conditions(Sessitsch et al. 2002). Furthermore, technological rigor must be applied to the formulation process, ensuring that the selected microbes are protected during storage and delivery (Khan et al. 2023).

16.4.2 Designer Consortia and Synthetic Ecology

To address the multifunctional demands of modern sustainable agriculture (which often requires simultaneous N-fixation, P-solubilisation, and stress tolerance), researchers are moving beyond single-strain inoculation towards the design of Synthetic Microbial Communities (SynComs). SynComs are intentionally engineered communities composed of multiple microbial taxa, selected for their complementary or synergistic traits, designed to mimic and optimise the functional benefits of natural, complex microbiomes (Tariq et al. 2025). SynComs offer enhanced resilience, particularly in systems prone to abiotic stress. For instance, consortia containing IAA-producing *Bacillus* and *Pseudomonas* species have shown enhanced water-use efficiency under drought, while the inclusion of phosphate-solubilising bacteria improves P uptake in nutrient-deficient settings (Benmrid et al. 2023). However, the design process is fraught with complexity. Synergy cannot be assumed; co-inoculation studies reveal that adding too many strains (e.g. three or four) can sometimes result in lower performance compared to dual inoculations, suggesting antagonistic or resource-competition effects that undermine collective function. Effective design requires rigorous evaluation of microbial interactions, moving beyond traditional, low-throughput microbiological assays (Negi et al. 2024). The deployment of SynComs is strategically relevant within integrated systems such as intercropping, where legumes are grown simultaneously with cereals. In intercropping systems, the combined application of synergistic PGPR and rhizobia consortia is highly advantageous, especially in soils with low native rhizobia populations (Chamkhi et al. 2022). This approach maximises the efficiency of the legume-rhizobium symbiosis, boosting BNF and the subsequent growth of both the legume and its intercropped partner.

TABLE 16.2
Leveraging Omics Technologies for Rational SynCom Design

Omics Technology	Application in SynCom Development	Goal/Insight Derived
Metagenomics (DNA)	Community structure analysis, genetic reservoir mapping	Identifying potential PGP traits, functional redundancy, and community membership
Meta transcriptomics (RNA)	Gene expression dynamics *in planta*	Determining microbial activity, resource utilisation, and signalling pathways during symbiosis
Metabolomics (Small Molecules)	Profiling root exudates and microbial secretions	Identifying communication molecules (e.g. Osmo protectants, antibiotics) and nutrient fluxes
Systems Biology/ Modelling (FBA)	Metabolic network prediction, flux balance analysis	Optimising metabolic interactions and predicting stability/synergy among strains

16.4.3 "Omics" Technologies for Microbiome-Guided Breeding

Integrated multi-omics approaches are crucial for the rational design of SynComs. Genomics and Metagenomics reveal the entire genetic blueprint, mapping the reservoir of potential plant growth–promoting (PGP) traits, such as genes encoding Osmo protectants (trehalose, betaine) and ion transporters necessary for stress resilience (Mishra et al. 2022). Transcriptomics and meta-transcriptomics provide insight into gene expression dynamics under specific environmental conditions, revealing which metabolic pathways are actively functioning, such as those related to nitrogen fixation or induced systemic resistance. Proteomics and Metabolomics profile the functional output identifying active proteins and small signalling molecules (volatile compounds, phytohormones) that mediate communication between the host and its symbionts (Nong et al. 2023). This data is then leveraged by Systems Biology, using computational methods such as Flux Balance Analysis (FBA) and constraint-based modelling.

Furthermore, the integration of these molecular tools enables a novel approach to crop improvement: microbiome-guided breeding. By combining Metagenomics with Genome-Wide Association Studies (GWAS), researchers can identify specific host genetic loci that influence the recruitment and maintenance of beneficial microbial communities in the rhizosphere. This capacity allows breeders to develop legume varieties that are genetically optimised to selectively recruit and sustain high-efficacy, rationally designed SynComs, advancing the goal of precision intervention in agriculture (Wang et al. 2022).

16.4.4 Integration into Sustainable Agricultural Practices

The successful adoption of microbiome technologies requires seamless integration into established sustainable farming systems. Bioinoculants amplify the benefits derived from practices such as crop rotation, where the nitrogen legacy of legumes enhances subsequent non-legume crops, and simultaneous intercropping systems. In conservation agriculture, bioinoculants contribute significantly to maintaining and enhancing soil quality, improving microbial activity, and sustaining soil organic carbon content (Hanif et al. 2024). The transition to bioinoculants is increasingly recognised as a vital mitigation strategy against the rising operational costs and environmental damage associated with high-energy chemical inputs required by conventional farming, particularly in the context of global warming and climate change. Bioinoculants are not merely nutrient supplements; they are multifunctional tools that promote growth, mobilise nutrients, and detoxify soil, supporting the resilience and profitability of sustainable agricultural systems (Kumar et al. 2022).

16.5 CONCLUSION: SYNTHESISING KNOWLEDGE AND FUTURE TRAJECTORIES

16.5.1 The Paradigm Shift: From Legume-to-Legume Holobiont Management

Microbial solutions provide a potent, environmentally sustainable alternative to traditional reliance on synthetic inputs. The collective evidence presented underscores the necessity of managing the legume not as a discrete organism but as a complex holobiont, where functional plasticity and stress resilience are conferred by synergistic microbial consortia. Harnessing the combined power of Rhizobia for Nitrogen Use Efficiency (NUE), AMF for Phosphorus Use Efficiency (PUE), and PGPB for growth promotion and biocontrol offers a pathway to achieve high yields and improve soil health simultaneously, minimising the negative environmental footprint of agriculture.

16.5.2 Future Trajectories for Global Legume Security

The future trajectory of legume microbiome research must prioritise a dedication to functional ecology and precision agricultural tools. This includes enhancing the resolution of 'Omics technologies, exploring epigenomics, and using advanced computational modelling to guide the systematic construction of highly adapted, robust SynComs. These advancements will enable precision microbial intervention tailored to specific crop-soil-climate contexts.

The research indicates a persistent disconnect between high-level scientific innovation (e.g. consortia) and farmer adoption rates. Constraints such as differential access to timely, high-quality seeds, reliable biofertilisers, affordable credit, and equitable agricultural information severely limit the utilisation of these sustainable technologies, particularly among smallholder farmers. Therefore, the path to global legume security requires integrated policy intervention focusing not just on enhancing the microbiological quality of bioinoculants, but on establishing robust, reliable, and economically accessible distribution channels to ensure that the scientific gains translate into tangible and sustained benefits at the field level (Katole, Chinchmalatpure, and More 2017).

REFERENCES

Abobatta, WF, RH Hegab, and EF El-Hashash. 2022. 'Challenges and opportunities for the global cultivation and adaptation of legumes B. Opportunities for increasing legumes production and availability', *Annals of Agriculture Crop & Sciences*, 7: 1107.

Behera, Subhransu Sekhar, Suchismita Nivedita, Pratyush Kumar Behera, Zahra Parwez, Seemon Giri, Sourav Ranjan Parida, and Lopamudra Ray. 2023. 'Streptomyces griseoincarnatus strain RB7AG promotes Oryza sativa (var. swarna) growth under salt stress: Mechanisms and potential applications', *Letters in Applied Microbiology*, 76: ovad125.

Benmrid, Bouchra, Cherki Ghoulam, Youssef Zeroual, Lamfeddal Kouisni, and Adnane Bargaz. 2023. 'Bioinoculants as a means of increasing crop tolerance to drought and phosphorus deficiency in legume-cereal intercropping systems', *Communications Biology*, 6: 1016.

Berlanga-Clavero, María Victoria, Carlos Molina-Santiago, Antonio de Vicente, and Diego Romero. 2020. 'More than words: The chemistry behind the interactions in the plant holobiont', *Environmental Microbiology*, 22: 4532–4544.

Cabello, Purificación, M Dolores Roldan, and Conrado Moreno-Vivian. 2004. 'Nitrate reduction and the nitrogen cycle in archaea', *Microbiology*, 150: 3527–3546.

Chamkhi, Imane, Said Cheto, Joerg Geistlinger, Youssef Zeroual, Lamfeddal Kouisni, Adnane Bargaz, and Cherki Ghoulam. 2022. 'Legume-based intercropping systems promote beneficial rhizobacterial community and crop yield under stressing conditions', *Industrial Crops and Products*, 183: 114958.

Compant, Stéphane, Christophe Clément, and Angela Sessitsch. 2010. 'Plant growth-promoting bacteria in the rhizo-and endosphere of plants: Their role, colonization, mechanisms involved and prospects for utilization', *Soil Biology and Biochemistry*, 42: 669–678.

Diagne, Nathalie, Mariama Ngom, Pape Ibrahima Djighaly, Dioumacor Fall, Valérie Hocher, and Sergio Svistoonoff. 2020. 'Roles of arbuscular mycorrhizal fungi on plant growth and performance: Importance in biotic and abiotic stressed regulation', *Diversity*, 12: 370.

Dobbelaere, Sofie, Jos Vanderleyden, and Yaacov Okon. 2003. 'Plant growth-promoting effects of diazotrophs in the rhizosphere', *Critical Reviews in Plant Sciences*, 22: 107–149.

Dutta, Jintu, and Utpal Bora. 2019. 'Role of PGPR for alleviating aluminum toxicity in acidic soil', in *Plant Growth Promoting Rhizobacteria for Sustainable Stress Management: Volume 1: Rhizobacteria in Abiotic Stress Management* (Springer).

Egamberdieva, Dilfuza, Stephan J Wirth, Abdulaziz A Alqarawi, Elsayed F Abd_Allah, and Abeer Hashem. 2017. 'Phytohormones and beneficial microbes: Essential components for plants to balance stress and fitness', *Frontiers in Microbiology*, 8: 2104.

Farhaoui, Abdelaaziz, Mohammed Taoussi, Salah-Eddine Laasli, Ikram Legrifi, Nizar El Mazouni, Abdelilah Meddich, Mohamed Hijri, and Rachid Lahlali. 2025. 'Arbuscular mycorrhizal fungi and their role in plant disease control: A state-of-the-art', *The Microbe*: 100438.

Gourion, Benjamin, Fathi Berrabah, Pascal Ratet, and Gary Stacey. 2015. 'Rhizobium–legume symbioses: The crucial role of plant immunity', *Trends in Plant Science*, 20: 186–194.

Hanif, Muhammad Sajid, Muhammad Tayyab, Elamin Hafiz Baillo, M Mominul Islam, Waqar Islam, and Xiaofang Li. 2024. 'Plant microbiome technology for sustainable agriculture', *Frontiers in Microbiology*, 15: 1500260.

Hocine, Abbaci, Bensidhoum Leila, Houali Karim, and Nabti Elhafid. 2020. 'Functional microbiome for crop improvement under a changing environment', *The Plant Microbiome in Sustainable Agriculture*: 101–118.

Katole, RT, Umesh R Chinchmalatpure, and GB More. 2017. 'Constraints faced by the farmers in adoption of biofertilizers', 628–633.

Kavadia, Athanasia, Michalis Omirou, Dionysia A Fasoula, Filippa Louka, Constantinos Ehaliotis, and Ioannis M Ioannides. 2021. 'Co-inoculations with rhizobia and arbuscular mycorrhizal fungi alters mycorrhizal composition and lead to synergistic growth effects in cowpea that are fungal combination-dependent', *Applied Soil Ecology*, 167: 104013.

Kebede, Erana. 2021. 'Contribution, utilization, and improvement of legumes-driven biological nitrogen fixation in agricultural systems', *Frontiers in Sustainable Food Systems*, 5: 767998.

Khan, Amir, Ajay Veer Singh, Shiv Shanker Gautam, Aparna Agarwal, Arjita Punetha, Viabhav Kumar Upadhayay, Bharti Kukreti, Vindhya Bundela, Arun Kumar Jugran, and Reeta Goel. 2023. 'Microbial bioformulation: A microbial assisted biostimulating fertilization technique for sustainable agriculture', *Frontiers in Plant Science*, 14: 1270039.

Khawula, Sindiswa, Augustine Innalegwu Daniel, Nkosingiphile Nyawo, Khethiwe Ndlazi, Sanele Sibiya, Siyabonga Ntshalintshali, Gcinokuhle Nzuza, Arun Gokul, Marshall Keyster, and Ashwil Klein. 2025. 'Optimizing plant resilience with growth-promoting Rhizobacteria under abiotic and biotic stress conditions', *Plant Stress*: 100949.

Kumar, Pawan, Surender Singh, Kumar Pranaw, Sandeep Kumar, Balkar Singh, and Vikram Poria. 2022. 'Bioinoculants as mitigators of multiple stresses: A ray of hope for agriculture in the darkness of climate change', *Heliyon*, 8.

Kumar, Rahul, Beatrice Farda, Amedeo Mignini, Rihab Djebaili, Leonard Koolman, Alivia Paul, Subhankar Mondal, Joy M Joel, Aditi Pandit, and Periyasamy Panneerselvam. 2025. 'Microbial solutions in agriculture: Enhancing soil health and resilience through bio-inoculants and bioremediation', *Bacteria*, 4: 28.

Kundan, Rishi, Garima Pant, Nitesh Jadon, and Pavan Kumar Agrawal. 2015. 'Plant growth promoting rhizobacteria: Mechanism and current prospective', *J Fertil Pestic*, 6: 9.

Laskar, Islamul Hoque, Udaya Kumar Vandana, Nandita Das, Piyush Pandey, and Pranab Bihari Mazumder. 2024. 'Role of microbial bio-inoculants in sustainable agriculture', in *Microbial Biotechnology for Sustainable Agriculture Volume 2* (Springer).

Mandal, Santi M, and Rabindranath Bhattacharyya. 2012. 'Rhizobium–legume symbiosis: A model system for the recovery of metal-contaminated agricultural land', in *Toxicity of Heavy Metals to Legumes and Bioremediation* (Springer).

Masson-Boivin, Catherine, and Joel L Sachs. 2018. 'Symbiotic nitrogen fixation by rhizobia – the roots of a success story', *Current Opinion in Plant Biology*, 44: 7–15.

Mirzaee, S, and A Mirzakhani Nafchi. 2025. 'Advancing global nitrogen use efficiency for environmental sustainability', *Land Degradation & Development*, 36(12): 4005–4016.

Mishra, Ajay Kumar, Naganeeswaran Sudalaimuthuasari, Khaled M Hazzouri, Esam Eldin Saeed, Iltaf Shah, and Khaled MA Amiri. 2022. 'Tapping into plant–microbiome interactions through the lens of multi-omics techniques', *Cells*, 11: 3254.

Negi, Rajeshwari, Babita Sharma, Tawseefa Jan, Tanvir Kaur, Sohini Chowdhury, Monit Kapoor, Sangram Singh, Anu Kumar, Ashutosh Kumar Rai, and Sarvesh Rustagi. 2024. 'Microbial consortia: Promising tool as plant bioinoculants for agricultural sustainability', *Current Microbiology*, 81: 222.

Nong, Qian, Mukesh Kumar Malviya, Manoj Kumar Solanki, Li Lin, Jinlan Xie, Zhanghong Mo, Zeping Wang, Xiupeng Song, Xin Huang, and Changning Li. 2023. 'Integrated metabolomic and transcriptomic study unveils the gene regulatory mechanisms of sugarcane growth promotion during interaction with an endophytic nitrogen-fixing bacteria', *BMC Plant Biology*, 23: 54.

Saleem, Muhammad, Muhammad Arshad, Sarfraz Hussain, and Ahmad Saeed Bhatti. 2007. 'Perspective of plant growth promoting rhizobacteria (PGPR) containing ACC deaminase in stress agriculture', *Journal of industrial Microbiology and Biotechnology*, 34: 635–648.

Sessitsch, Angela, JG Howieson, X Perret, H Antoun, and Esperanza Martínez-Romero. 2002. 'Advances in Rhizobium research', *Critical Reviews in Plant Sciences*, 21: 323–378.

Smýkal, Petr, Clarice J Coyne, Mike J Ambrose, Nigel Maxted, Hanno Schaefer, Matthew W Blair, Jens Berger, Stephanie L Greene, Matthew N Nelson, and Naghmeh Besharat. 2015. 'Legume crops phylogeny and genetic diversity for science and breeding', *Critical Reviews in Plant Sciences*, 34: 43–104.

Tariq, Arneeb, Shengzhi Guo, Fozia Farhat, and Xihui Shen. 2025. 'Engineering synthetic microbial communities: Diversity and applications in soil for plant resilience', *Agronomy*, 15: 513.

Vincze, Éva-Boglárka, Annamária Becze, Éva Laslo, and Gyöngyvér Mara. 2024. 'Beneficial soil microbiomes and their potential role in plant growth and soil fertility', *Agriculture*, 14: 152.

Wang, Yayu, Xiaolin Wang, Shuai Sun, Canzhi Jin, Jianmu Su, Jinpu Wei, Xinyue Luo, Jiawen Wen, Tong Wei, and Sunil Kumar Sahu. 2022. 'GWAS, MWAS and mGWAS provide insights into precision agriculture based on genotype-dependent microbial effects in foxtail millet', *Nature Communications*, 13: 5913.

Yan, Zhengjun, Jie Zhou, Chunyan Liu, Rong Jia, Kevin Z Mganga, Lei Yang, Yadong Yang, Leanne Peixoto, Huadong Zang, and Zhaohai Zeng. 2023. 'Legume-based crop diversification reinforces soil health and carbon storage driven by microbial biomass and aggregates', *Soil and Tillage Research*, 234: 105848.

Yasmeen, Tahira, Waqas Haidar, Afira Syrish, Muhammad Rizwan, Muhammad Saleem Arif, Sher Muhammad Shahzad, and Mohsin Tariq. 2024. 'Endophytic fungi: Plant growth-promoting phytohormones and their potential application', in *Fungal Secondary Metabolites* (Elsevier).

17 Legume Metagenomics
Methods and Protocols

Alok Ranjan, Shruti Kaushik,
Piyush Kumar Rai, and Anil Kumar Singh

17.1 INTRODUCTION

Metagenomics refers to the direct examination of genetic material extracted from environmental samples, enabling the study of microbial communities without the need for their cultivation. Metagenomics began almost two decades ago with cloning environmental DNA and screening it for function (Handelsman et al., 1998). It quickly advanced to random shotgun sequencing of environmental samples (Tyson et al., 2004), which confirmed the method's potential and revealed vast gene diversity in microbial communities. Plants harbour a rich diversity of microorganisms that play vital roles in promoting growth, enhancing resilience, and improving tolerance to abiotic stressors. Studies on rhizosphere microbiota have demonstrated their influence on plant development, stress adaptation, and pathogen defence (Venturi & Keel, 2016). Understanding plant physiology increasingly requires examining the associated microbial communities. However, most microbes remain difficult to culture in laboratory. Advances in DNA sequencing and bioinformatics have enabled the exploration of genetic diversity within these host-associated communities, especially the uncultured fraction. Among the available techniques, amplicon sequencing is widely used to assess microbial diversity. This method involves extracting DNA from environmental samples such as soil or plant tissue followed by PCR amplification of a taxonomica genomic marker. The resulting sequences are analysed to determine microbial composition and relative abundance. The 16S rRNA gene is frequently used to study bacteria and archaea because of its strong taxonomic and phylogenetic significance (Brescia et al., 2023; Brown et al., 2020; Zhang et al., 2023). The 16S amplicon sequencing has revealed extensive microbial diversity across environments like *Arabidopsis* roots (Lundberg et al., 2012), hot springs (Kochetkova et al., 2020), and acidic and alkaline soils (Lopes et al., 2021; Y. Shi et al., 2021). Comparative analysis of these profiles helps uncover patterns of microbial diversity and host microbe interactions, offering insights into disease mechanisms. Recent metagenomic studies have identified several unculturable rhizosphere microbes, including *Acidobacteria* (nutrient cycling; Goel et al., 2018), Planctomycetes (nitrogen cycling; Wiegand et al., 2020), *Verrucomicrobia* (carbon and nitrogen cycling; Dash et al., 2020), *Gemmatimonadetes* (organic matter degradation), and *Chloroflexi* (organic decomposition; Mehrshad et al., 2018). One challenge in plant metagenomics is the large size of plant genomes, which can obscure

 DOI: 10.1201/9781003733324-18

microbial signals. Improved lab techniques, such as physically separating microbes from plant tissues, help to overcome this barrier and enhance the efficiency and accuracy of metagenomic analysis in plant systems. Overall, metagenomics contributes to soil health, ecosystem management, diagnostics, and biotechnology, with applications spanning agriculture, waste management, and food production. This chapter provides a comprehensive overview of legume-associated metagenomics, focusing specifically on the key stages of a standard sequence-based metagenomic workflow (Figure 17.1). It outlines and examines critical components, including sample preparation, sequencing platforms, data assembly, genome binning, functional annotation, study design, statistical evaluation, and strategies for data management and dissemination. The integration of insights from existing legume metagenomic datasets can greatly enhance the interpretation of new data (Table 17.1).

17.2 SAMPLING STRATEGIES AND EXPERIMENTAL DESIGN FOR METAGENOMICS STUDY IN LEGUMES

Metagenomics has revolutionised our understanding of plant–microbe interactions, enabling researchers to uncover the diversity and functional potential of microbial communities associated with legumes. A well-planned sampling and experimental

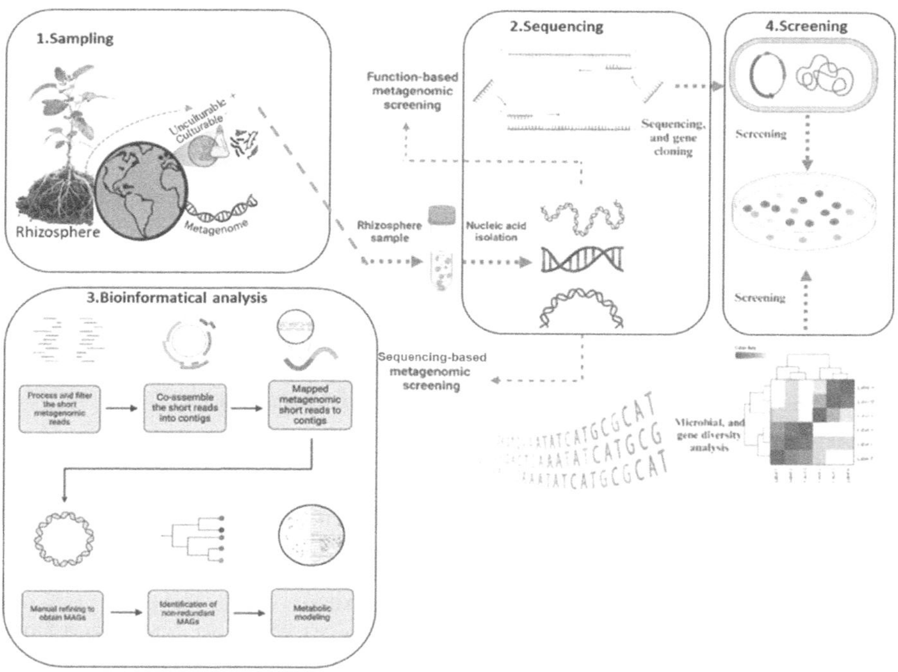

FIGURE 17.1 Metagenomics experimental work flow

TABLE 17.1A
Metagenomics Database

S.No	Data base	Description	References
1.	MAGdatabase https://magdb.nanhulab.ac.cn	High quality microbial metagenome-assemble genomes	Ye et al. (2025)
2.	GOLD:Genomes OnLine Database;https://gold.jgi-psf.org	Genomic catalog of Earth's microbiomes	Nayfach et al. (2021)
3.	GSC: Genomic Standards Consortium http://gensc.org	Information about bacterial and archaeal genome sequences	Nayfach et al. (2021)
4.	MGnify Genomes https://www.ebi.ac.uk/metagenomics/browse/genomes	Biome-specific Microbial Genome Catalogues	Gurbich et al. (2023)
5.	gcMeta https://gcmeta.wdcm.org	Metagenomics standardisation and analysis of microbiome data	W. Shi et al. (2019)

TABLE 17.1B Metagenomics Study of Major Legume Crops

S.No	Plant Species	Sample	Methods	reference
1.	*Trifolium pratense*	Rhizospheric soil, Plant Root (5cm fragment)	Culture-dependent and independent methods. 16SRNA Illumina MiSeq	Hartman et al. (2017)
2.	*Glycine max*	Rhizosphere soil	16SRNA Illumina MiSeq	Liu et al. (2019)
3.	*Medicago sativa*	Soils	Miseq-Sequencing of 16S rRNA	Zhou et al. (2017)
4.	*Bambara groundnut*	Rhizosphere/soil	Paired-end Illumina-Miseq sequencing of 16S rRNA	Ajilogba and Babalola (2020)
5.	*Medicago truncatula*	Root endosphere, nodule endosphere, and leaf endosphere	16S rRNA gene amplicon and shotgun metagenomic sequencing	Brown et al. (2020)
6.	*Cicer arietinum*	Rhizosphere Soil	Paired-end Illumina-Miseq sequencing of 16S rRNA	Zhang et al. (2023)
7.	*Lens culinaris*	Bulk soil, Rhizosphere Soil	16S rRNA metagenomic sequencing	Brescia et al. (2023)
8.	*Arachis hypogaea*	Composite soil samples	16SRNA Illumina MiSeq	Bigatton et al. (2024)
9.	*Pisum sativum*	Pea seeds	16S rRNA (V3-V4 regions)	Chartrel et al. (2021)
10.	*Cajanus cajan*	Soil, Nodules	16S rRNA metagenomic sequencing	Basu et al. (2025)

design is fundamental to obtaining reliable and biologically meaningful data (Table 17.2, Figure 17.2). This section outlines the essential considerations for sampling strategies in legume metagenomics, focusing on the selection of plant species and growth conditions, sampling sites, replication and controls, and sample preservation and transport.

TABLE 17.2
Comprehensive Overview of the Main Techniques Used in Metagenomics

Sampling and DNA Extraction Techniques		
Steps	**Common Methods**	**Remarks**
Sample selection	Rhizosphere soil, root endosphere, nodules, leaves, bulk soil	Different compartments have distinct microbial communities. Use sterile tools and consistent depths for soil sampling.
DNA extraction	Commercial kits (e.g. Qiagen PowerSoil, NucleoSpin Soil), CTAB (Cetyltrimethylammonium Bromide)-based methods	Nodules and root tissues often need extra steps for tough cell walls and polysaccharide removal.
DNA Quality check	Nanodrop, Qubit, agarose gel electrophoresis	High molecular weight DNA is needed for long-read or shotgun sequencing.
Metagenomics approach for sequencing		
Types	**Description**	**Output**
Amplicon-based sequencing (Marker gene sequencing)	PCR amplification of conserved marker genes (16S rRNA for bacteria, ITS for fungi, 18S for eukaryotes)	Provides who is their taxonomic composition. Common in legume rhizosphere/nodule microbiome studies.
Shotgun (Whole-genome) metagenomics	Sequencing of all DNA from a sample without amplification of a single marker	Reveals who is there and what they can do (functional genes, metabolic pathways)
Sequencing platform		
Platform	**Read length**	**Application**
Illumina MiSeq/ NovaSeq	150–300 bp (short reads)	Most common for 16S/ITS amplicon or shotgun metagenomics.
PacBio SMRT/ Oxford Nanopore	>10 kb (long reads)	Used for long-read metagenomics, improved assembly of MAGs (Metagenome-Assembled Genomes).
Ion Torrent/BGISEQ platforms	Alternative short-read platforms	Used in some plant–microbiome studies with budget constraints.

(Continued)

TABLE 17.2 (*Continued*)

Comprehensive Overview of the Main Techniques Used in Metagenomics

Bioinformatics Techniques

Stage	Tools/Pipelines	Purpose
Quality control	FastQC, Trimmomatic, Cutadapt	Remove low-quality reads, adapters.
OTU/ASV picking (for amplicon)	QIIME2, Mothur, DADA2	Identify and classify microbial taxa.
Taxonomic annotation (for shotgun)	Kraken2, MetaPhlAn, Kaiju	Assign taxonomy directly from reads.
Assembly & binning	MEGAHIT, MetaSPAdes, MaxBin, MetaBAT	Assemble reads into contigs; group into MAGs.
Functional annotation	PROKKA, KEGG, eggNOG, HUMAnN3	Predict functional genes and metabolic pathways.
Diversity & statistics	R packages (phyloseq, vegan)	Alpha/beta diversity, PCoA, PERMANOVA, network analysis.
Visualisation	Krona plots, Circos, heatmaps, phylogenetic trees	For graphs, table

Complementary Validation Techniques

Technique	Purpose
qPCR (quantitative PCR)	Validate abundance of specific functional genes (e.g. nifH, phoD).
FISH (Fluorescent in situ hybridisation)	Visualise specific microbes in plant tissues or nodules.
Stable isotope probing (SIP)	Identify active microbial taxa involved in nutrient cycles (e.g. ^15N fixation).
Culture-independent single-cell genomics	Recovery of genomes from unculturable microbes; advanced but growing.

17.2.1 Selection of Legume Species and Growth Conditions

The first step in designing a metagenomics study in legumes is the careful selection of plant species and growth conditions. Legumes are a diverse plant family (Fabaceae) with over 22,500 species, including important food, feed, and cover crops such as soybean (*Glycine max*), chickpea (*Cicer arietinum*), pea (*Pisum sativum*), lentil (*Lens culinaris*), cowpea (*Vigna unguiculata*), and alfalfa (*Medicago sativa*) (Hughes et al., 2025). The choice of species should align with the study's objective whether to understand model legume–microbiome interactions (e.g. *Medicago truncatula*), improve crop performance under stress (e.g. *Cicer arietinum* under drought), or explore microbial diversity in under-studied legumes. Environmental and agronomic conditions significantly influence microbial community assembly. Thus, controlled greenhouse experiments offer advantages such as reproducibility and environmental consistency, allowing the isolation of host or soil effects. However, field studies capture realistic microbial diversity and dynamic plant–soil interactions. Ideally, researchers should combine controlled and field environments or use factorial designs to disentangle the effects of soil type, climate, genotype, and management practices. When selecting growth conditions, it is essential to document parameters such as soil physicochemical properties (pH, organic matter, texture, nutrient content), prior crop history, irrigation regime, and fertilisation. These variables affect both microbial composition and function. Moreover, legumes form symbiotic associations with rhizobia and mycorrhizal fungi; therefore, inoculation practices should be standardised or clearly described. If metagenomic comparisons involve inoculated versus uninoculated plants, the rhizobial strain identity and inoculation method must be consistent. Growth stage is another critical consideration sampling at vegetative, flowering, and pod-filling stages often reveals shifts in microbiome structure related to plant phenology and nutrient demand.

17.2.2 Sampling Sites: Soil, Rhizosphere, Nodules, Roots, and Leaves

Microbial communities associated with legumes inhabit distinct niches, including the surrounding soil, rhizosphere, rhizoplane, root endosphere, nodules, and even aerial tissues. Each niche harbours unique microbial assemblages shaped by nutrient gradients, oxygen levels, and plant exudates. Therefore, defining sampling sites precisely and adopting appropriate collection methods are crucial for metagenomic accuracy and comparability.

17.2.3 Bulk Soil

Bulk soil samples represent the microbial background independent of direct root influence (Daniel, 2005). Sampling typically occurs 5–10 cm away from the root zone and at the same depth as the root system. It serves as a baseline to compare with root-associated microbiomes. Soils should be collected using sterile spatulas or augers, avoiding contamination from plant residues. Homogenisation of soil subsamples from multiple plants improves representativeness.

17.2.4 Rhizosphere

The rhizosphere soil tightly adhering to roots is the hotspot of plant–microbe interactions (Ling et al., 2022). To collect rhizosphere samples, roots are gently shaken to remove loosely attached soil, and the remaining soil is brushed or vortexed off in a sterile saline or phosphate buffer. The suspension is then centrifuged to pellet microbial biomass for DNA extraction. In legume studies, rhizosphere sampling must consider the developmental stage and nitrogen fixation status, as root exudation patterns and nodule formation alter microbial community composition.

17.2.5 Root Endosphere and Rhizoplane

To access the root endosphere microbiome, roots are surface-sterilised (usually with 70% ethanol and sodium hypochlorite) to remove external microorganisms. The sterilised tissue is then rinsed thoroughly with sterile water and macerated using liquid nitrogen or mechanical homogenisers. This compartment provides insights into endophytic bacteria and fungi residing within plant tissues, often involved in nutrient cycling, stress tolerance, and plant growth promotion (Ling et al., 2022).

17.2.6 Nodules

Nodules are specialised root organs that host nitrogen-fixing rhizobia(Kay et al., 2014). Their microbiome is of particular interest because, in addition to rhizobia, diverse non-rhizobial bacteria inhabit nodules and may influence symbiotic efficiency. For nodule sampling, healthy nodules are collected, surface-sterilised, and immediately frozen or processed for DNA extraction. If functional metagenomics is intended, care should be taken to collect nodules at the active nitrogen fixation stage (visible by pink leghemoglobin colour). It is often advisable to separate large and small nodules or to sample from different plant zones for comparative analysis.

17.2.7 Leaves and Aerial Tissues

Although most metagenomics work on legumes focuses on belowground compartments, phyllosphere microbiomes are gaining attention (Fadiji & Babalola, 2020). Leaf surfaces harbour distinct bacterial and fungal taxa that contribute to disease suppression and stress tolerance. Leaves should be handled with sterile gloves and tools, and surface sterilisation protocols must be optimised depending on whether epiphytic or endophytic microbes are targeted. Across all sampling sites, consistent spatial and temporal sampling is essential. Variations in depth, plant age, or diurnal cycles can lead to unwanted heterogeneity.

17.2.8 Replication and Controls

Replication and appropriate controls are critical for statistical robustness and biological interpretation in metagenomics. Biological replication using multiple independent plants or plots is necessary to capture natural variation and ensure reproducibility. For soil and rhizosphere studies, at least three to five biological replicates per treatment are recommended. Technical replication (e.g. DNA extraction duplicates) helps assess methodological consistency, especially when dealing with complex matrices like soil or nodules.

Controls should include: *Negative controls* (e.g. sterile buffer, blank DNA extraction controls) to identify contamination during processing. *Positive controls* such as mock microbial communities or reference soils to evaluate sequencing accuracy. *Environmental controls* for example, bulk soil not influenced by plant roots, or uninoculated plants, to assess host- or inoculum-driven effects. Randomisation of sampling order and spatial arrangement minimises bias due to environmental gradients. In field experiments, a randomised complete block design (RCBD) or split-plot layout helps account for spatial heterogeneity in soil conditions. Metadata collection is equally important recording soil temperature, moisture, plant phenology, and environmental parameters enables downstream correlation analyses. Replication should be planned considering sequencing depth and cost; fewer samples with deeper sequencing may be preferable to many shallow samples if the goal is to detect low-abundance taxa or functional genes.

17.2.9 Sample Preservation and Transport Considerations

Microbial community structure and nucleic acid integrity can change rapidly after sampling. Proper preservation and transport protocols are vital to avoid bias before DNA extraction. For most metagenomic applications, flash freezing in liquid nitrogen followed by storage at −80°C is the gold standard (Mordant & Kleiner, 2021). This halts enzymatic degradation and preserves nucleic acids. When liquid nitrogen is unavailable in field conditions, samples can be temporarily stored on dry ice or at −20°C for short periods (less than 24 hours). Commercial DNA/RNA preservation reagents such as LifeGuard Soil Preservation Solution, RNAlater, or DNA/RNA Shield are effective alternatives for stabilising microbial DNA and RNA, especially during transport from remote sites. These reagents prevent nucleic acid degradation without the need for freezing and can maintain sample integrity for several days.

17.3 DNA EXTRACTION AND QUALITY ASSESSMENT

Metagenomic studies of legume microbiomes depend on extracting clean, high-quality DNA from complex samples such as soil, roots, and nodules. This is challenging because these materials contain inhibitors such as humic acids and plant metabolites that interfere with sequencing. Careful sample preparation, effective DNA extraction methods, and thorough quality checks are essential to remove contaminants and obtain reliable data. When done properly, these steps ensure accurate analysis of the microbes associated with legumes and their roles in plant health.

17.4 METAGENOMIC SEQUENCING APPROACHES

Metagenomics has revolutionised the study of complex microbial communities by enabling direct analysis of genetic material recovered from environmental samples without the need for culturing individual microorganisms. In the context of legume-associated microbiomes, metagenomics provides valuable insights into rhizosphere ecology, symbiotic nitrogen fixation, plant growth–promoting rhizobacteria (PGPR), and the molecular mechanisms driving plant–microbe interactions. Various sequencing approaches and protocols are available, each offering specific advantages depending on the research objective, sequencing platform, and desired resolution (Figure 17.2). The following sections describe the major metagenomic sequencing strategies and methodological considerations relevant to legume microbiome studies.

17.4.1 Amplicon-Based Metagenomics (16S rRNA, ITS, *nifH*)

Amplicon-based metagenomics is a targeted sequencing approach focusing on specific conserved marker genes to assess microbial diversity, community composition, and relative abundance (Eloe-Fadrosh et al., 2016). This method is particularly cost-effective and suitable for large sample sets, such as rhizospheric or endophytic communities in legumes.

17.4.2 16S rRNA Gene Sequencing

The 16S ribosomal RNA (rRNA) gene is widely used for bacterial and archaeal community profiling. The gene contains conserved and variable regions (V1–V9), allowing taxonomic discrimination across different phylogenetic levels. For legume-associated microbiota, commonly targeted regions include V3–V4 or V4–V5, which provide an optimal balance between read length and taxonomic resolution. Universal primers such as 341F/806R or 515F/806R are employed for PCR amplification, followed by sequencing on Illumina MiSeq or similar platforms(Johnson et al., 2019).

17.4.3 ITS Sequencing

For fungal communities, the Internal Transcribed Spacer (ITS) region serves as the standard barcode marker(Asthana et al., 2025). ITS1 and ITS2 regions provide high species-level resolution and are suitable for studying legume-associated mycorrhizal fungi, endophytes, and pathogenic communities. Amplicon sequencing of ITS regions facilitates comparative analysis of fungal communities across legume rhizospheres, nodules, and non-legume control soils.

17.4.4 *nifH* Gene Sequencing

The nifH gene, encoding the nitrogenase reductase enzyme, is a key functional marker for nitrogen-fixing microorganisms(Delmont et al., 2018). Targeted amplification and sequencing of nifH allows exploration of diazotrophic diversity in legume nodules and rhizospheres. Degenerate primers such as PolF/PolR or Ueda19F/Ueda407R are typically employed. This approach provides insights into the functional potential of symbiotic and free-living nitrogen fixers contributing to legume productivity.

Amplicon-based metagenomics offers the advantages of simplicity, reproducibility, and established bioinformatics pipelines (e.g. QIIME2, Mothur, DADA2). However, it is limited by PCR amplification biases, inability to capture non-target genes, and relatively low functional resolution.

17.5 SHOTGUN METAGENOMICS

Shotgun metagenomics involves random sequencing of all genomic DNA extracted from environmental samples, allowing comprehensive characterisation of taxonomic diversity and functional potential (Quince et al., 2017). This approach is increasingly preferred for legume metagenome studies aiming to uncover metabolic pathways, gene clusters, and microbial interactions. In shotgun metagenomics, total DNA is fragmented and subjected to high-throughput sequencing without prior amplification of specific genes. The resulting reads represent a mixture of microbial genomes, including bacteria, archaea, fungi, viruses, and host-associated DNA. Bioinformatic analyses such as taxonomic binning, gene annotation, and metabolic reconstruction provide detailed insights into community structure and function. Shotgun metagenomics provides higher resolution and functional insights compared to amplicon-based methods, enabling genome assembly and discovery of novel microbial taxa. However, it is relatively expensive and requires higher DNA quality and sequencing depth. Host DNA contamination, especially from root tissues, can also complicate data analysis.

17.6 LONG-READ VERSUS SHORT-READ SEQUENCING PLATFORMS (ILLUMINA, PACBIO, OXFORD NANOPORE)

The choice of sequencing platform is a critical consideration in metagenomic studies, influencing read length, throughput, error rates, and assembly quality (Xu et al., 2022). Modern sequencing technologies can be broadly classified into short-read (e.g. Illumina) and long-read (e.g. PacBio, Oxford Nanopore) platforms, each offering distinct advantages for legume metagenome analyses.

17.6.1 Illumina Sequencing

Illumina platforms (MiSeq, HiSeq, NovaSeq) are widely used due to their high accuracy and cost-effectiveness for both amplicon and shotgun metagenomics. They generate short reads (100–300 bp) with low error rates (<0.1%), enabling reliable taxonomic profiling and gene quantification. For legume rhizosphere samples, paired-end 2×250 bp sequencing provides sufficient overlap for 16S rRNA and ITS amplicon reconstruction.

17.6.2 PacBio Sequencing

Pacific Biosciences' (PacBio) Single Molecule Real-Time (SMRT) sequencing produces long reads (>10 kb) with improved consensus accuracy (HiFi reads). This platform is valuable for metagenome-assembled genomes (MAGs) and full-length 16S rRNA sequencing, providing higher phylogenetic resolution. PacBio sequencing aids in resolving repetitive regions and reconstructing complete operons such as nif and nod clusters in symbiotic bacteria.

17.6.3 Oxford Nanopore Sequencing

Oxford Nanopore Technologies (ONT) platforms, including MinION and PromethION, offer real-time, portable sequencing with ultra-long reads (up to >100 kb). Although ONT has a higher raw error rate compared to Illumina, continuous improvements in base-calling algorithms and hybrid assembly methods have enhanced accuracy. ONT is particularly advantageous for in-field legume microbiome studies, rapid community profiling, and plasmid or phage genome reconstruction. Short-read technologies are ideal for high-throughput community profiling, whereas long-read platforms excel in genome assembly and resolving structural variants. Hybrid sequencing strategies combining Illumina and PacBio/ONT data are increasingly adopted to achieve accurate and contiguous metagenome assemblies in legume-associated microbial studies.

17.7 LIBRARY PREPARATION AND SEQUENCING DEPTH CONSIDERATIONS

Successful metagenomic analysis depends critically on optimal library preparation and sequencing depth, as these factors determine data quality, coverage, and interpretability. High-quality, intact DNA is essential for reliable sequencing outcomes. For Illumina platforms, libraries are prepared using tagmentation (Nextera XT) or fragmentation ligation workflows to generate insert sizes between 300–500 bp. PacBio and ONT libraries, in contrast, require high-molecular-weight DNA with minimal fragmentation to yield long reads. Library quantification and normalisation ensure balanced representation of samples in multiplexed sequencing runs. Optimal sequencing depth depends on community complexity and study objectives. For amplicon sequencing, 20,000–50,000 reads per sample generally suffice to capture bacterial diversity in rhizosphere communities. In shotgun metagenomics, deeper sequencing (5–20 Gb per sample) is recommended to achieve comprehensive gene coverage and assembly quality. For highly complex soil microbiomes, greater depth improves rare taxa detection and functional annotation reliability.

17.8 BIOINFORMATICS AND DATA ANALYSIS PIPELINES

Metagenomic sequencing generates large volumes of complex data that require systematic computational analysis to extract biologically meaningful insights. Bioinformatics pipelines are therefore essential for processing raw sequencing reads, performing taxonomic classification, predicting gene functions, and visualising microbial community structures. In legume metagenomics, these analyses reveal the composition, diversity, and metabolic potential of microbial consortia inhabiting the rhizosphere, nodules, and endosphere, contributing to our understanding of nitrogen fixation, nutrient cycling, and plant–microbe interactions. This section provides a detailed overview of the major computational steps and commonly used bioinformatics tools for metagenomic data analysis, including quality control, taxonomic profiling, functional annotation, comparative metagenomics, and data visualisation.

17.8.1 Quality Control and Preprocessing (e.g. FastQC, Trimmomatic)

High-throughput sequencing produces millions of short reads that may contain errors, adapter sequences, and low-quality regions. These artifacts can significantly affect downstream analyses, leading to false taxonomic assignments or inaccurate gene predictions. Therefore, quality control (QC) and data preprocessing constitute the first critical stage in metagenomic analysis.

17.8.2 FastQC

FastQC is a widely used open-source tool for initial assessment of raw sequencing data. It provides detailed visual summaries of sequence quality scores, GC content, duplication levels, overrepresented sequences, and adapter contamination (Andrews,

2010). Typical QC metrics include: Per-base quality score: Indicates read accuracy at each nucleotide position. Sequence length distribution: Ensures consistent fragment sizes. GC content: Detects anomalies or contamination. Adapter contamination: Identifies residual adapter sequences from library preparation. The results are displayed in HTML format, allowing quick identification of problematic datasets before trimming or filtering steps.

17.8.3 Trimmomatic

Trimmomatic is a versatile tool used to remove low-quality bases, adapters, and artifacts from Illumina reads (Bolger et al., 2014). It allows both single-end and paired-end data processing using customisable parameters. Key operations include: Sliding window trimming: Removes low-quality regions within a defined window size. Adapter removal: Eliminates sequencing adapters using built-in or user-supplied adapter files. Length filtering: Discards reads shorter than a specified threshold after trimming.

Typical command-line options include:

ILLUMINACLIP: adapters.fa:2:30:10 SLIDINGWINDOW:4:20 MINLEN:100

These parameters remove low-quality bases (quality <20) and discard short reads (<100 bp). After preprocessing, another round of FastQC analysis verifies the improvement in data quality.

17.9 OTHER TOOLS

Additional tools such as Cutadapt, Sickle, and PRINSEQ provide similar functionalities. For long-read data (PacBio, Oxford Nanopore), specialised QC tools such as NanoPlot and Filtlong are recommended. Clean, high-quality reads serve as the foundation for accurate taxonomic and functional analyses.

17.9.1 Taxonomic Profiling Tools (QIIME2, DADA2, Kraken2, MetaPhlAn)

Taxonomic profiling identifies the microbial composition of metagenomic samples, enabling researchers to understand community structure, diversity, and ecological relationships. Depending on sequencing type (amplicon or shotgun), different pipelines and algorithms are employed.

17.9.1.1 QIIME2

QIIME2 (Quantitative Insights Into Microbial Ecology 2) is one of the most comprehensive frameworks for analysing amplicon-based metagenomic data, such as 16S rRNA or ITS sequences (Bolyen et al., 2019). The important features of QIIMF2 include import and demultiplexing of raw reads, denoising using DADA2 or Deblur to correct sequencing errors, taxonomic classification via reference databases such as SILVA, Greengenes, or UNITE, alpha and beta diversity analyses to assess within and between-sample variation.

17.9.1.2 DADA2

DADA2 (Divisive Amplicon Denoising Algorithm 2) is both a standalone R package and a QIIME2 plugin that performs high-resolution sequence variant detection(Callahan et al., 2016). Unlike traditional OTU (Operational Taxonomic Unit) clustering, DADA2 identifies Amplicon Sequence Variants (ASVs) at single-nucleotide resolution, providing improved accuracy and reproducibility.

17.9.1.3 Kraken2

For shotgun metagenomics, Kraken2 provides ultrafast taxonomic classification based on exact k-mer matching against reference genomes(Wood et al., 2019). It constructs a database of k-mers associated with known taxa and assigns reads by determining the lowest common ancestor. Kraken2 is computationally efficient, enabling large-scale analyses of soil or root metagenomes.

17.9.1.4 MetaPhlAn

MetaPhlAn (Metagenomic Phylogenetic Analysis) uses clade-specific marker genes instead of entire genomes for taxonomic profiling(Segata et al., 2012). This approach reduces computational demands and improves accuracy in complex communities. MetaPhlAn outputs species-level relative abundances and is particularly suitable for human and plant microbiomes. Its integration with HUMAnN enables combined taxonomic and functional profiling, providing a comprehensive view of microbial ecology.

17.10 FUNCTIONAL ANNOTATION AND GENE PREDICTION (PROKKA, KEGG, EGGNOG, COG)

Functional annotation aims to identify genes and predict metabolic capabilities from metagenomic assemblies. This step is crucial in understanding how microbial communities contribute to nutrient cycling, nitrogen fixation, and plant growth promotion in legumes.

17.10.1 Prokka

Prokka is a rapid prokaryotic genome annotation tool designed for assembled contigs or metagenome-assembled genomes (MAGs)(Seemann, 2014). It predicts open reading frames (ORFs), tRNAs, rRNAs, and other genomic features using embedded databases such as UniProt and RefSeq. The output includes annotated GenBank, GFF, and FASTA files compatible with downstream visualisation tools. Prokka is often the first step after assembly in shotgun metagenome pipelines.

17.10.2 Kyoto Encyclopedia of Genes and Genomes (KEGG)

The KEGG database provides a comprehensive resource for mapping genes to metabolic pathways(Kanehisa & Goto, 2000). After annotation, functional genes are assigned to KEGG Orthology (KO) identifiers, enabling pathway reconstruction. In legume microbiome studies, KEGG analysis can reveal enrichment in pathways related to: nitrogen fixation (nif, fix, nod genes), phosphate solubilisation, siderophore biosynthesis, phytohormone production (IAA, cytokinin biosynthesis). Tools such as KAAS (KEGG Automatic Annotation Server) or GhostKOALA automate KEGG mapping from protein sequences. eggNOG and COG eggNOG (evolutionary genealogy of genes: Non-supervised Orthologous Groups) and COG (Clusters of Orthologous Groups) databases classify genes into orthologous clusters for functional categorisation (Huerta-Cepas et al., 2019). eggNOG-mapper uses DIAMOND or HMMER-based alignment for large datasets, providing annotation in terms of Gene Ontology (GO) terms, KEGG pathways, and COG categories. COG analysis aids in identifying gene functions across broad functional classes such as energy production, amino acid metabolism, or signal transduction. Together, these annotation frameworks elucidate the ecological roles and biotechnological potential of legume-associated microbes.

17.11 COMPARATIVE METAGENOMICS AND DIVERSITY ANALYSES

Comparative metagenomics involves the quantitative comparison of microbial communities across different samples, environments, or host species. In legume systems, it helps identify microbial signatures associated with specific legume species, soil types, or symbiotic efficiencies.

17.11.1 Alpha and Beta Diversity

Alpha diversity measures within-sample diversity using indices such as Shannon, Simpson, and Chao1, reflecting species richness and evenness (Daniel, 2005). Beta diversity assesses differences between samples using dissimilarity measures such as Bray–Curtis, Jaccard, or UniFrac distances. Principal Coordinate Analysis (PCoA) or Non-metric Multidimensional Scaling (NMDS) visualisations are employed to observe clustering patterns.

For example, comparing the rhizosphere microbiome of chickpea and lentil may reveal distinct bacterial communities enriched in nitrogen-fixing taxa or stress-tolerant microbes.

17.11.2 Differential Abundance Analysis

Statistical frameworks such as DESeq2, ALDEx2, or LEfSe (Linear Discriminant Analysis Effect Size) identify taxa or functions significantly enriched in particular conditions (e.g. nodulated versus non-nodulated soils). These analyses pinpoint indicator species and functional pathways crucial for plant performance.

17.11.3 Metagenome-Assembled Genomes (MAGs)

Comparative genome reconstruction identifies MAGs shared or unique among legume systems(Bowers et al., 2017). This allows tracking of beneficial microbial populations and evaluating genome-level adaptations, such as nitrogenase gene cluster organisation or stress-response pathways.

17.11.4 Integration with Environmental Metadata

Combining metagenomic data with soil physicochemical parameters, plant growth data, or geographic metadata provides holistic insights into microbial ecology. Multivariate analyses (Canonical Correspondence Analysis, Redundancy Analysis) reveal correlations between community composition and environmental variables.

17.11.5 Statistical and Visualisation Tools (R, Python, Cytoscape)

Effective statistical analysis and visualisation are vital for interpreting complex metagenomic datasets and communicating results clearly. A variety of programming languages and visualisation tools support these tasks.

17.11.6 R Programming

R is the most widely used statistical environment for metagenomic data analysis due to its robust ecosystem of packages:

17.11.7 Phyloseq

Integrates taxonomic, phylogenetic, and environmental data for diversity analyses.

17.11.8 Vegan

Performs community ecology statistics (ordination, diversity indices, permutation tests).

17.11.9 ggplot2

Generates publication-quality graphics for data visualisation.

17.11.10 Cytoscape

Cytoscape is a powerful platform for network visualisation and analysis. It represents microbial associations as networks, where nodes denote taxa or genes and edges represent co-occurrence or correlation relationships. In legume metagenome studies, Cytoscape helps visualise: microbial co-occurrence networks in nodules or rhizospheres, functional interaction networks linking metabolic pathways and gene clusters. Plugins such as Clustermaker, CytoHubba, and STRINGApp enhance its functionality by identifying hub species or key metabolic interactions.

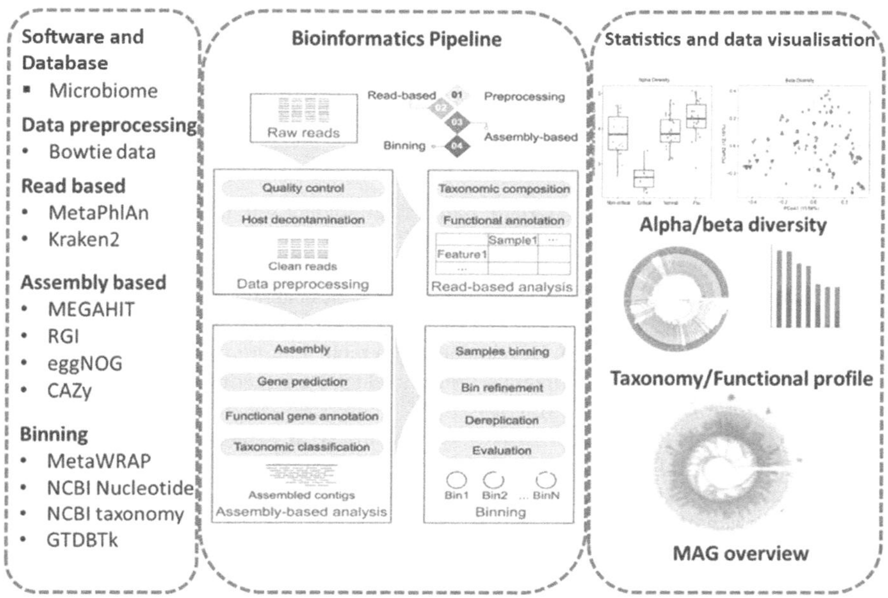

FIGURE 17.2 Metagenome Bioinformatics pipeline. The pipelines contain three key components, including software and database installation, data analysis pipeline, statistics and visualisation.

17.12 FUTURE TRENDS IN LEGUME METAGENOMICS AND SYNTHETIC MICROBIOMES

Legumes are ideal plants for studying plant–microbe partnerships because of their close link with nitrogen-fixing bacteria and their importance as food and fodder crops. Research is now shifting from simply identifying which microbes are present to understanding what they do and how they support plant growth, nutrition, and stress tolerance. By combining multi-omics tools, advanced sequencing, and AI-based analysis, scientists can link microbial functions to real field outcomes like yield and resilience. At the same time, synthetic microbial communities are being developed to improve nodulation, nutrient use, and stress resistance in legumes, offering more reliable alternatives to single-strain inoculants. Progress in formulation, delivery, regulation, and farmer participation will be key to success. Overall, integrating metagenomics, computation, and practical field application promises more resilient legumes, reduced chemical inputs, and more sustainable agriculture.

17.13 CONCLUSION

Legumes are valuable models for studying plant–microbe interactions because of their role in nitrogen fixation and agriculture. Research is now moving beyond identifying microbes to understanding how they function and support plant health, stress tolerance, and yield. By combining advanced omics tools, AI, and designed microbial communities, scientists aim to develop more resilient legume crops, reduce fertiliser use, and promote sustainable farming.

REFERENCES

Ajilogba, C. F., & Babalola, O. O. (2020). Bambara groundnut soil metagenomics data. *Data in Brief.* https://doi.org/10.1016/j.dib.2020.105542

Andrews, S. (2010). FastQC: A quality control tool for high throughput sequence data. *Babraham Bioinformatics.* http://www.bioinformatics.babraham.ac.uk/projects/fastqc/.

Asthana, V., Nieves, E. M., Bugga, P., Smith, C., Dunn, T., Narayanasamy, S., Dickson, R. P., & VanEpps, J. S. (2025). Development of a rapid, culture-free, universal microbial identification system using internal transcribed spacer targeting primers. *The Journal of Infectious Diseases, 231*(5), 1155–1164.

Basu, A., Chalasani, D., Sarma, P., Uikey, S., Chenna, V. R., Choudhari, P. L., & Podile, A. R. (2025). Influence of genotype, nodule position, and edaphic factors on microbial diversity and assembly of pigeonpea (Cajanus cajan) root nodules in Indian soils. *Environmental Microbiome, 20*(1), 41.

Bigatton, E. D., Verdenelli, R. A., Haro, R. J., Ayoub, I., Barbero, F. M., Mart\'\in, M. P., Dubini, L. E., Jorr\'\in Novo, J. V, Lucini, E. I., & Castillejo, M. Á. (2024). Metagenomic analysis to assess the impact of plant growth-promoting rhizobacteria on peanut (Arachis hypogaea L.) crop production and soil enzymes and microbial diversity. *Journal of Agricultural and Food Chemistry, 72*(40), 22385–22397.

Bolger, A. M., Lohse, M., & Usadel, B. (2014). Trimmomatic: A flexible trimmer for Illumina sequence data. *Bioinformatics.* https://doi.org/10.1093/bioinformatics/btu170

Bolyen, E., Rideout, J. R., Dillon, M. R., Bokulich, N. A., Abnet, C. C., Al-Ghalith, G. A., Alexander, H., Alm, E. J., Arumugam, M., Asnicar, F., Bai, Y., Bisanz, J. E., Bittinger, K., Brejnrod, A., Brislawn, C. J., Brown, C. T., Callahan, B. J., Caraballo-Rodríguez, A. M., Chase, J., . . . Caporaso, J. G. (2019). Reproducible, interactive, scalable and extensible microbiome data science using QIIME 2. *Nature Biotechnology.* https://doi.org/10.1038/s41587-019-0209-9

Bowers, R. M., Kyrpides, N. C., Stepanauskas, R., Harmon-Smith, M., Doud, D., Reddy, T. B. K., Schulz, F., Jarett, J., Rivers, A. R., Eloe-Fadrosh, E. A., Tringe, S. G., Ivanova, N. N., Copeland, A., Clum, A., Becraft, E. D., Malmstrom, R. R., Birren, B., Podar, M., Bork, P., . . . Woyke, T. (2017). Minimum information about a single amplified genome (MISAG) and a metagenome-assembled genome (MIMAG) of bacteria and archaea. *Nature Biotechnology.* https://doi.org/10.1038/nbt.3893

Brescia, F., Sillo, F., Franchi, E., Pietrini, I., Montesano, V., Marino, G., Haworth, M., Zampieri, E., Fusini, D., Schillaci, M., Papa, R., Santamarina, C., Vita, F., Chitarra, W., Nerva, L., Petruzzelli, G., Mennone, C., Centritto, M., & Balestrini, R. (2023). The 'microbiome counterattack': Insights on the soil and root-associated microbiome in diverse chickpea and lentil genotypes after an erratic rainfall event. *Environmental Microbiology Reports.* https://doi.org/10.1111/1758-2229.13167

Brown, S. P., Grillo, M. A., Podowski, J. C., & Heath, K. D. (2020). Soil origin and plant genotype structure distinct microbiome compartments in the model legume Medicago truncatula. *Microbiome.* https://doi.org/10.1186/s40168-020-00915-9

Callahan, B. J., McMurdie, P. J., Rosen, M. J., Han, A. W., Johnson, A. J. A., & Holmes, S. P. (2016). DADA2: High-resolution sample inference from Illumina amplicon data. *Nature Methods.* https://doi.org/10.1038/nmeth.3869

Chartrel, V., Dugat-Bony, E., Sarthou, A. S., Huchette, S., Bonnarme, P., & Irlinger, F. (2021). The microbial community associated with pea seeds (Pisum sativum) of different geographical origins. *Plant and Soil*. https://doi.org/10.1007/s11104-021-04856-6

Daniel, R. (2005). The metagenomics of soil. *Nature Reviews Microbiology*. https://doi.org/10.1038/nrmicro1160

Dash, B., Nayak, S., Pahari, A., & Nayak, S. K. (2020). Verrucomicrobia in soil: An agricultural perspective. *Frontiers in Soil and Environmental Microbiology*. https://doi.org/10.1201/9780429485794-4

Delmont, T. O., Quince, C., Shaiber, A., Esen, Ö. C., Lee, S. T., Rappé, M. S., MacLellan, S. L., Lücker, S., & Eren, A. M. (2018). Nitrogen-fixing populations of Planctomycetes and Proteobacteria are abundant in surface ocean metagenomes. *Nature Microbiology*. https://doi.org/10.1038/s41564-018-0176-9

Eloe-Fadrosh, E. A., Ivanova, N. N., Woyke, T., & Kyrpides, N. C. (2016). Metagenomics uncovers gaps in amplicon-based detection of microbial diversity. *Nature Microbiology*. https://doi.org/10.1038/NMICROBIOL.2015.32

Fadiji, A. E., & Babalola, O. O. (2020). Metagenomics methods for the study of plant-associated microbial communities: A review. *Journal of Microbiological Methods*. https://doi.org/10.1016/j.mimet.2020.105860

Goel, R., Kumar, V., Suyal, D. C., Narayan, N., & Soni, R. (2018). Toward the unculturable microbes for sustainable agricultural production. *Role of Rhizospheric Microbes in Soil: Stress Management and Agricultural Sustainability*. https://doi.org/10.1007/978-981-10-8402-7_4

Gurbich, T. A., Almeida, A., Beracochea, M., Burdett, T., Burgin, J., Cochrane, G., Raj, S., Richardson, L., Rogers, A. B., Sakharova, E., Salazar, G. A., & Finn, R. D. (2023). MGnify genomes: A resource for biome-specific microbial genome catalogues: MGnify biome-specific genome catalogues. *Journal of Molecular Biology*. https://doi.org/10.1016/j.jmb.2023.168016

Handelsman, J., Rondon, M. R., Brady, S. F., Clardy, J., & Goodman, R. M. (1998). Molecular biological access to the chemistry of unknown soil microbes: A new frontier for natural products. *Chemistry and Biology*. https://doi.org/10.1016/S1074-5521(98)90108-9

Hartman, K., van der Heijden, M. G. A., Roussely-Provent, V., Walser, J. C., & Schlaeppi, K. (2017). Deciphering composition and function of the root microbiome of a legume plant. *Microbiome*. https://doi.org/10.1186/s40168-016-0220-z

Huerta-Cepas, J., Szklarczyk, D., Heller, D., Hernández-Plaza, A., Forslund, S. K., Cook, H., Mende, D. R., Letunic, I., Rattei, T., Jensen, L. J., Von Mering, C., & Bork, P. (2019). EggNOG 5.0: A hierarchical, functionally and phylogenetically annotated orthology resource based on 5090 organisms and 2502 viruses. *Nucleic Acids Research*. https://doi.org/10.1093/nar/gky1085

Hughes, C., Ringelberg, J. J., & Bruneau, A. (2025). Legumes. *Current Biology, 35*(9), R323–R328.

Johnson, J. S., Spakowicz, D. J., Hong, B. Y., Petersen, L. M., Demkowicz, P., Chen, L., Leopold, S. R., Hanson, B. M., Agresta, H. O., Gerstein, M., Sodergren, E., & Weinstock, G. M. (2019). Evaluation of 16S rRNA gene sequencing for species and strain-level microbiome analysis. *Nature Communications*. https://doi.org/10.1038/s41467-019-13036-1

Kanehisa, M., & Goto, S. (2000). KEGG: Kyoto Encyclopedia of Genes and Genomes. *Nucleic Acids Research*. https://doi.org/10.1093/nar/28.1.27

Kay, G. L., Sergeant, M. J., Giuffra, V., Bandiera, P., Milanese, M., Bramanti, B., Bianucci, R., & Pallen, M. J. (2014). Recovery of a medieval Brucella melitensis genome using shotgun metagenomics. *MBio*. https://doi.org/10.1128/mBio.01337-14

Kochetkova, T. V., Toshchakov, S. V., Zayulina, K. S., Elcheninov, A. G., Zavarzina, D. G., Lavrushin, V. Y., Bonch-Osmolovskaya, E. A., & Kublanov, I. V. (2020). Hot in cold: Microbial life in the hottest springs in permafrost. *Microorganisms*. https://doi.org/10.3390/microorganisms8091308

Ling, N., Wang, T., & Kuzyakov, Y. (2022). Rhizosphere bacteriome structure and functions. *Nature Communications*. https://doi.org/10.1038/s41467-022-28448-9

Liu, F., Hewezi, T., Lebeis, S. L., Pantalone, V., Grewal, P. S., & Staton, M. E. (2019). Soil indigenous microbiome and plant genotypes cooperatively modify soybean rhizosphere microbiome assembly. *BMC Microbiology*. https://doi.org/10.1186/s12866-019-1572-x

Lopes, L. D., Hao, J., & Schachtman, D. P. (2021). Alkaline soil pH affects bulk soil, rhizosphere and root endosphere microbiomes of plants growing in a Sandhills ecosystem. *FEMS Microbiology Ecology*. https://doi.org/10.1093/femsec/fiab028

Lundberg, D. S., Lebeis, S. L., Paredes, S. H., Yourstone, S., Gehring, J., Malfatti, S., Tremblay, J., Engelbrektson, A., Kunin, V., Rio, T. G. Del, Edgar, R. C., Eickhorst, T., Ley, R. E., Hugenholtz, P., Tringe, S. G., & Dangl, J. L. (2012). Defining the core Arabidopsis thaliana root microbiome. *Nature*. https://doi.org/10.1038/nature11237

Mehrshad, M., Salcher, M. M., Okazaki, Y., Nakano, S., Šimek, K., Andrei, A.-S., & Ghai, R. (2018). Hidden in plain sight – highly abundant and diverse planktonic freshwater Chloroflexi. *Microbiome*. https://doi.org/10.1186/s40168-018-0563-8

Mordant, A., & Kleiner, M. (2021). Evaluation of sample preservation and storage methods for metaproteomics analysis of intestinal microbiomes. *Microbiology Spectrum*. https://doi.org/10.1128/spectrum.01877-21

Nayfach, S., Roux, S., Seshadri, R., Udwary, D., Varghese, N., Schulz, F., Wu, D., Paez-Espino, D., Chen, I. M., Huntemann, M., Palaniappan, K., Ladau, J., Mukherjee, S., Reddy, T. B. K., Nielsen, T., Kirton, E., Faria, J. P., Edirisinghe, J. N., Henry, C. S., . . . Eloe-Fadrosh, E. A. (2021). A genomic catalog of Earth's microbiomes. *Nature Biotechnology*. https://doi.org/10.1038/s41587-020-0718-6

Quince, C., Walker, A. W., Simpson, J. T., Loman, N. J., & Segata, N. (2017). Shotgun metagenomics, from sampling to analysis. *Nature Biotechnology*. https://doi.org/10.1038/nbt.3935

Seemann, T. (2014). Prokka: Rapid prokaryotic genome annotation. *Bioinformatics*. https://doi.org/10.1093/bioinformatics/btu153

Segata, N., Waldron, L., Ballarini, A., Narasimhan, V., Jousson, O., & Huttenhower, C. (2012). Metagenomic microbial community profiling using unique clade-specific marker genes. *Nature Methods*. https://doi.org/10.1038/nmeth.2066

Shi, W., Qi, H., Sun, Q., Fan, G., Liu, S., Wang, J., Zhu, B., Liu, H., Zhao, F., Wang, X., Hu, X., Li, W., Liu, J., Tian, Y., Wu, L., & Ma, J. (2019). GcMeta: A global catalogue of metagenomics platform to support the archiving, standardization and analysis of microbiome data. *Nucleic Acids Research*. https://doi.org/10.1093/nar/gky1008

Shi, Y., Li, Y., Yang, T., & Chu, H. (2021). Threshold effects of soil pH on microbial co-occurrence structure in acidic and alkaline arable lands. *Science of the Total Environment*. https://doi.org/10.1016/j.scitotenv.2021.149592

Tyson, G. W., Chapman, J., Hugenholtz, P., Allen, E. E., Ram, R. J., Richardson, P. M., Solovyev, V. V., Rubin, E. M., Rokhsar, D. S., & Banfield, J. F. (2004). Community structure and metabolism through reconstruction of microbial genomes from the environment. *Nature*. https://doi.org/10.1038/nature02340

Venturi, V., & Keel, C. (2016). Signaling in the rhizosphere. *Trends in Plant Science*. https://doi.org/10.1016/j.tplants.2016.01.005

Wiegand, S., Jogler, M., Boedeker, C., Pinto, D., Vollmers, J., Rivas-Marín, E., Kohn, T., Peeters, S. H., Heuer, A., Rast, P., Oberbeckmann, S., Bunk, B., Jeske, O., Meyerdierks, A., Storesund, J. E., Kallscheuer, N., Lücker, S., Lage, O. M., Pohl, T., . . . Jogler, C. (2020). Cultivation and functional characterization of 79 planctomycetes uncovers their unique biology. *Nature Microbiology*. https://doi.org/10.1038/s41564-019-0588-1

Wood, D. E., Lu, J., & Langmead, B. (2019). Improved metagenomic analysis with Kraken 2. *Genome Biology*. https://doi.org/10.1186/s13059-019-1891-0

Xu, G., Zhang, L., Liu, X., Guan, F., Xu, Y., Yue, H., Huang, J. Q., Chen, J., Wu, N., & Tian, J. (2022). Combined assembly of long and short sequencing reads improve the efficiency of exploring the soil metagenome. *BMC Genomics*. https://doi.org/10.1186/s12864-021-08260-3

Ye, G., Hong, H., Li, T., Li, J., Wu, J.-Q., Jiang, S., Meng, Z.-T., Yuan, H.-T., Xue, W., Li, A.-L., & others. (2025). MAGdb: A comprehensive high quality MAGs repository for exploring microbial metagenome-assemble genomes. *Genome Biology*, *26*(1), 276.

Zhang, J., Wang, N., Li, S., Wang, J., Feng, Y., Wang, E., Li, Y., Yang, T., & Chen, W. (2023). The effect of different rhizobial symbionts on the composition and diversity of rhizosphere microorganisms of chickpea in different soils. *Plants*. https://doi.org/10.3390/plants12193421

Zhou, Y., Zhu, H., Fu, S., & Yao, Q. (2017). Variation in soil microbial community structure associated with different legume species is greater than that associated with different grass species. *Frontiers in Microbiology*. https://doi.org/10.3389/fmicb.2017.01007

18 The Bioinformatics Lens to Legume Metagenomic Data Analysis

Tikshana Yadav, Varsha Singh, Nikita Srivastava, Pooja Sharma, and Surendra Pratap Singh

18.1 INTRODUCTION

With the advent of research on plant–microbial interactions, particularly those involving rhizobia and arbuscular mycorrhizal fungi, the diversity of microbes identified in the plant rhizosphere has increased. Numerous studies have elucidated the evolutionary dynamics and communication mechanisms between these microbes and their plant hosts (Tsiknia et al., 2021). Legume trees play a vital role in nitrogen-fixing symbioses, boosting plant growth through nutrient cycling facilitated by the soil microbiome community. Metagenomic and marker analyses reveal an increased microbial capacity for anaerobic iron reduction (Epihov et al., 2021). To thrive, leguminous plants undergo a process called "nodulation," during which specialised structures called "nodules" harbour leghemoglobin, which provides essential oxygen for rhizobia – the bacteria that partner with these plants. Rhizobia respond to signals known as lipochitooligosaccharides, produced by plants to initiate nodulation. The uptake of chemical signals by rhizosphere transporters indicates the existence of highly coordinated, light-dependent processes in the legume root system (Tkacz & Poole, 2021). The metagenome of legumes is important because it includes a variety of rhizobia endosymbionts found in the nodules of leguminous plants (Johnson, 2019). The genetic profile of resistant strains is determined by extracting DNA from root nodules using amplicon-based methods, such as 16S ribosomal DNA sequencing and multi-locus sequence analysis. These approaches are vital for unravelling the rich tapestry of nodule rhizobia diversity (Weir et al., 2004). Within a single legume nodule, populations of diverse rhizobial and non-rhizobial species coexist; thus, isolating and sequencing via in vitro techniques may inadvertently elevate the prominence of the most dominant taxa. In contrast, direct DNA sequencing of whole nodules not only minimises bias towards specific taxa but also offers a more authentic glimpse into the complex symbiotic relationships within (Eaton et al., 2024). With the advent of deep sequencing, large-scale sequencing libraries are prepared by extracting high-quality DNA from entire nodule tissue (Healey et al., 2014). Legume metagenome refers to the total genomic DNA extracted from environmental samples taken from nodules of leguminous plants. Metagenomics, first described by Handelsman et al.

 DOI: 10.1201/9781003733324-19

(1998), is a major research tool in microbial ecology that studies these metagenomes. This approach helps researchers gain a deeper understanding of the ecological roles, metabolic functions, and evolutionary histories of microbes across different ecosystems by analysing DNA directly collected from the environment (Visconti et al., 2018). Metagenomics provides a latitudinal gradient of biodiversity, through taxonomic distinction, which refers to the classification of organisms into groups based on shared characteristics. In contrast, functional distinction involves categorising organisms by their roles or activities within an ecosystem (Yadav et al., 2024). By studying both taxonomic and functional differences among biomes, metagenomics reveals how microbial communities vary and adapt across different environments (Coban et al., 2022). In this chapter, we have discussed databases such as RDP, SILVA, Greengenes, EzTaxon-e, MG-RAST, IMG/M, and tools such as MEGAN, SEPP, QIIME, and AMPHORA that can be used for legume metagenomics, along with the pipelines. The pipeline for legume metagenomics discussed in this chapter is illustrated in Figure 18.1 and describes stepwise processes such as quality control, sequence analysis, functional pathway annotation, interpretation, and validation. The respective tools have been described in detail in this chapter.

18.2 DATABASES FOR LEGUME METAGENOMICS

There are plenty of databases that can be used for analysing legume metagenome, a few are described below, and many more are mentioned in Table 18.1 along with their functions.

18.3 RIBOSOMAL DATABASE PROJECT (RDP)

The RDP provides aligned and annotated rRNA gene sequences and analysis tools across various fields, including human health, microbial ecology, and phylogenetics. It includes fungal large-subunit rRNA genes and bacterial and archaeal small-subunit rRNA genes. High-throughput sequencing has advanced the characterisation of environmental microbial populations, resulting in larger datasets. RDP offers tools for taxonomic classification, nearest-neighbour searches, primer-probe testing, and tree building, specifically designed for handling high-volume amplicon sequence data. Sequences are aligned and classified using RDP's aligner and classifier, with classifications supported by a bootstrap of 80% or higher. The Infernal Aligner incorporates secondary structure information, and RDP browsers provide interactive interfaces for exploring the sequence collection. Users can analyse subsets of sequences from various categories, including type and non-type strains, as well as partial or high-quality sequences. The Genome Browser organises rRNA sequences from genome projects, offering details on rRNA copy number and genome size, along with links to additional information (Cole et al., 2014; https://github.com/rdpstaff/classifier).

18.4 SILVA

The SILVA database is a comprehensive web resource that provides quality-controlled databases of aligned ribosomal RNA (rRNA) gene sequences from the

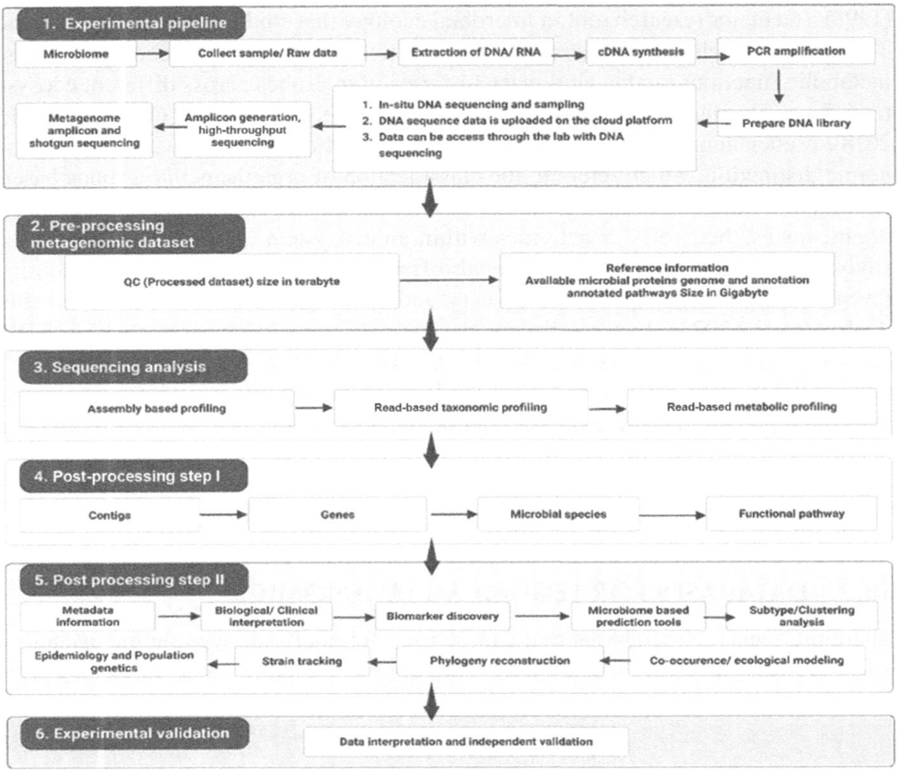

FIGURE 18.1 A comprehensive workflow for legume metagenomic data analysis. It states a pipeline and steps to design an experiment, quality control, sequence analysis, post-processing, and validation.

Bacteria, Archaea, and Eukaryota domains. It includes sequences from both the small and large subunits of rRNA. The database also provides sequence-related information, including taxonomic classification, multiple sequence alignments, type strain information, and the latest valid nomenclature. These datasets serve as reference data in next-generation sequencing (NGS) analysis pipelines such as MOTHUR, QIIME, and MG-RAST. The selection of rRNA gene candidate sequences is enhanced by hidden Markov model-based rRNA gene prediction. It has improved and fine-tuned the quality criteria to ensure that only reliable sequence information is included in its databases. Users can access the Taxonomy Browser and Search pages, where the Browser provides a hierarchical view of the database contents, like a file browser, allowing visualisation of the various taxonomies included in SILVA. Users can upload FASTA files containing up to 1,000 sequences, and the sequences will be aligned using the same reference alignments utilised to prepare the SILVA databases. Signature sequences are essential for studying microbial communities. However, since these signatures are derived from previously characterised sequences, their accuracy is limited by the diversity covered by the available data at the time of their design. The probe matches evaluation tool tests and visualises the *in silico* target group coverage of rRNA gene-targeting probes and single primers, with the option to include ambiguity codons against the SILVA datasets. The tool can be configured to allow up to five mismatches between the probe and target sequences, and these mismatches can also be weighted. Additionally, the TestPrime tool enables users to search for sequences within the SILVA databases that are targeted by a pair of primers. Users can configure the maximum number of allowed base mismatches and set a 'zero tolerance zone' at the 3' end of primers (Quast et al., 2012; https://www.arb-silva.de/).

18.5 GREENGENES

This database for the 16S rRNA gene focuses on chimera screening, standard alignment, and taxonomic classification using various established taxonomies. It offers a standardised array of descriptive fields, taxonomic designations, chimera screening, and compatibility with ARB. It ensures a uniform multiple-sequence alignment for both archaeal and bacterial 16S small-subunit rRNA genes to aid in taxonomic categorisation. A key feature of the Greengenes data assembly process is thorough chimera evaluation. Each sequence is evaluated for its potential to be chimeric, a breakpoint is determined, and the parent sequences are identified. It provides an illustration and examination of the data assembly process, along with an assessment of the reliability of prokaryotic candidate phyla; we aligned and assessed chimeras in over 90,000 publicly available 16S small subunit rRNA gene sequences. The database provides enhanced control over probe selection input to the probe selection software, which should help prevent inaccurate predictions of terminal restriction fragment length polymorphism patterns. It serves as a functional workbench for analysing user-generated 16S rRNA gene sequences, and batches of sequencing reads can be uploaded for quality-based trimming and the creation of multiple-sequence alignments (DeSantis et al., 2006; https://www.lcsciences.com/documents/sample_data/16S_sequencing/src/html/top2.html).

18.6 EZTAXON-E

The EzTaxon-e database contains 34,607 representative 16S rRNA gene sequences from prokaryotes. It is widely used for the routine identification of prokaryotic isolates, as well as sequences from uncultured prokaryotes. All sequences in the database have been taxonomically assigned to create a complete hierarchical system based on 16S rRNA gene sequence similarity and phylogenetic relationships. For formal taxonomic descriptions, it is recommended to use 16S rRNA gene sequences with completeness values greater than 95%. Incomplete or partial sequences can hinder various taxonomic analyses, including BLAST searches and phylogenetic tree analyses. According to phylogenetic relationships, sequences belonging to the major phyla of Bacteria and Archaea form clusters that support taxonomic assignments (Kim et al., 2012; https://eztaxon-e.ezbiocloud.net/).

18.7 MG-RAST

This platform has served as a public resource for the annotation and analysis of metagenomic sequence data, currently housing over 150,000 datasets totalling more than 60 terabases, with over 23,000 datasets available to the public. It facilitates the study of microbiota through assays such as amplicon sequencing, shotgun metagenomics, and metatranscriptomics. Providing large quantities of data analysed using standardised procedures enables the comparison of annotated sequence data and functional reconstructions. The platform centralises analysis tools and services to support data processing development, while also archiving and providing access to originally submitted sequence data and their corresponding results (Wilke et al., 2016; https://www.mg-rast.org/, https://github.com/MG-RAST/MG-RAST-Tools).

18.8 IMG/M

The integrated microbial genomes and metagenomes system provides a comprehensive framework for analysing microbial community aggregate genomes, referred to as metagenomes. It combines datasets from isolated microbial genomes with assembled or unassembled metagenomic data generated using 454 or Illumina sequencing platforms. The IMG/M annotation pipeline is employed to process these datasets. For unassembled reads, a quality control step is implemented that includes quality trimming, detection of low-complexity regions, and masking and removal of technical replicates. Sequences are annotated by the same pipeline that identifies CRISPR repeats, non-coding RNAs, and protein-coding genes. Predictions for RNAs are conducted using tRNAscan-SE for tRNAs, while in-house developed HMM models are used for rRNAs. Protein-coding sequences (CDSs) are identified through ab initio gene prediction tools. Additionally, sequences ranging from 100 to 800 bp are compared to the IMG non-redundant protein database using BLASTx to identify CDSs that might have been overlooked by ab initio tools. The analysis of metagenomes addresses questions about phylogenetic composition and the functional or metabolic potential of individual microbiomes. This system also facilitates the integration of metagenomic datasets with isolated microbial genomes, using NCBI RefSeq as the source of sequence data. It incorporates draft and complete microbial genomes from bacteria, archaea, and eukaryotes, as well as plasmids and viruses. Furthermore, it records primary sequence information for both isolate genomes and metagenomes in scaffolds and/or contigs, along with computationally predicted protein-coding and RNA-coding sequences (Markowitz et al., 2012; https://img.jgi.doe.gov/cgi-bin/m/main.cgi).

18.9 TOOLS FOR LEGUME METAGENOMICS

There are plenty of tools that can be used for analysing legume metagenome; a few are described below, and many more are mentioned in Table 18.2 – that is, stepwise tools along with their functions.

TABLE 18.1
List of Databases with Their Usage in Legume Metagenomics

Database	Brief information/function	Links	Reference
16S rRNA database (SILVA)	Library of high-quality rRNA sequences for bacteria, archaea, and eukarya.	https://www.arb-silva.de/	Quast et al. (2013)
Complete genome database (RefSeq)	Non-redundant reference for genomic research.	https://www.ncbi.nlm.nih.gov/refseq	O'Leary et al. (2016)
Pathway database (MetaCyc)	Library of enzymes and metabolic pathways.	https://metacyc.org/	Caspi et al. (2018)
Gene families database (UniRef90)	Metagenomic analysis, functional annotation, and sequence similarity searches.	https://www.uniprot.org/	Suzek et al. (2007)
VOGDB	Viral genome annotation.	https://vogdb.org/	Grazziotin et al. (2017)

18.10 MEGAN

The metagenome analyser handles large data sets. After preprocessing, it estimates and explores the dataset's taxonomic content. It uses a simple algorithm that assigns each read to the lowest common ancestor (LCA) of the set of taxa, reflecting the distance: species-specific sequences are assigned to taxa near the leaves of the NCBI tree, while widely conserved sequences are assigned to higher taxa closer to the root. Analysis is initiated by opening the output file of any member of the BLAST family. This is a typical processing pipeline for an initial analysis of a metagenomic sample. It involves collecting shotgun reads, then comparing them against one or more databases of known sequences. The results of these comparisons are processed to gather all hits and assign a taxon ID to each sequence according to NCBI taxonomy. Finally, the user interacts with the program to run the lowest common ancestor (LCA) algorithm, analyse the data, inspect the assignment of individual reads to taxa based on their hits, and produce summaries of the results at different levels of NCBI taxonomy (Huson et al., 2007; http://www-ab.informatik.uni-tuebingen.de/software/megan).

18.11 SEPP

This method serves as a meta-approach for two-step phylogenetic placement, employing the dataset decomposition technique found in SATé-2, which simultaneously estimates sequence alignments and phylogenies. It begins by generating an initial alignment and a maximum likelihood (ML) tree, then uses these results to compute updated alignments and subsequently estimate a new ML tree based on those alignments. If the new alignment/tree pair yields a lower ML score, realignment is performed using an alternative decomposition strategy. The sizes of the alignment and placement subsets influence the accuracy, computational time, and memory usage of SEPP (Mirarab et al., 2012).

18.12 QUANTITATIVE INSIGHTS INTO MICROBIAL ECOLOGY (QIIME)

It performs microbial community analysis. It analyses and interprets nucleic acid sequence data from fungal, viral, bacterial, and archaeal communities. Sequencer data from microbial communities are analysed. It processes data from a high throughput 16S rRNA sequencing study, from multiplexed sequence reads to taxonomic and phylogenetic profiles, and to sample comparisons. It can run across many environments and support parallel computation for long-running processes. Firstly, it is performed on 16S DNA sequence data from microbial communities. The sequence is then quality checked and mapped. The multiplexed reads are then assigned to their nucleotide barcode. The Operational Taxonomic Units (OTUs) are based on sequence similarity within the reads, and a representative sequence is picked from each OTU. To assign taxonomic identities using reference databases, align OUT sequences, create a phylogenetic tree, and construct an OYU table to represent the abundance of each OUT in each microbial sample. Lastly, computing within the community diversity (Kuczynski et al., 2012; http://bmf.colorado.edu/QIIME/QIIME-1.3.0-amd64.vdi.gz).

18.13 AMPHORA2

It relies on a core phylogenetic marker database to identify marker genes in the input sequences. It handles marker genes from the bacterial and archaeal domains. Users can add markers from a choice to build their personalised phylogenetic marker database using the Zorro program. It uses RAxML and Pplacer to perform parsimony- or likelihood-based phylogenetic analyses. It places sequences into the NCBI taxonomic hierarchy and assigns a confidence score at each taxonomic rank. It does not require annotation of sequences, thus the tool can be applied directly to the metagenomic reads as it uses HMMER3 as its sequence search engine, which accurately detects homologous sequences (Wu & Scott, 2012; http://wolbachia.biology.virginia.edu/WuLab/Software.html).

18.14 WORKFLOW AND PIPELINES

For each step of metagenomics analysis algorithms and assembly methods have been mentioned in Table 18.3.

18.15 CLOVR-METAGENOMICS

It employs several well-known tools and protocols for the analysis of metagenomic whole-genome shotgun (WGS) sequence datasets. It accepts multiple FASTA files and a metadata file that specifies features associated with samples. It clusters all

TABLE 18.2
List of Tools with Their Function for Legume Metagenomics

Steps	Tools	Function	Reference
General Software	FastQC	Quality parameters	Andrews et al., (2010)
	Microbiome	exploring, analysing, and visualising microbiome data.	Lahti et al. (2023)
	Microbiomeutilities	data preparation, visualisation, and statistical analysis of microbiome datasets.	Shetty and Lahti (2023)
	MMUPHin	For meta-analysis and batch-effect correction of microbiome data.	Ma et al. (2022)
	SIAMCAT	Feature selection, model training, and performance assessment.	Wirbel et al. (2021)
	Vegan	Analyse ecological patterns and species abundance data.	Oksanen et al. (2022)
Amplicon data analysis software	PICRUSt2	16S rRNA gene sequences	Douglas et al. (2020)
	DADA2	High-resolution analysis of amplicon sequencing data	Callahan et al. (2016)
Whole Metagenome Data analysis software	MetaPhlAn4	Clade-specific marker genes	Beghini et al. (2023)
	MetaBAT 2	Sequence composition and coverage data to bin metagenomic contigs into draft genomes.	Kang et al. (2019)
	VirSorter2	Recognise and categorise viral sequences.	Guo et al. (2021)
	CheckV	Evaluate the completeness, quality, and contamination of viral genomes.	Nayfach et al. (2021)
	DRAM-v	Annotates and characterises viral genomes.	Shaffer et al. (2020)
	VirStrain	Identifies RNA viruses.	Xia et al. (2020)
	MicrobiomeAnalyst2.0	Meta-analysis of microbiome data.	Chong et al. (2020)
Meta-transcriptome Data Analysis Software	SortMeRNA	Filter, map, and classify rRNA sequences.	Kopylova et al. (2012)

DNA sequences using UCLUST. Any non-representative sequence in a cluster that shares a prefix of length eight is determined from sequencing a replicate. Then, DNA sequence clusters are searched against the RefSeq database of finish prokaryotic genomes using BLASTN. Furthermore, all DNA sequences from UCLUST are searched against the COG database of orthologous gene groups (Fricke et al., 2011).

18.16 METAMOS

The system is built around a collection of publicly available assembly and analysis tools, utilising the lightweight workflow system Ruffus. It is broadly divided into three main sections: preprocessing, identification of genomic repeats, and the post-scaffolding stage. To generate the initial high-confidence contigs, these were analysed with Bambus 2, which is provided by MetAMOS. The feature response curve technique tracks the cumulative size of the assembly and the number of errors detected within it. The contigs are sorted by decreasing sequence length, with the number of errors and cumulative contig size plotted along the x- and y-axes, respectively. MetAMOS encompasses independent preprocessing and scaffolding routines, and the performance of both Velvet and MetaVelvet is enhanced within this framework (Treangen et al., 2013).

18.17 YAMP WORKFLOW

This user-friendly workflow facilitates the analysis of whole shotgun metagenomic data by utilising containerisation to ensure computational reproducibility and promote collaborative research. It can be executed on UNIX systems and extends seamless support for various job schedulers. This ready-to-use containerised workflow incorporates state-of-the-art tools to process raw shotgun metagenomic sequencing data for taxonomic and functional annotation. The workflow is structured into three components: quality control, quality assessment, and community characterisation. Developed using Nextflow, a workflow management system allows for the easy development, deployment, and execution of complex distributed computational workflows. Accompanied by a Docker container that defines a platform-independent, virtualised, and lightweight operating system, it also includes a Singularity container for transferring features to high-performance computing (HPC) systems, supporting a multi-container scenario. Quality control is performed using several tools from the BBmap suite. Taxonomic binning and profiling are conducted with MetaPhlAn2,

TABLE 18.3
Legume Metagenome Analysis Algorithms and Assembly Methods

Type	Sub-type	Tool	Reference
Algorithm	Denoising	PyroNoise	Quince et al. (2009)
		Denoiser	Reeder and Knight (2010)
	Chimera Detection	UCHIME	Edgar et al. (2011)
		ChimeraSlayer	Haas et al. (2011)
		Perseus	Tyanova et al. (2016)
	OTU Clustering	UCLUST	Edgar et al. (2011)
		CLOTU	Lindahl et al. (2013)
		CrunchCluster	Portik and Wiens (2020)
Assembly	Referenced assembly	AMOS	Treangen et al. (2011)
	Denovo assembly	SOAPdenovo	Luo et al. (2012)
		Meta-Velvet	Namiki et al. (2012)

while the functional capabilities of the microbial community are assessed using the HUMAnN2 pipeline. Additionally, the identified taxonomic profile is used by YAMP to evaluate various alpha-diversity measures via the alpha_diversity.py function in the widely recognised QIIME pipeline (Visconti et al., 2018).

18.18 IMPORTANCE OF LEGUME MICROBIOME

The legume microbiome is essential for soil fertility, aiding in nitrogen fixation, disease protection, and nutrient uptake. It mainly consists of Proteobacteria, Actinobacteria, and Bacteroidetes, which have been studied extensively in plants such as *Arabidopsis*, maize, and rice (Sahil et al., 2025). This microbiome includes rhizobia, arbuscular mycorrhizal fungi, nitrogen-fixing bacteria, and disease-suppressive bacteria, enhancing nutrient uptake and defending against pathogens. Genotype variability influences the rhizosphere microbiome depicted in Figure 18.2. For example, in transgenic *Arabidopsis*, glucosinolate production alters the microbial

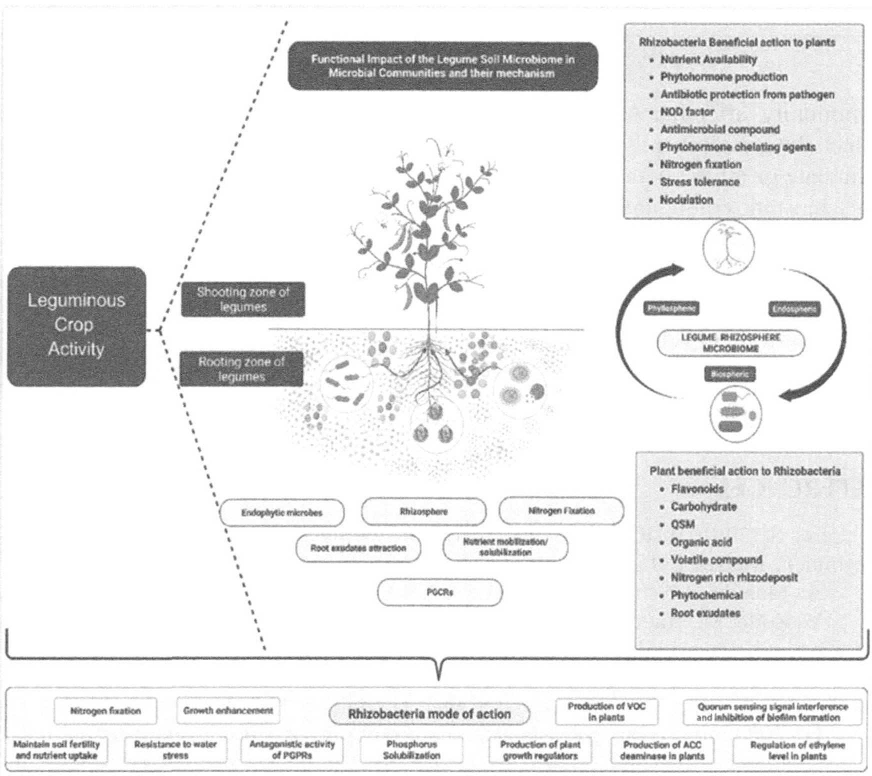

FIGURE 18.2 The functional impact of the legume microbiome illustrating crop activity, microbial interaction, mechanisms of action, and microbiome composition.

TABLE 18.4
Summarising the Importance and Function of the Legume Microbiome

Aspect	Description	Reference
Soil fertility and nitrogen fixation	Enhances nitrogen fixation.	Schaedel et al. (2021)
Climate change mitigation	Reduces greenhouse gas emissions.	Buernor et al. (2022)
Non-rhizobial microbes	Supports stress tolerance, and disease resistance.	Hassen et al. (2025)
Stage-specific associations	Microbial composition and function shift.	Sahil et al. (2025)
Genotype and environmental interplay	Plant genotypes and environmental factors modulate rhizosphere microbial diversity.	Lanzavecchia et al. (2024)
Microbiome engineering	Crop improvement.	Patra et al. (2025)
Plant–microbe signalling	Root exudates and quorum sensing.	Jain et al. (2024)
Nutritional and health impact	Improves the nutritional quality of seeds.	Özdemir and Buyuktuncer (2024)

community, affecting Alphaproteobacteria and fungi (Bressan et al., 2009). Plants affect the rhizosphere's composition through compounds they exude, which can stimulate or inhibit pathogens. Successful competitors among microorganisms produce biostatic compounds and activate the plant immune system. Additionally, some plant-associated bacteria produce N-acyl-homoserine lactones (AHLs) to regulate gene expression through quorum sensing (Pudasaini et al., 2023).

18.19 ACKNOWLEDGEMENTS

We are grateful to the Council of Science & Technology, UP (CSTUP), Government of Uttar Pradesh, for the project grant (Sanction No. CST/D/Biotech-1271).

REFERENCES

Andrews, S. (2010). FastQC: A quality control tool for high throughput sequence data.

Beghini, F., McIver, L. J., Blanco-Míguez, A., Dubois, L., Asnicar, F., Maharjan, S., Mailyan, A., Manghi, P., Scholz, M., Thomas, A. M., Valles-Colomer, M., Weingart, G., Zhang, Y., Zolfo, M., Huttenhower, C., Segata, N., & Franzosa, E. A. (2023). Integrating taxonomic, functional, and strain-level profiling of diverse microbial communities with MetaPhlAn4. *Nature Biotechnology*, *41*, 1140–1151.

Bressan, M., Roncato, M. A., Bellvert, F., Comte, G., Haichar, F. E. Z., Achouak, W., & Berge, O. (2009). Exogenous glucosinolate produced by Arabidopsis thaliana has an impact on microbes in the rhizosphere and plant roots. *The ISME Journal*, *3*(11), 1243–1257.

Buernor, A. B., Kabiru, M. R., Bechtaoui, N., Jibrin, J. M., Asante, M., Bouraqqadi, A., . . . & Jemo, M. (2022). Grain legume yield responses to rhizobia inoculants and phosphorus supplementation under Ghana soils: A meta-synthesis. *Frontiers in Plant Science*, *13*, 877433.

Callahan, B. J., McMurdie, P. J., Rosen, M. J., Han, A. W., Johnson, A. J. A., & Holmes, S. P. (2016). DADA2: High-resolution sample inference from Illumina amplicon data. *Nature Methods*, *13*(7), 581–583.

Caspi, R., Billington, R., Ferrer, L., Foerster, H., Fulcher, C. A., Keseler, I. M., Kothari, A., Krummenacker, M., Latendresse, M., Mueller, L. A., Ong, Q., Paley, S., Subhraveti, P., Weaver, D. S., Weerasinghe, D., Zhang, P., & Karp, P. D. (2018). The MetaCyc database of metabolic pathways and enzymes. *Nucleic Acids Research*, *46*(D1), D633–D639.

Chong, J., Liu, P., Zhou, G., & Xia, J. (2020). Using MicrobiomeAnalyst 2.0 for comprehensive statistical, functional, and meta-analysis of microbiome data. *Nature Protocols*, *16*, 479–502.

Coban, O., De Deyn, G. B., & van der Ploeg, M. (2022). Soil microbiota as game-changers in restoration of degraded lands. *Science*, *375*(6584), abe0725.

Cole, J. R., Wang, Q., Fish, J. A., Chai, B., McGarrell, D. M., Sun, Y., . . . & Tiedje, J. M. (2014). Ribosomal database project: Data and tools for high throughput rRNA analysis. *Nucleic Acids Research*, *42*(D1), D633–D642.

DeSantis, T. Z., Hugenholtz, P., Larsen, N., Rojas, M., Brodie, E. L., Keller, K., . . . & Andersen, G. L. (2006). Greengenes, a chimera-checked 16S rRNA gene database and workbench compatible with ARB. *Applied and Environmental Microbiology*, *72*(7), 5069–5072.

Douglas, G. M., Maffei, V. J., Zaneveld, J., Yurgel, S. N., Brown, J. R., Taylor, C. M., Huttenhower, C., & Langille, M. G. I. (2020). PICRUSt2 for prediction of metagenome functions.

Eaton, W. D., Hamilton, D. A., Chen, W., Lemenze, A., & Soteropoulos, P. (2024). Use of high throughput DNA analysis to characterize the nodule-associated bacterial community from four ages of Inga punctata trees in a Costa Rican cloud forest. *AIMS Microbiology*, *10*(3), 572.

Edgar, R. C., Haas, B. J., Clemente, J. C., Quince, C., & Knight, R. (2011). UCHIME improves sensitivity and speed of chimera detection. *Bioinformatics*, *27*(16), 2194–2200.

Epihov, D. Z., Saltonstall, K., Batterman, S. A., Hedin, L. O., Hall, J. S., van Breugel, M., . . . & Beerling, D. J. (2021). Legume–microbiome interactions unlock mineral nutrients in regrowing tropical forests. *Proceedings of the National Academy of Sciences*, *118*(11), e2022241118.

Grazziotin, A. L., Koonin, E. V., & Kristensen, D. M. (2017). Prokaryotic Virus Orthologous Groups (pVOGs): A resource for comparative genomics and protein family annotation. *Nucleic Acids Research*, *45*(D1), D491–D498.

Guo, J., Bolduc, B., Zayed, A. A., Varsani, A., Dominguez-Huerta, G., Delmont, T. O., Pratama, A. A., Gazitúa, M. C., Vik, D., Sullivan, M. B., & Roux, S. (2021). VirSorter2: A multi-classifier, expert-guided approach to detect diverse DNA and RNA viruses. *Microbiome*, *9*, 37.

Haas, B. J., Gevers, D., Earl, A. M., Feldgarden, M., Ward, D. V., Giannoukos, G., Ciulla, D., Tabbaa, D., Highlander, S. K., Sodergren, E., Methe, B. A., DeSantis, T. Z., Petrosino, J. F., Andersen, G. L., & Relman, D. A. (2011). ChimeraSlayer: Improved chimera detection for amplicon sequencing. *Nucleic Acids Research*, *39*(14), e92.

Handelsman, J., Rondon, M. R., Brady, S. F., Clardy, J., & Goodman, R. M. (1998). Molecular biological access to the chemistry of unknown soil microbes: A new frontier for natural products. *Chemistry & Biology*, *5*(10), R245–R249.

Hassen, A. I., Muema, E. K., Diale, M. O., Mpai, T., & Bopape, F. L. (2025). Non-rhizobial endophytes (NREs) of the nodule microbiome have synergistic roles in beneficial tripartite plant–microbe interactions. *Microorganisms*, *13*(3), 518.

Healey, A., Furtado, A., Cooper, T., & Henry, R. J. (2014). Protocol: A simple method for extracting next-generation sequencing quality genomic DNA from recalcitrant plant species. *Plant Methods*, *10*(1), 21.

Huson, D. H., Auch, A. F., Qi, J., & Schuster, S. C. (2007). MEGAN analysis of metagenomic data. *Genome Research*, *17*(3), 377–386.

Jain, A., Sarsaiya, S., Singh, R., Gong, Q., Wu, Q., & Shi, J. (2024). Omics approaches in understanding the benefits of plant-microbe interactions. *Frontiers in Microbiology*, *15*, 1391059.

Johnson, G. (2019). High-throughput DNA extraction of legume root nodules for rhizobial metagenomics. *AMB Express*, *9*(1), 47.

Kang, D. D., Li, F., Kirton, E., Thomas, A., Egan, R., An, H., & Wang, Z. (2019). MetaBAT 2: An adaptive binning algorithm for robust and efficient genome reconstruction from metagenome assemblies. *PeerJ*, *7*, e7359.

Kim, O. S., Cho, Y. J., Lee, K., Yoon, S. H., Kim, M., Na, H., . . . & Chun, J. (2012). Introducing EzTaxon-e: A prokaryotic 16S rRNA gene sequence database with phylotypes that represent uncultured species. *International Journal of Systematic and Evolutionary Microbiology*, *62*(Pt_3), 716–721.

Kopylova, E., Noé, L., & Touzet, H. (2012). SortMeRNA: Fast and accurate filtering of ribosomal RNAs in metatranscriptomic data. *Bioinformatics*, *28*(24), 3211–3217.

Kuczynski, J., Stombaugh, J., Walters, W. A., González, A., Caporaso, J. G., & Knight, R. (2012). Using QIIME to analyze 16S rRNA gene sequences from microbial communities. *Current Protocols in Microbiology*, *27*(1), 1E-5.

Lahti, L., Shetty, S., Blake, T., & Salojärvi, J. (2023). Microbiome: Microbiome analytics. R package version 1.22.0.

Lanzavecchia, G., Frascarelli, G., Rocchetti, L., Bellucci, E., Bitocchi, E., Di Vittori, V., . . . & Papa, R. (2024). Genotype combinations drive variability in the microbiome configuration of the rhizosphere of maize/bean intercropping system. *International Journal of Molecular Sciences*, *25*(2), 1288.

Lindahl, B. D., Nilsson, R. H., Tedersoo, L., Abarenkov, K., Carlsen, T., Kjøller, R., Kõljalg, U., Pennanen, T., Rosendahl, S., Stenlid, J., Kauserud, H., & Lindahl, B. D. (2013). CLOTU: A flexible and user-friendly pipeline for clustering 454 amplicon sequences into OTUs. *BMC Bioinformatics*, *14*, 314.

Luo, R., et al. (2012). SOAPdenovo2: An empirically improved memory-efficient short-read de novo assembler. *Gigascience*, *1*(1), 18.

Ma, S., Shungin, D., Mallick, H., Schirmer, M., Nguyen, L. H., Kolde, R., Franzosa, E. A., Vlamakis, H., Xavier, R. J., Huttenhower, C., & Li, H. (2022). MMUPHin: Meta-analysis methods for studying microbial community heterogeneity across environments and datasets. *Genome Biology*, *23*(1), 100.

Markowitz, V. M., Chen, I. M. A., Chu, K., Szeto, E., Palaniappan, K., Grechkin, Y., . . . & Kyrpides, N. C. (2012). IMG/M: The integrated metagenome data management and comparative analysis system. *Nucleic Acids Research*, *40*(D1), D123–D129.

Mirarab, S., Nguyen, N., & Warnow, T. (2012). SEPP: SATé-enabled phylogenetic placement. In *Biocomputing 2012* (pp. 247–258).

Namiki, T., Hachiya, T., Tanaka, H., & Sakakibara, Y. (2012). MetaVelvet: An extension of Velvet assembler to de novo metagenome assembly from short sequence reads. *Nucleic Acids Research*, *40*(20), e155.

Nayfach, S., Camargo, A. P., Schulz, F., Eloe-Fadrosh, E., Roux, S., & Kyrpides, N. C. (2021). CheckV assesses the quality and completeness of metagenome-assembled viral genomes. *Nature Biotechnology*, *39*, 578–585.

Oksanen, J., Blanchet, F. G., Friendly, M., Kindt, R., Legendre, P., McGlinn, D., Minchin, P. R., O'Hara, R. B., Simpson, G. L., Solymos, P., Stevens, M. H. H., Szoecs, E., & Wagner, H. (2022). Vegan: Community ecology package. R package version 2.6–4.

O'Leary, N. A., Wright, M. W., Brister, J. R., Ciufo, S., Haddad, D., McVeigh, R., Rajput, B., Robbertse, B., Smith-White, B., Ako-Adjei, D., Astashyn, A., Badretdin, A., Bao, Y., Blinkova, O., Brover, V., Chetvernin, V., Choi, J., Cox, E., Ermolaeva, O., . . . & Pruitt, K. D. (2016). Reference sequence (RefSeq) database at NCBI: Current status, taxonomic expansion, and functional annotation. *Nucleic Acids Research*, *44*(D1), D733–D745.

Özdemir, A., & Buyuktuncer, Z. (2024). Dietary legumes and gut microbiome: A comprehensive review. *Critical Reviews in Food Science and Nutrition*, 1–15.

Patra, D., Mishra, S., Mohanty, J., Subhadarsani, S., & Sahoo, J. P. (2025). Harnessing the plant microbiome: A genomic and metagenomic perspective for sustainable agriculture. *Academia Molecular Biology and Genomics*, *2*(3).

Portik, D. M., & Wiens, J. J. (2020). SuperCRUNCH: A flexible and robust toolkit for phylogenetic dataset assembly. *Methods in Ecology and Evolution*, *11*(11), 1368–1374.

Pudasaini, R., Hewedy, O. A., & Raizada, M. N. (2023). Improving field legume nodulation by crushing nodules onto seeds: Implications for small-scale farmers. *Frontiers in Agronomy*, *5*, 1161978.

Quast, C., Pruesse, E., Yilmaz, P., Gerken, J., Schweer, T., Yarza, P., . . . & Glöckner, F. O. (2012). The SILVA ribosomal RNA gene database project: Improved data processing and web-based tools. *Nucleic Acids Research*, *41*(D1), D590–D596.

Quince, C., Lanzén, A., Curtis, T. P., Davenport, R. J., Hall, N., Head, I. M., Read, L. F., & Sloan, W. T. (2009). Accurate determination of microbial diversity from 454 pyrosequencing data. *Nature Methods*, *6*(9), 639–641.

Reeder, J., & Knight, R. (2010). Rapid denoising of pyrosequencing amplicon data: Exploiting the rank-abundance distribution. *Nature Methods*, *7*(9), 668–669.

Sahil, R., Pal, V., Kharat, A. S., & Jain, M. (2025). A multi-omics meta-analysis of rhizosphere microbiome reveals growth-promoting marker bacteria at different stages of legume development. *Plant, Cell & Environment*.

Schaedel, M., Hidrobo, G., & Grossman, J. (2021). From microns to meters: Exploring advances in legume microbiome diversity for agroecosystem benefits. *Frontiers in Sustainable Food Systems*, *5*, 668195.

Shaffer, M., Borton, M. A., McGivern, B. B., Zayed, A. A., La Rosa, S. L., Solden, L. M., Liu, P., Narrowe, A. B., Rodríguez-Ramos, J., Bolduc, B., Gazitúa, M. C., Daly, R. A., Smith, G. J., Vik, D., Pope, P. B., Sullivan, M. B., & Wrighton, K. C. (2020). DRAM for distilling microbial metabolism to automate the curation of microbiome function. *Nucleic Acids Research*, *48*(16), 8883–8900.

Shetty, S., & Lahti, L. (2023). Microbiomeutilities: An R package for simplified microbiome analysis and visualization. R package version 1.00.15.

Suzek, B. E., Huang, H., McGarvey, P., Mazumder, R., & Wu, C. H. (2007). UniRef: Comprehensive and non-redundant UniProt reference clusters. *Bioinformatics*, *23*(10), 1282–1288.

Tkacz, A., & Poole, P. (2021). The plant microbiome: The dark and dirty secrets of plant growth. *Plants, People, Planet*, *3*(2), 124–129.

Treangen, T. J., Koren, S., Sommer, D. D., Liu, B., Astrovskaya, I., Ondov, B., . . . & Pop, M. (2013). MetAMOS: A modular and open source metagenomic assembly and analysis pipeline. *Genome Biology*, *14*(1), R2.

Tsiknia, M., Tsikou, D., Papadopoulou, K. K., & Ehaliotis, C. (2021). Multi-species relationships in legume roots: From pairwise legume-symbiont interactions to the plant–microbiome–soil continuum. *FEMS Microbiology Ecology*, *97*(2), fiaa222.

Tyanova, S., Temu, T., & Cox, J. (2016). The Perseus computational platform for comprehensive analysis of (prote)omics data. *Nature Methods*, *13*, 731–740.

Visconti, A., Martin, T. C., & Falchi, M. (2018). YAMP: A containerized workflow enabling reproducibility in metagenomics research. *Gigascience*, *7*(7), giy072.

Weir, B. S., Turner, S. J., Silvester, W. B., Park, D. C., & Young, J. M. (2004). Unexpectedly diverse Mesorhizobium strains and Rhizobium leguminosarum nodulate native legume genera of New Zealand, while introduced legume weeds are nodulated by Bradyrhizobium species. *Applied and Environmental Microbiology*, *70*(10), 5980–5987.

Wilke, A., Bischof, J., Gerlach, W., Glass, E., Harrison, T., Keegan, K. P., . . . & Meyer, F. (2016). The MG-RAST metagenomics database and portal in 2015. *Nucleic Acids Research*, *44*(D1), D590–D594.

Wirbel, J., Zych, K., Essex, M., Karcher, N., Kartal, E., Salazar, G., Bork, P., & Zeller, G. (2021). SIAMCAT: An R package for statistical inference of associations between microbial communities and host phenotypes. *Nature Methods*, *18*(9), 1076–1078.

Wu, M., & Scott, A. J. (2012). Phylogenomic analysis of bacterial and archaeal sequences with AMPHORA2. *Bioinformatics*, *28*(7), 1033–1034.

Xia, Y., Li, F., & Li, J. (2020). VirStrain: A strain-level identification tool for RNA viruses. *Bioinformatics*, *36*(18), 5043–5044.

Yadav, T., Singh, V., Sharma, P., & Singh, S. P. (2024). Deciphering microbe-driven remediation of environmental pollutants: An omics perspective. In *Microbiome-Based Decontamination of Environmental Pollutants* (pp. 147–166). Academic Press.

Index

Note: Page numbers in *italics* indicate a figure, and page numbers in **bold** indicate a table on the corresponding page.

G

H

I

J

K

L

M

N

O

P